Basic Phylogenetic Combinatorics

Phylogenetic combinatorics is a branch of discrete applied mathematics concerned with the combinatorial description and analysis of phylogenetic trees and related mathematical structures such as phylogenetic networks and tight spans. Based on a natural conceptual framework, the book focuses on the interrelationship between the principal options for encoding phylogenetic trees: split systems, quartet systems, and metrics. Such encodings provide useful options for analyzing and dealing with phylogenetic trees and networks, and are at the basis of much of phylogenetic data processing. The book highlights how each one provides a unique perspective for viewing and perceiving the combinatorial structure of a phylogenetic tree and is, simultaneously, a rich source for combinatorial analysis and theory building. It is dedicated to Manfred Eigen who inspired many of the results presented in this book.

Graduate students and researchers in mathematics and computer science will enjoy exploring this fascinating new area, and learn how mathematics may be used to help solve topical problems arising in evolutionary biology.

ANDREAS DRESS works currently as a scientific advisor at infinity^3 GmbH, Bielefeld, Germany.

KATHARINA T. HUBER is a Lecturer in the School of Computing Sciences at the University of East Anglia, UK.

JACOBUS KOOLEN is an Associate Professor in the Department of Mathematics at Pohang University of Science and Technology (POSTECH), South Korea.

VINCENT MOULTON is a Professor in the School of Computing Sciences at the University of East Anglia, UK.

ANDREAS SPILLNER is an Assistant Professor in the Department of Mathematics and Computer Science at the University of Greifswald, Germany.

Basic Phylogenetic Combinatorics

ANDREAS DRESS (德乐思)
infinity[3] GmbH, Bielefeld, Germany
and
CAS-MPG Partner Institute for Computational Biology, Shanghai Institutes for Biological Sciences

KATHARINA T. HUBER
University of East Anglia

JACOBUS KOOLEN
Pohang University of Science and Technology (POSTECH), South Korea

VINCENT MOULTON
University of East Anglia

ANDREAS SPILLNER
University of Greifswald

CAMBRIDGE
UNIVERSITY PRESS

CAMBRIDGE
UNIVERSITY PRESS

Shaftesbury Road, Cambridge CB2 8EA, United Kingdom

One Liberty Plaza, 20th Floor, New York, NY 10006, USA

477 Williamstown Road, Port Melbourne, VIC 3207, Australia

314–321, 3rd Floor, Plot 3, Splendor Forum, Jasola District Centre, New Delhi – 110025, India

103 Penang Road, #05–06/07, Visioncrest Commercial, Singapore 238467

Cambridge University Press is part of Cambridge University Press & Assessment, a department of the University of Cambridge.

We share the University's mission to contribute to society through the pursuit of education, learning and research at the highest international levels of excellence.

www.cambridge.org
Information on this title: www.cambridge.org/9780521768320

First published 2012

A catalogue record for this publication is available from the British Library

Library of Congress Cataloging-in-Publication data
Basic phylogenetic combinatorics / Andreas Dress . . . [et al.].
p. cm.
ISBN 978-0-521-76832-0 (Hardback)
1. Branching processes. 2. Combinatorial analysis. I. Dress, Andreas.
QA274.76.B37 2011
511′.6–dc23

2011043264

ISBN 978-0-521-76832-0 Hardback

We dedicate this book to Manfred Eigen whose
questions concerning the evolution of RNA and DNA sequences
inspired many of the early results in
phylogenetic combinatorics that ultimately led
to the work presented in this book.

Contents

Preface

More than one and a half centuries have passed since Charles Darwin presented his theory on the origin of species asserting that all organisms are related to each other by common descent via a "tree of life". Since then, biologists have been able to piece together a great deal of information concerning this tree — relying in particular in more recent times on the advent of ever cheaper and faster DNA sequencing technologies. Even so, there remain many fascinating open problems concerning the tree of life and the evolutionary processes underlying it, problems that often require sophisticated techniques from areas such as mathematics, computer science, and statistics.

Phylogenetic combinatorics can be regarded as a branch of discrete applied mathematics concerned with the combinatorial description and analysis of *phylogenetic* or *evolutionary trees* and related mathematical structures such as phylogenetic networks, complexes, and tight spans. In this book, we present a *systematic* approach to phylogenetic combinatorics based on a natural conceptual framework that, simultaneously, allows and forces us to encompass many classical as well as a good number of new pertinent results.

More specifically, this book concentrates on the interrelationship between the three principal ways commonly used for **encoding** phylogenetic trees: *Split systems*, *metrics*, and *quartet systems* (see Figure 1). Informally, for X some finite set, a split system over X is a collection of bipartitions of X, a quartet system is a collection of two-versus-two bipartitions of subsets of X of size four, and a metric is a bivariate function assigning a "distance" to any pair of elements in X.

Such encodings provide useful options for analyzing and manipulating phylogenetic trees with leaves labeled by X, and are at the basis of much of phylogenetic data processing. Indeed, they arise naturally from the various types of data from which phylogenetic trees are typically (re-)constructed: Comparative sequence analysis of genes or genomes may lead to metrics, character tables as

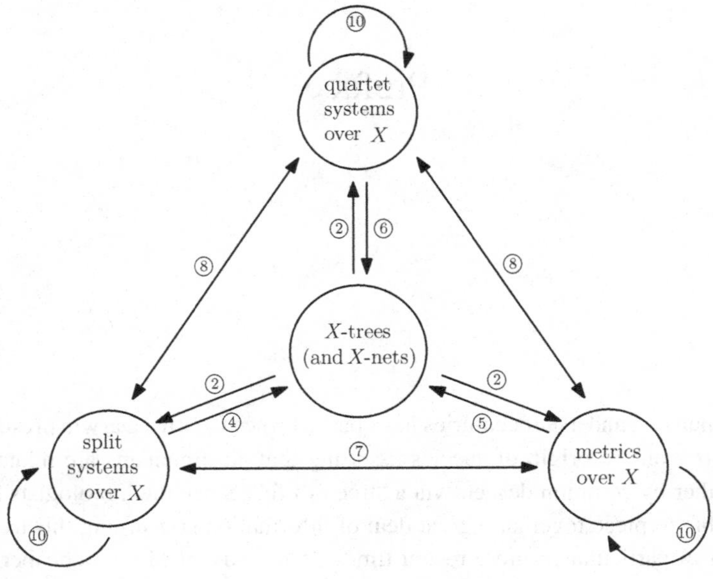

Figure 1 In this figure, we indicate the manifold relationships between various combinatorial objects relevant in phylogenetic analysis that will be studied in this book.

well as single nucleotide polymorphisms give rise to splits (and metrics), and quartet systems arise from restricting attention in phylogenetic data analysis to just four taxa at a time to avoid reconstruction algorithms becoming overwhelmed by the sheer number of taxa that need to be treated simultaneously. All three types of encodings require a solid theoretical foundation and provide, at the same time, a rich source for combinatorial analysis and theory building.

This book aims to highlight how each of the three types of encodings provides a unique perspective for viewing the combinatorial structure of a phylogenetic tree, for assessing the suitability of given data for tree reconstruction, and, if suitable, for recovering such trees from such data. And it will, of course, also discuss how split systems, metrics, and quartet systems are related to one another.

Here is an outline of the contents: After presenting some basic definitions and concepts that will be used throughout the book in Chapter 1, we introduce the formal concept of a phylogenetic tree or, a bit more generally, an X-tree in Chapter 2. We then define split systems, metrics, and quartet systems, and show that X-trees may indeed be uniquely encoded in terms of such combinatorial objects. In Chapter 3, we then proceed to identify which split systems, metrics, or quartet systems are induced by — and thus encode — an X-tree: That is, we characterize the split systems, metrics, or quartet systems in the "image" of the "maps" labeled ② in Figure 1 in terms of some simple, yet instructive and enlightening conditions.

In Chapters 4, 5, and 6, we move on to the problem of deciding how to *decode* a given tree-encoding using appropriate constructions corresponding, respectively, to the maps labeled ④, ⑤, ⑥ in Figure 1. In other words, given a split system, a metric, or a quartet system in the image of the maps labeled ②, we consider how to find that (unique!) X-tree encoded by them. We will also explain how, when applied to data sets that do not encode a tree, these constructions can produce *networks* (as opposed to trees) and discuss some pertinent consequences.

In Chapters 7 and 8, we investigate the *recoding* problem: How can we compute the split system, metric, or quartet system encoding a tree from its other two encodings, i.e., how can we define maps (as indicated by the arrows labeled ⑦ and ⑧, in either direction) so that the resulting triangular subdiagrams in Figure 1 are commutative? Generally, the pertinent constructions lead to pairs of maps between any two of the three classes of objects that appear to be of some independent interest in themselves.

"Rooting" an X-tree is rather important for a realistic interpretation of X-trees in terms of evolutionary history. Correspondingly, we consider in Chapter 9 how the previously mentioned maps and constructions can be modified so as to apply to *rooted X-trees*, giving rise to *cluster systems* (rather than split systems), *triplet systems* (rather than quartet systems), and *hierarchical dissimilarities* and *ultrametrics* (rather than metrics). Mathematically speaking, this can be regarded as taking an *affine* (more concrete) versus a *projective* (more elegant) approach to working with phylogenetic trees.

In the final chapter, Chapter 10, we address the problem of how to measure and remove "inconsistencies" in split systems, metrics, and quartet systems. In other words, given one such structure that does not encode an X-tree, we explore how we may find another one in its "neighborhood" that does. As we shall see, this not only yields some interesting mathematical results, but also new ways to analyze and understand phylogenetic data.

A major feature of this book is that full proofs are provided for all of the fundamental results, thus giving the motivated reader a chance to get to the forefront of the field of phylogenetic combinatorics without having to spend too much time seeking references (see also Figure 2 for a *Leitfaden*, i.e., an overview of chapter dependencies). It also includes various new results and proofs that have not been published previously, and it attempts to introduce most topics in an elementary way. Overall, we hope that the reader will be

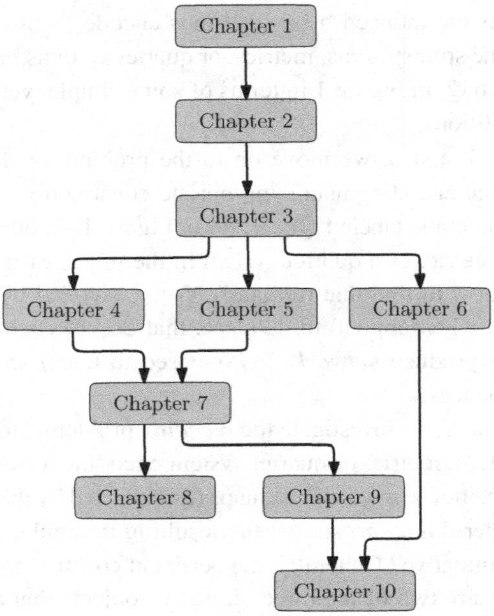

Figure 2 A diagram depicting the main dependencies between chapters in this book.

motivated by the book to explore this interesting area of mathematics whilst, at the same time, having the opportunity of seeing how mathematics may be used to help with solving topical problems that arise outside mathematics.

Finally, we would like to thank the Isaac Newton Institute for Mathematical Sciences, Cambridge, for giving us the opportunity to jointly draft and work on this book there, and also the UK Engineering and Physical Sciences Research Council and Royal Society, the Basic Science Research Program through the National Research Foundation of Korea (NRF) (grant number 2010-0008138), the DFG and the Max Planck Society, Germany, and the Chinese Academy for Sciences for financial support. We also thank our friends and colleagues and, in particular, David Bryant, Stefan Grünewald, Mike Hendy, Daniel Huson, Saitou Naruya, David Penny, LI Qiang, Charles Semple, Mike Steel, and WU Yaokun for many stimulating, critical, as well as encouraging discussions and comments. In addition, we thank students and colleagues at Bangalore, Bandar Lampung, Christchurch, Manila, New York, Paris, Pohang, and Shanghai, for their helpful feedback in courses where early versions of this text were presented. And last but not least, we all thank our families and, especially, Christiana, Keiko, Therese, Eugen, Jacky, and Robin, for their patience and support.

1
Preliminaries

We begin by reviewing some simple concepts regarding set systems, graphs, metric spaces, and computational complexity which will be used throughout this book. For more information on these topics see, for example, [48, 49, 79].

1.1 Sets, set systems, and partially ordered sets

In this section, we introduce useful terminology regarding sets, set systems, and partially ordered sets.

A finite set V of cardinality n will also be called an *n-set* and the n-set $\{1, 2, \ldots, n\}$ will be denoted by $\langle n \rangle$. A *set system* (over V) is a subset $\mathcal{V}$ of the *power set* $\mathbb{P}(V)$ of V, i.e., the set consisting of *all* subsets of V. The subsets in $\mathcal{V}$ are often also called the *clusters* in $\mathcal{V}$. For any non-negative integer $k \in \mathbb{N}_{\geq 0}$, the set system consisting of all k-subsets of V will also be denoted by $\binom{V}{k}$, and the set system consisting of all subsets of V of cardinality at least/at most k will also be denoted by $\mathbb{P}_{\geq k}(V)$ or $\mathbb{P}_{\leq k}(V)$, respectively. Given a subset A of a set V and an element $x \in V$, we denote the union $A \cup \{x\}$ also by $A + x$ and the difference $A \setminus \{x\} = \{a \in A : a \neq x\}$ also by $A - x$. Also, given two subsets A, B of V, we may write $A - B$ for $A \setminus B$.

Set systems are special instances of *partially ordered sets*, i.e., sets U together with a binary relation "$\preceq$" defined on U such that

$$u_1 \preceq u_2 \quad \text{and} \quad u_2 \preceq u_3 \Rightarrow u_1 \preceq u_3$$

and

$$u_1 \preceq u_2 \quad \text{and} \quad u_2 \preceq u_1 \iff u_1 = u_2$$

holds for all $u_1, u_2, u_3 \in U$ in which case the binary relation "$\preceq$" — or as well the (also transitive!) binary relation "$\prec$" defined by "$u \prec u' \iff u \preceq u'$ and $u \neq u'$" — is called a *partial order*. For any partially ordered set U, we denote by

$$\max(U) = \max_{\preceq}(U) := \{u \in U : \forall_{u' \in U}\ u \preceq u' \Rightarrow u = u'\}$$

the set of maximal elements in U (relative to the partial order "$\preceq$") and by

$$\min(U) = \min_{\preceq}(U) := \{u \in U : \forall_{u' \in U}\ u' \preceq u \Rightarrow u = u'\}$$

the set of minimal elements in U (relative to the partial order "$\preceq$"), we denote, for any $u \in U$, by $U_{\preceq u}$ the set of all $u' \in U$ with $u' \preceq u$ and by $U_{\prec u}$ the set of all $u' \in U_{\preceq u}$ that are distinct from u. We also consider any subset U' of a partially ordered set $U = (U, \preceq)$ as being itself a partially ordered set relative to the restriction of the binary relation $\preceq$ to U' which we keep denoting by $\preceq$ as long as no misunderstanding can arise. In particular, we denote by $U'_{\preceq u}$ the set $U_{\preceq u} \cap U'$ and by $U'_{\prec u}$ the set $U_{\prec u} \cap U'$. Furthermore, the elements in $\max(U_{\prec u})$ will sometimes also be called the *children* of u, and we will therefore denote the set $\max(U_{\prec u})$ also by $\text{chld}_U(u)$.

In particular, we denote, for any set system $\mathcal{V} \subseteq \mathbb{P}(V)$ over a set V and any subset L of V, by $\mathcal{V}_{\subseteq L}$ the set of all $U \in \mathcal{V}$ with $U \subseteq L$ and by $\mathcal{V}_{\subset L}$ the set of all $U \in \mathcal{V}_{\subseteq L}$ with $U \subsetneq L$. We will also denote by $\bigcup \mathcal{V}$ the union $\bigcup_{U \in \mathcal{V}} U$ of all clusters in a set system $\mathcal{V}$ and by $\bigcap \mathcal{V}$ the intersection $\bigcap_{U \in \mathcal{V}} U$ of all clusters in $\mathcal{V}$.

Of particular significance will be partitions and hierarchies. A set system $\mathcal{V} \subseteq \mathbb{P}(V)$ is defined to be a *partition* if it is contained in the set $\mathbb{P}_{\geq 1}(V) := \{U \subseteq V : U \neq \emptyset\}$ consisting of all non-empty subsets of V and $U_1 \cap U_2 = \emptyset$ holds for any two distinct clusters U_1, U_2 in $\mathcal{V}$, it is called a *partition of* V if, in addition, $\bigcup \mathcal{V} = V$ holds, it is called a *bipartition* or a *split* (of V) if it is a partition (of V) and contains exactly two distinct clusters, and every cluster in a partition will also be called a *part* of that partition. Often, we will also refer to splits by the letter S and denote a split S of the form $\{A, B\}$ by $A|B$. We will not distinguish between $A|B$ and $B|A$ as both terms stand for the same split $\{A, B\}$. Given a split $S = A|B$, the number $\min\{|A|, |B|\}$ will also be called its *size* and denoted by $\|S\|$ or, as well, by $\|A|B\|$. A split of size 1 is also called *trivial*, and a split of size k a *k-split*. And, given an element $x \in X$ and a split $S = A|B$ with $x \in A \cup B$, we denote that subset, A or B, in S that contains the element x by $S(x)$ and its complement in $A \cup B$ by $\overline{S}(x)$.

Clearly, $\bigcup \mathcal{U}_1 \cup \bigcup \mathcal{U}_2 = \bigcup(\mathcal{U}_1 \cup \mathcal{U}_2)$ holds for any two subcollections $\mathcal{U}_1, \mathcal{U}_2$ of a set system $\mathcal{V} \subseteq \mathbb{P}(V)$ while a set system $\mathcal{V} \subseteq \mathbb{P}_{\geq 1}(V)$ is a partition if and only if $\bigcup \mathcal{U}_1 \cap \bigcup \mathcal{U}_2 = \bigcup(\mathcal{U}_1 \cap \mathcal{U}_2)$ holds for any two subcollections $\mathcal{U}_1, \mathcal{U}_2$ of $\mathcal{V}$.

Further, a set system $\mathcal{H} \subseteq \mathbb{P}(V)$ is defined to be a *hierarchy* (over V) if it is contained in $\mathbb{P}_{\geq 1}(V)$ and $H_1 \cap H_2 \in \{\emptyset, H_1, H_2\}$ holds for any two clusters $H_1, H_2 \in \mathcal{H}$. Clearly, $\mathrm{chld}_{\mathcal{H}}(H)$ must be a partition for every H in a hierarchy $\mathcal{H}$: Indeed, if H_1 and H_2 are two distinct children of some cluster H in a hierarchy $\mathcal{H}$, we must have $H_1 \cap H_2 = \emptyset$ as neither $H_1 \cap H_2 = H_1$ nor $H_1 \cap H_2 = H_2$ can hold.

It is easy to see that, conversely, a set system $\mathcal{V} \subseteq \mathbb{P}_{\geq 1}(V)$ must be a hierarchy if $\mathrm{chld}_{\mathcal{V}}(U)$ is a partition for every U in $\mathcal{V}$ provided V is finite and a member of $\mathcal{V}$: Indeed, if this holds and if U_1 and U_2 are any two clusters in $\mathcal{V}$, there exists — in view of $V \in \mathcal{V}$ — an inclusion-minimal cluster U in $\mathcal{V}$ containing $U_1 \cup U_2$. If $U = U_1$ or $U = U_2$ holds, we have $U_1 \cap U_2 \in \{U_1, U_2\}$. Otherwise, there exist largest proper subsets U_1', U_2' of U in $\mathcal{V}$ that contain U_1 and U_2, respectively, and we must have $U_1' \neq U_2'$ by the choice of U (as, otherwise, $U_1' = U_2'$ would be a smaller cluster than U in $\mathcal{V}$ that contains $U_1 \cup U_2$). So, U_1' and U_2' must be distinct members of the partition $\mathrm{chld}_{\mathcal{V}}(U)$ and, therefore, disjoint, implying that also $U_1 \cap U_2 \subseteq U_1' \cap U_2' = \emptyset$ must be empty. So, $U_1 \cap U_2 \in \{U_1, U_2, \emptyset\}$ must indeed hold for any two clusters U_1 and U_2 in $\mathcal{V}$.

It is also easy to see that every hierarchy $\mathcal{H}$ over an n-set contains at most $2n - 1$ clusters: Indeed, this clearly holds in case $n := 1$, and if it holds for any hierarchy over any proper subset of an n-set V, then it holds for $\mathcal{H}$, too, in view of $\mathcal{H} \subseteq V + \dot{\bigcup}_{H \in \mathrm{chld}_{\mathcal{H}}(V)} \mathcal{H}_{\subseteq H}$ and, hence,

$$|\mathcal{H}| \leq 1 + \sum_{H \in \mathrm{chld}_{\mathcal{H}}(V)} |\mathcal{H}_{\subseteq H}|,$$

the fact that $\sum_{U \in \mathcal{V}} |U| \leq n$ must hold for every partition $\mathcal{V} \subseteq \mathbb{P}(V)$, and that $\mathcal{H}_{\subseteq H}$ is a hierarchy over H for every $H \in \mathcal{H}$. So,

$$|\mathcal{H}| \leq 1 + \sum_{H \in \mathrm{chld}_{\mathcal{H}}(V)} (2\,|H| - 1) \leq 1 + 2n - |\mathrm{chld}_{\mathcal{H}}(V)| \leq 2n - 1$$

must hold in case $2 \leq |\mathrm{chld}_{\mathcal{H}}(V)|$. And it must hold in case $|\mathrm{chld}_{\mathcal{H}}(V)| < 2$ as this implies that even $1 + \sum_{U \in \mathrm{chld}_{\mathcal{H}}(V)} (2\,|U| - 1) \leq 1 + 2\,(n-1) - 1 = 2n - 2$ must hold.

In particular, we have $|\mathcal{H}| = 2n - 1$ if and only if $V \in \mathcal{H}$ holds, $\mathrm{chld}_{\mathcal{H}}(V)$ is a split of V, and $|\mathcal{H}_{\subseteq U}| = 2\,|U| - 1$ holds for both clusters $U \in \mathrm{chld}_{\mathcal{H}}(V)$ — so, by recursion, this holds if and only if $V \in \mathcal{H}$ holds and $\mathrm{chld}_{\mathcal{H}}(U)$ is a split of U for every cluster $U \in \mathcal{H}$ with $|U| \geq 2$.

More generally, the following fact is well known and easy to see:

Lemma 1.1 *Given a hierarchy $\mathcal{H}$ over a finite set V of cardinality n, the following assertions are equivalent:*

(i) $|\mathcal{H}| = 2n - 1$ *holds.*
(ii) $\mathcal{H}$ *contains* V *and* $\mathrm{chld}_{\mathcal{H}}(H)$ *is a split of* H *for every cluster* $H \in \mathcal{H}$ *with* $|H| \geq 2$.
(iii) $\mathcal{H}$ *is a maximal hierarchy over* V, *i.e.,* $U \in \mathcal{H}$ *holds for every subset* U *of* V *with* $U \cap H \in \{U, H, \emptyset\}$ *for every cluster* $H \in \mathcal{H}$.
(iv) $\mathcal{H}$ *contains* V *and all one-element subsets of* V, *and* $H_2 - H_1 \in \mathcal{H}$ *holds for any two subsets* $H_1, H_2 \in \mathcal{H}$ *with* $H_1 \in \mathrm{chld}_{\mathcal{H}}(H_2)$.

Proof We have seen already that (i) $\iff$ (ii) holds. And it is also clear that (i) $\Rightarrow$ (iii) holds: Otherwise, there would exist a hierarchy over V containing more than $2n-1$ clusters. And (iii) $\Rightarrow$ (iv) holds as $U \cap H \in \{U, H, \emptyset\}$ holds for every cluster $H \in \mathcal{H}$ for $U := V$ or U a one-element subset of V. It also holds for $U := H_2 - H_1 \in \mathcal{H}$ in case $H_2 \in \mathcal{H}$ holds and H_1 is a child of H_2: Indeed, $U \subseteq H$ holds in case $H_2 \subseteq H$, $H \subseteq U$ holds in case $H \subseteq H_2$ and $H \cap H_1 = \emptyset$, and $H \cap U = \emptyset$ holds in case $H \cap H_2 = \emptyset$ and in case $H \subseteq H_1$. Finally, if neither $H_2 \subseteq H$ nor $H \cap H_1 = \emptyset$ nor $H \subseteq H_1$ holds, we would necessarily have $H \subsetneq H_2$ (in view of $H \cap H_2 \neq H_2, \emptyset$) and $H_1 \subsetneq H$ (in view of $H \cap H_1 \neq H, \emptyset$) in contradiction to our assumption that H_1 is a child of H_2 and that, therefore, $\{U \in \mathcal{H} : H_1 \subsetneq U \subsetneq H_2\} = \emptyset$ holds.

Finally, we have (iv) $\Rightarrow$ (ii) as $\mathrm{chld}_{\mathcal{H}}(H)$ must be a partition of H for every $H \in \mathcal{H}$ whenever $\mathcal{H}$ contains all one-element subsets of V, and it must, of course, be a bipartition of H if $H - H' \in \mathcal{H}$ holds for any $H' \in \mathrm{chld}_{\mathcal{H}}(H)$. □

Note that hierarchies over a set V are sometimes required to also contain V or the empty set or, as well, all one-element subsets of V — see e.g., [28] where it was shown that a hierarchy $\mathcal{H}$ over an arbitrary set V is a maximal hierarchy over V if and only if $\mathcal{H}$ satisfies the condition (iv) and, in addition, $\bigcup \mathcal{C}, \bigcap \mathcal{C} \in \mathcal{H}$ holds for any "chain" $\mathcal{C}$ of clusters contained in $\mathcal{H}$ (i.e., any subset $\mathcal{C}$ of $\mathcal{H}$ with $C_1 \cap C_2 \in \{C_1, C_2\}$ for all $C_1, C_2 \in \mathcal{C}$) with $\bigcap \mathcal{C} \neq \emptyset$.

1.2 Graphs

A *graph* is a pair $G = (V, E)$ consisting of a non-empty set V, the *vertex set* of G, and a subset E of $\binom{V}{2}$, the *edge set* of G. G is called *finite* if its vertex

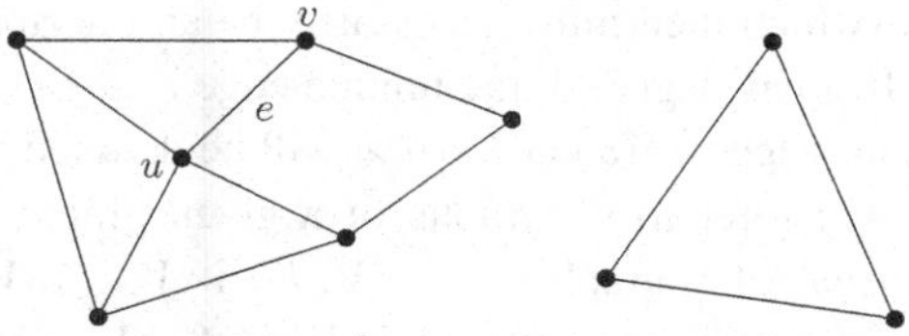

Figure 1.1 A (non-connected) graph with nine vertices and 12 edges.

set — and, hence, also its edge set — is finite. The elements of V and E are also called the *vertices* and the *edges* of G, respectively. Two graphs $G = (V, E)$ and $G' = (V', E')$ are called *isomorphic* if and only if there exists a bijective map $\iota : V \to V'$ with $\{u, v\} \in E \Longleftrightarrow \{\iota(u), \iota(v)\} \in E'$ for all $u, v \in V$. Clearly, graphs can be viewed as particularly simple set systems, that is, set systems $\mathcal{V}$ for which any cluster $e \in \mathcal{V}$ has cardinality 2. In Figure 1.1, we present a (drawing of a) graph: Vertices are represented by dots, and edges by straight line segments.

Two vertices u and v of a graph G are called *adjacent* if $\{u, v\}$ is an edge of G. For any edge $e = \{u, v\}$ of G, we call the vertices u and v the *endpoints* of e, and we will say that an edge $e \in E$ and a vertex $v \in V$ are *incident* if (and only if) $v \in e$ holds. The vertices that are adjacent to a vertex v of G are also called the *neighbors* of v in G, the set of neighbors of v in G is denoted by $N_G(v)$ or just $N(v)$, and the set of edges that are incident to v by $E_G(v)$ or just $E(v)$. The number of edges that are incident to a vertex v — or, equivalently, the number of neighbors of v — is called its *degree*, denoted by $\deg(v)$ or, more specifically, by $\deg_G(v)$.

For instance, referring to Figure 1.1, the vertex u has degree 4 and is adjacent to the vertex v, and the edge e is incident to both, u and v.

A vertex of degree 1 is called a *leaf* (of G), and the unique edge e of G that is incident to a leaf a is denoted by $e_G(a)$. Any such edge is also called a *pendant edge* while the unique vertex in $e_G(a)$ distinct from a is denoted by $v_G(a)$.

Every vertex that is not a leaf is called an *interior vertex* of G, and every edge that is not a pendant edge is called an *interior edge*. We denote the set of interior vertices and edges of G by $V_{int}(G)$ and $E_{int}(G)$, respectively. Clearly, "plucking off" all of the leaves and pendant edges from a graph $G = (V, E)$ yields a graph with vertex set $V_{int}(G)$ and edge set $E_{int}(G)$ that we dub the graph *derived* from G and denote, for short, by ∂G.

A pair of distinct leaves a, b is said to form a *cherry* (in G) — or, just, to be a cherry (of G) — if $v_G(a) = v_G(b)$ holds, i.e., if both leaves are adjacent to the same vertex (which then must necessarily be an interior vertex, having degree at least 2). If v has degree 3, the unique edge $e \in E(v)$ that is distinct from the two pendant edges $e_G(a)$ and $e_G(b)$ will be denoted by $e_G(a, b)$.

Frequently, we will refer to subgraphs of a given graph: A graph $G' = (V', E')$ is a *subgraph* of a graph $G = (V, E)$ if $V' \subseteq V$ and $E' \subseteq E$ holds, and it is *the subgraph of G induced on V′ by G*, also denoted by $G[V']$, if — in addition — $E' = E_{V'} := E \cap \binom{V'}{2}$ holds, that is, if and only if G' is the largest subgraph of G with vertex set V'.

A *path* $\mathbf{p}$ in a graph $G = (V, E)$ is a sequence $v_0, v_1, \dots, v_\ell$ of consecutively adjacent vertices of G, i.e., with $e_i := \{v_{i-1}, v_i\} \in E$ for all $i = 1, \dots, \ell$, such that $v_{i-1} \neq v_{i+1}$ holds for all $i \in \{1, \dots, \ell - 1\}$ — more specifically, any such sequence $v_0, v_1, \dots, v_\ell$ will be called a *path of length* ℓ while the vertices $v_0, v_1, \dots, v_\ell$ and the edges $e_1 = \{v_0, v_1\}, \dots, e_\ell = \{v_{\ell-1}, v_\ell\}$ will be called the *vertices* and the *edges of* $\mathbf{p}$ or, as well, the vertices and edges that are *passed* by $\mathbf{p}$, and the sets $\{v_0, v_1, \dots, v_\ell\}$ and $\{e_1, \dots, e_\ell\}$ will also be denoted by $V(\mathbf{p})$ and $E(\mathbf{p})$, respectively. The vertex v_0 is also called the *starting point*, and the vertex v_ℓ the *end point* of $\mathbf{p}$ (though sometimes also both vertices, v_0 and v_ℓ, may be referred to as its endpoints), and $\mathbf{p}$ is also called a path *from* v_0 *to* v_ℓ.

A path $\mathbf{p}$ is called *proper* if all of its vertices except perhaps its starting and its end point are distinct, i.e., if $v_i \neq v_j$ holds for all $i, j \in \{0, 1, \dots, \ell\}$ with $i \neq j$ and $\{i, j\} \neq \{0, \ell\}$, and it is called a (*cyclically*) *closed path* if its starting and its end point coincide, i.e., if $v_0 = v_\ell$ holds, its length is positive and, hence, exceeds 2, and also $v_1 \neq v_{\ell-1}$ holds.

In Figure 1.1, there is exactly one proper path of length 1, 2, 4, and 5, respectively, from u to v, and two such paths of length 3.

A graph $G = (V, E)$ is *connected* if there exists, for any two vertices $u, v \in V$ of G, a path in G with endpoints u and v. More generally, a subset $U \subseteq V$ of the vertex set V of a graph $G = (V, E)$ is *connected* (relative to G) if the associated induced subgraph $G[U]$ is connected. And a subset $F \subseteq E$ of the edge set E of a graph $G = (V, E)$ is connected (relative to G) if the graph $\left(\bigcup F, F\right)$ is connected.

A *connected component* of a graph $G = (V, E)$ is an inclusion-maximal connected subset $U \subseteq V$ of V or, equivalently, an inclusion-minimal nonempty subset U of V for which $e \subseteq U$ holds for all $e \in E$ with $e \cap U \neq \emptyset$. So, the graph in Figure 1.1, for example, "contains" exactly two distinct connected components.

Clearly, any two connected components of a graph G either coincide or have an empty intersection. We denote the set of connected components of a graph

$G = (V, E)$ by $\pi_0(G)$, and we denote the (unique!) connected component of G containing a given vertex $v \in V$ by $G(v)$.

It is also obvious that the set system $\pi_0(G)$ forms a partition of the vertex set V of a graph $G = (V, E)$ and that, if G is a connected graph with at least one interior vertex, the set $V_{int}(G)$ of its interior vertices is a connected subset of G and the associated induced — and necessarily connected — subgraph $G[V_{int}(G)]$ coincides with the derived graph ∂G. More precisely, a graph G with $V_{int}(G) \neq \emptyset$ is connected if and only if the derived graph ∂G is connected and G contains no *isolated vertices* or *isolated edges*, i.e., vertices or edges that form a connected component of G.

We will say that an edge $e \in E$ *separates* a vertex $v \in V$ from a vertex $u \in V$ if $G(v) = G(u)$ holds while the two connected components $G^{(e)}(u)$ and $G^{(e)}(v)$ of the graph $G^{(e)} := (V, E - e)$ containing u and v, respectively, are distinct — that is, if there is a path in G connecting u and v, but every such path passes e. The set of all edges of G separating the vertices u and v will be denoted by $E_G(u|v)$ or simply $E(u|v)$. And any edge $e = \{u, v\} \in E$ that separates its two endpoints u and v will be called a *bridge*.

More generally, we call a subset E' of E an *edge-cutset* of G if $G(v)$ differs from $(V, E - E')(v)$ for at least one vertex $v \in V$. Analogously, a subset $U \subseteq V$ is a *vertex-cutset* of G if there exist two vertices $u, u' \in V - U$ with $G(u) = G(u')$, but $G[V - U](u) \neq G[V - U](u')$. In particular, a vertex $v \in V$ such that $\{v\}$ is a vertex-cutset of G is called a *cut vertex* of G.

A *cycle* is a finite connected graph all of whose vertices have degree 2. Clearly, a graph $G = (V, E)$ is a cycle if and only if it is finite and we can label its vertices as $v_1, v_2, \dots, v_\ell$ ($\ell := |V|$) so that E coincides with $\{\{v_1, v_2\}, \dots, \{v_{\ell-1}, v_\ell\}, \{v_\ell, v_1\}\}$ in which case the sequence $v_0 := v_\ell, v_1, v_2, \dots, v_\ell$ forms a proper closed path in G that encompasses all vertices and edges of G. A *cycle in a graph G* is a subgraph of G that is a cycle. The graph in Figure 1.1 contains exactly four cycles of length 3 and 5, and three of length 4 and 6, respectively.

Clearly, an edge e in a finite graph G is contained in a cycle in G if and only if it is contained in $\partial^k G$ for every natural number $k \in \mathbb{N}_{\geq 0}$ (where $\partial^k G$ is, of course, defined recursively by $\partial^0 G := G$ and $\partial^{k+1} G := \partial(\partial^k G)$ for every $k \in \mathbb{N}_{\geq 0}$).

A graph $T = (V, E)$ is a *tree* if it is connected and contains no cycles or equivalently, as every "shortest" closed path "is" a cycle, no closed path. A subgraph $T' = (V', E')$ of a tree T that is connected is called a *subtree* of T in which case it must coincide with the induced subgraph $T[V']$ of T with vertex set V'. An example of a tree is given in Figure 1.2.

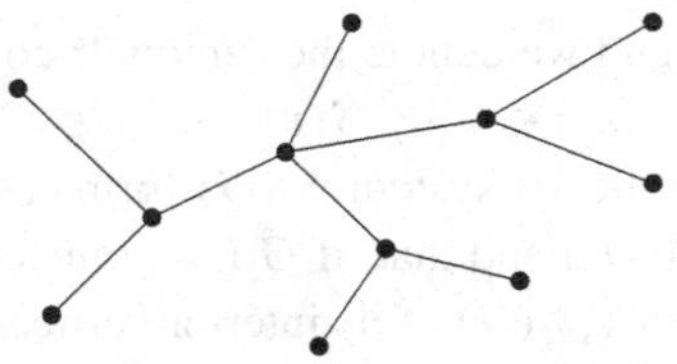

Figure 1.2 An example of a tree.

A tree $T = (V, E)$ is called *binary* if every interior vertex has degree 3, and it is called a *star tree* if it has precisely one interior vertex which then must be necessarily of degree $|V| - 1$. The unique interior vertex of a star tree will also be called the *central vertex* of that tree.

The tree in Figure 1.2 has one interior vertex of degree 4 and is, therefore, not binary. There are three cherries in this tree.

Note that, for any two distinct vertices u and v in a tree $T = (V, E)$, there is a unique edge $e^{v \to u} \in E(v)$ in the intersection $E_T(u|v) \cap E_v$. In consequence, there is precisely one path in T from u to v for any two distinct vertices u and v of T that we will denote by $\mathbf{p}_T(u, v)$ or simply by $\mathbf{p}(u, v)$, while its vertex set $V(\mathbf{p}_T(u, v))$ will be denoted by $V_T[u, v]$ and its edge set (that actually coincides with $E_T(u|v)$) also by $E_T[u, v]$. Clearly, a subset U of V is connected if and only if $V_T[u, v] \subseteq U$ holds for all $u, v \in U$ implying that, given any subset U of V, there exists a unique smallest connected subset of V that contains U, *viz.* the subset $V_T[U] := \bigcup_{u,v \in U} V_T[u, v]$. And we have $e^{v \to u} \neq e^{v \to u'}$ for three distinct vertices u, u', v of T if and only if $v \in V_T[u, u']$ holds.

Note that, for any three vertices u, v, w in a tree $T = (V, E)$, there is a unique vertex $m \in V$ that is contained in the intersection $V_T[u, v] \cap V_T[v, w] \cap V_T[w, u]$, called the *median* of u, v, w in T and denoted by $med(u, v, w) = med_T(u, v, w)$.

Note also that, for every edge $e = \{u, v\}$ of a tree $T = (V, E)$, the subgraph $T^{(e)} = (V, E - e)$ of T has precisely two connected components, *viz.* $T^{(e)}(u)$, the one containing u, and $T^{(e)}(v)$, the one containing v. Note also that $e \in E_T[u', v']$ holds for some edge $e \in E$ and any two vertices $u', v' \in V$ if and only if $T^{(e)}(u') \neq T^{(e)}(v')$ or, equivalently, $\pi_0(T^{(e)}) = \{T^{(e)}(u'), T^{(e)}(v')\}$ holds.

In particular, one has

$$E_T[u, w] = E_T[u, v] \,\Delta\, E_T[v, w] \subseteq E_T[u, v] \cup E_T[v, w] \tag{1.1}$$

for any three vertices u, v, w of a tree $T = (V, E)$ (where $A \Delta B$ denotes, for any two sets A, B, their *symmetric difference* $A \cup B - A \cap B$) as $T^{(e)}(u) \neq T^{(e)}(v)$ holds for any edge $e \in E$ and any two vertices $u, v \in V$ if and only

$T^{(e)}(u) = T^{(e)}(w) \neq T^{(e)}(v)$ or $T^{(e)}(u) \neq T^{(e)}(w) = T^{(e)}(v)$ holds for any further vertex $w \in V$.

Further, a graph that, whether connected or not, at least contains no cycles is called a *forest*. Clearly, a graph $F = (V, E)$ is a forest if and only if the induced graph $F[U]$ is a tree for every connected component U of F and, hence, just as well for every connected subset U of V. Note that a graph $G = (V, E)$ is a forest if and only if every edge $e = \{u, v\} \in E$ is a bridge.

In the context of graphs and trees, we will also follow popular practice and freely use the term *network* instead of the term *graph*, in particular when referring to connected graphs that are not (necessarily) trees.

A surjective map $\psi : V \to U$ from the vertex set V of a graph $G = (V, E)$ onto another set U is called a *contracting map* (for G) if all subsets of V of the form $\psi^{-1}(u)$ $(u \in U)$ are connected. Clearly, given an equivalence relation $\sim$ on V, the canonical map $V \to V/\sim$ from V onto the set $V/\sim$ of $\sim$-equivalence classes is a contracting map if and only if all $\sim$-equivalence classes are connected.

Further, given a graph $G = (V, E)$ and a contracting map $\psi : V \to U$ for G, we denote by ψG the graph with vertex set U and edge set

$$\psi E := \{\{\psi(u), \psi(v)\} : \{u, v\} \in E, \psi(u) \neq \psi(v)\}.$$

Note that ψG is a tree whenever G is a tree T, and that the map

$$\psi_\star : V \to \{\star\}\dot{\cup}(V - V_{int}(G),) : v \mapsto \begin{cases} \star & \text{if } v \in V_{int}(G), \\ v & \text{otherwise}, \end{cases}$$

from V onto the disjoint union of the set $V - V_{int}(G)$ of leaves of G and just one additional element $\star$ not yet involved in G is a contracting map if and only if $V_{int}(G)$ is a connected subset of V. So, this holds in particular whenever G is connected in which case the resulting graph $\psi_\star G$ is a star tree.

We will say that a graph $G' = (V', E')$ results from a graph $G = (V, E)$ by the *contraction of an edge* $e \in E$ if $G' = \psi G$ holds for some contracting map $\psi : V \to V'$ that only contracts the edge e, i.e., for which all but one of the subsets of V of the form $\psi^{-1}(v')$ $(v' \in V')$ have cardinality 1 while the unique remaining subset of V of that form has cardinality 2 and is actually the edge e.

Clearly, such a contracting map exists for every edge $e \in E$. For example, the canonical map $\psi_e : V \to V/\sim_e$ from V onto the set $V/\sim_e$ of equivalence classes $V/\sim_e$ of V relative to the equivalence relation $\sim_e$ defined by

$$u \sim_e v \iff u = v \quad \text{or} \quad \{u, v\} = e$$

is a contracting map that just contracts e. And it is also obvious that every contracting map is a concatenation of such "elementary" contracting maps.

And if, even more specifically, e is a pendant edge containing exactly one interior vertex v, we can choose this vertex as a canonical representative of the $\sim_e$-equivalence class $e \subseteq V$ of v in $V/\sim_e$ and replace the map $\psi_e : V \to V/\sim_e$ by the map

$$\psi^e : V \to V - v : w \mapsto \begin{cases} v & \text{if } w \in e, \\ w & \text{else.} \end{cases}$$

Note that, denoting the unique leaf in e by u, the graph $\psi^e G$ resulting from contracting the edge e in this way coincides with the subgraph

$$G^e := (V - u, E - e)$$

of G obtained by eliminating the pendant edge e and the leaf u in e.

Next, we state (without proof) some well-known simple facts that we will use later in this book.

Lemma 1.2 (i) *Given any finite graph $G = (V, E)$, one has*

$$2|E| = |\{(v, e) \in V \times E : v \in e\}| = \sum_{i \geq 1} i \; |V^{(i)}|$$

where $V^{(i)}$ denotes, for every $i \in \mathbb{N}_{\geq 0}$, the set

$$V^{(i)} := \{v \in V : \deg(v) = i\}$$

of vertices of degree i in G.

(ii) *A finite graph $G = (V, E)$ is a tree if and only if it is connected and $|V| = |E| + 1$ holds.*

(iii) *For every finite tree $T = (V, E)$ with at least two vertices, one has $\sum_{i \geq 1}(2 - i)|V^{(i)}| = 2$ or, equivalently,*

$$|V^{(1)}| = 2 + |V^{(3)}| + 2|V^{(4)}| + 3|V^{(5)}| + \cdots$$

and, denoting by $V^{(i|j)}$ the set of vertices of degree i that are adjacent to exactly j leaves (which clearly is empty if $j < 0$ holds), one also has

$$\sum_{i>1} |V^{(i|i-1)}| = 2 + \sum_i |V^{(i|i-3)}| + 2\sum_i |V^{(i|i-4)}| + 3\sum_i |V^{(i|i-5)}| + \cdots$$

provided T is not a star tree: Indeed, if T is not a star tree, the graph $\partial T := (V_{int}(T), E_{int}(T))$ derived from T by deleting all of its leaves and pendant edges is a tree with at least two vertices, exactly $\sum_{i>1} |V^{(i|i-1)}|$ leaves, and, $\sum_i |V^{(i|i-j)}|$ vertices of degree j in ∂T.

In particular, if $T = (V, E)$ *is a binary tree with exactly* n *leaves and* $n \geq 2$ *holds, one has* $n = |V^{(1)}| = 2 + |V^{(3)}| = 2 + |V_{int}(T)|$, *and, therefore, also* $|V| = |V^{(1)}| + |V_{int}(T)| = n + (n-2) = 2n - 2$, $|E| = |V| - 1 = 2n - 3$, *and* $|E_{int}(T)| = |E| - n = n - 3$.

And if $n \geq 4$ *holds, the number* $|V^{(3|2)}|$ *of its cherries exceeds the number* $|V^{(3|0)}|$ *of those interior vertices all of whose neighbors are also interior vertices, by exactly 2.*

(iv) *Every finite tree with at least two vertices has at least two leaves, and every finite tree with at least five vertices and without any vertices of degree 2 contains at least two disjoint cherries (possibly attached to the same interior vertex).*

Next, we present two important operations to obtain trees from trees. The subtree of T *induced* by a subset $U \subseteq V$ consists of all those vertices and edges of T that lie on some path between two vertices in U. The *restriction* $T|_U$ of T to U — or *the tree obtained by restricting* T *to* U — is the tree $T' = (V|_U, E')$ with vertex set $V|_U$ consisting of those vertices $v \in V$ that are medians of vertices in U, while two such vertices $u, v \in V|_U$ form an edge of T' if the path $\mathbf{p}_T(u, v)$ does not contain any vertex in $V|_U - \{u, v\}$. If U denotes the union of the left-most and the right-most cherry in the tree T depicted in Figure 1.2, then the subtree of T induced by U contains exactly seven vertices while the restriction $T|_U$ of T to U contains six vertices.

A graph $G = (V, E)$ is called *bipartite* if there exists a partition of V into two subsets V_1 and V_2 such that every edge of G has an endpoint in both V_1 and V_2. Clearly, there exists at most one such bipartition if G is connected. It is also well known that bipartite graphs can be characterized by the property that they do not contain a cycle of odd length as a subgraph. From this characterization, it follows immediately that every tree, containing just no cycles as subgraphs, is bipartite. An important example of a bipartite graph that we will encounter later is a *hypercube*, i.e., a graph that is isomorphic, for some non-negative integer k, to the graph $H_k = (V_k, E_k)$ with vertex set $V_k = \{0, 1\}^k$ and edge set E_k consisting of those pairs $\{(x_1, \ldots, x_k), (y_1, \ldots, y_k)\}$ of vertices in $\{0, 1\}^k$ that differ at exactly one index $i \in \{1, 2, \ldots, k\}$ or, equivalently, for which $\sum_{i=1}^{k} |x_i - y_i| = 1$ holds. We will sometimes refer to H_k as the *k-dimensional hypercube* (cf. Figure 1.3(a)).

As a direct generalization of bipartite graphs, a graph $G = (V, E)$ is called *multi-partite* or, more specifically *k-partite* for some integer $k \geq 2$, if V can be partitioned into k disjoint subsets $V_1, \ldots, V_k$ such that no edge of G has both endpoints in the same subset. A k-partite graph $G = (V, E)$ is *complete*

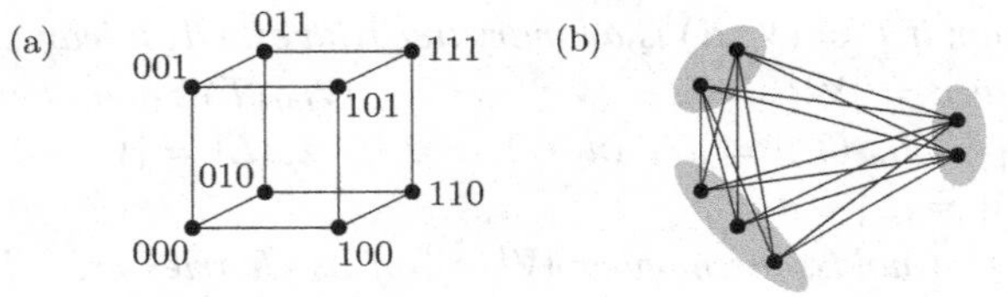

Figure 1.3 (a) The hypercube H_3. (b) A complete 3-partite graph.

if adding an arbitrary edge yields a graph that is not k-partite. An example of a complete 3-partite graph is given in Figure 1.3(b).

Occasionally, we will also use the concept of a *directed graph*, that is, a pair $G = (V, A)$ consisting of a non-empty set V and a subset $A \subseteq V \times V$. The elements in V are called the vertices of G and the elements in A the *arcs* of G. The *out-degree* of a vertex $v \in V$ is the number of arcs in A of the form (v, u), and the *in-degree* of v is the number of arcs in A of the form (u, v).

Furthermore, a sequence $v_0, v_1, \ldots, v_\ell$ of vertices of G is called a *directed path* in G if $(v_{i-1}, v_i) \in A$ holds for all $i = 1, \ldots, \ell$. The sequence is called a *(cyclically) closed directed path* if, in addition, $v_0 = v_\ell$, and a *directed cycle* if, furthermore, $|\{v_1, \ldots, v_\ell\}| = \ell$ holds.

Clearly, there exists, for any directed graph $G = (V, A)$, its *underlying (undirected) graph*

$$G^\circ = (V, A^\circ) := \left(V, \left\{\{u, v\} \in \binom{V}{2} : (u, v) \in A\right\}\right).$$

Whenever this cannot lead to confusion, we will freely use the terms introduced above also for directed graphs G presuming that they refer to the underlying undirected graph G°.

A directed graph (V, A) is called a (finite) *rooted tree* if the following holds:

(rt 1) There is no vertex $v \in V$ of in-degree larger than 1.

(rt 2) There is no pair u, v of vertices $u, v \in V$ with $(u, v) \in A$ and $(v, u) \in A$ — that is, (V, A) contains no loop and no directed cycle of length 2.

(rt 3) The associated undirected graph (V, A°) is a (finite) tree.

It is well-known folklore and easy to see that, given a finite rooted tree $T = (V, A)$, there exists precisely one vertex $\mathfrak{r} = \mathfrak{r}_T$ of in-degree 0, also called the *root* of T. Moreover, there exists a directed path from a vertex $u \in V$ to a vertex $v \in V$ if and only if one has $u \in V_{T^\circ}[\mathfrak{r}_T, v]$ in which case we will also write $u \preceq_T v$ so that, given any edge $e = \{u, v\}$ of the associated undirected tree $T^\circ = (V, A^\circ)$, the pair (u, v) forms an arc in A if and only if $u \preceq_T v$ holds.

In consequence, there is a canonical one-to-one correspondence between rooted and *augmented finite trees* — that is, pairs consisting of a finite tree and a distinguished vertex of that tree. This correspondence is given by associating, to any finite rooted tree $T = (V, A)$, the pair consisting of the associated undirected tree $T^\circ = (V, A^\circ)$ and the root $\mathfrak{r}_T$ of T.

We close this section by recalling some concepts and results that will be used in Chapter 6. Let V be an arbitrary finite set. Then, we may view the collection $\mathbb{P}(V)$ of all subsets of V as a vector space of dimension $|V|$ over the two-element field $\mathbb{F}_2 = \{0, 1\}$ — the sum of two vectors $V' \subseteq V$ and $V'' \subseteq V$ in $\mathbb{P}(V)$ being the symmetric difference $V' \Delta V''$ of V' and V''. The *cycle space* $\mathcal{C}(G)$ of a finite graph $G = (V, E)$ is defined as the subspace of the $\mathbb{F}_2$-space $\mathbb{P}(E)$ whose elements are all those subsets E' of E for which every vertex in the subgraph (V, E') has even degree. It is easy to check that $\mathcal{C}(G)$ is indeed a subspace of $\mathbb{P}(E)$, that the edge set of any cycle in G is contained in $\mathcal{C}(G)$, and that a subset E' of E is a vector in $\mathcal{C}(G)$ if and only if it is a sum of the edge sets of cycles in G. The dimension $c(G)$ of $\mathcal{C}(G)$ is called the *cyclomatic number* of G. We will also say that a set of cycles is *linearly (in)dependent* if the corresponding set of vectors in $\mathcal{C}(G)$ is linearly (in)dependent.

For later use, we state the following result, a concise proof of which can be found, for example, in [18].

Theorem 1.3 *Let $G = (V, E)$ be a graph with n vertices, m edges, and k connected components. Then, mapping each edge $e = \{u, v\}$, considered as a vector in $\mathbb{P}(E)$ (and, actually, a member of the canonical basis of $\mathbb{P}(E)$ consisting of all one-element subsets of E), onto the 2-subset $\{u, v\}$ considered as a vector in $\mathbb{P}(V)$, induces an $\mathbb{F}_2$-linear map from $\mathbb{P}(E)$ into $\mathbb{P}(V)$ whose kernel is $\mathcal{C}(G)$ while the co-kernel of this map can be identified with the $\mathbb{F}_2$-space $\mathbb{P}(\pi_0(G))$ associated with the set $\pi_0(G)$ consisting of all connected components of G. In particular, one has $k - c(G) = n - m$.*

1.3 Metric spaces

Given a set X, a (real-valued) *bivariate map* on X is a map $D : X \times X \to \mathbb{R}$ from the Cartesian product $X \times X$ of the set X with itself into the set $\mathbb{R}$ of real numbers. Such a map D is *symmetric* if $D(x, y) = D(y, x)$ holds for all $x, y \in X$. A symmetric bivariate map D on X is called a *dissimilarity* (on

X) if $D(x, x) \leq D(x, y)$ holds for all $x, y \in X$, and a *metric* (on X) if (i) $D(x, x) = 0$ holds for all $x \in X$, and (ii) the so-called *triangle inequality*

$$D(x, y) \leq D(x, z) + D(y, z) \tag{1.2}$$

holds for all $x, y, z \in X$. In this case, the pair (X, D) is also called a *metric space*. Further, D is called a *proper* metric (on X) if, in addition, $D(x, y) = 0$ implies $x = y$ for all $x, y \in X$ in which case the pair (X, D) is also called a *proper* metric space. Clearly, any bivariate map $D : X \times X \to \mathbb{R}$ that satisfies (i) and (ii) for all $x, y, z \in X$ is necessarily symmetric (put $z := x$ in (ii)) and a dissimilarity (put $y := x$ and $z := y$ in (ii)).

The point set X of a metric space (X, D) will also always be viewed as a topological space relative to the standard topology induced on X by D. A metric space (X', D') is called a *subspace* of a metric space (X, D) if X' is a subset of X and D' coincides with the *restriction* of D to $X' \times X'$ that is denoted by $D|_{X'}$. And a metric space (X, D) will be called *finite* if X is finite.

Next, given a metric space (X, D), we denote, for any two points $x, y \in X$, by $[x, y] = [x, y]_D := \{z \in X : D(x, y) = D(x, z) + D(z, y)\}$ the set of all elements z in X for which equality holds in (1.2). Clearly, we have

$$b \in [a, c]_D \quad \text{and} \quad c \in [a, d]_D \iff b \in [a, d]_D \quad \text{and} \quad c \in [b, d]_D \tag{1.3}$$

for any four points a, b, c, d in (X, D) as both assertions are equivalent to the assertion "$D(a, d) = D(a, b) + D(b, c) + D(c, d)$". So, we must also have $[a, b]_D \subseteq [a, c]_D$ for any three points a, b, c in a metric space (X, D) with $b \in [a, c]_D$.

Next, let us consider two metric spaces (X, D) and (X', D'). Two maps ρ and ρ' from (X, D) into (X', D') are called *homotopic* if there exists a continuous map $\tau : X \times [0, 1] \to X'$ such that $\tau(x, 0) = \rho(x)$ and $\tau(x, 1) = \rho'(x)$ holds for every $x \in X$. Such a map τ is called a *homotopy* from ρ to ρ'. An injective map $\iota : X \to X'$ is called an *isometric embedding* of (X, D) into (X', D') if $D(x, y) = D'(\iota(x), \iota(y))$ holds for all $x, y \in X$. And a metric space (X, D) is defined to be a *geodesic metric space* if, for all $x, y \in X$, there exists an isometric embedding $\iota : ([0, D(x, y)], L) \to (X, D)$ such that $\iota(0) = x$ and $\iota(D(x, y)) = y$ hold where L denotes the standard metric defined on the interval $[0, D(x, y)]$ by $L(a, b) := |a - b|$ for all $a, b \in [0, D(x, y)]$. Any such isometric embedding ι is called a *geodesic* from x to y.

A map $\kappa : X \to X'$ from a metric space (X, D) into a metric space (X', D') is a *contraction* if $D'(\kappa(x), \kappa(y)) \leq D(x, y)$ holds for all $x, y \in X$. Moreover, a subspace (X', D') of a metric space (X, D) is called a (metric) *retract* of (X, D) if there exists a surjective contraction κ of (X, D) to (X', D') and a homotopy $\tau : X \times [0, 1] \to X$ of the identity map on X to κ such that $\tau(x, t) = x$ holds for all $x \in X'$ and all $t \in [0, 1]$, and $D\left(\tau(x, t'), \tau(y, t')\right) \geq D\left(\tau(x, t''), \tau(y, t'')\right)$ for all $x, y \in X$ and $t', t'' \in [0, 1]$ with $t' \leq t''$. To emphasize that, for a given contraction $\kappa : X \to X'$, there exists a homotopy $\tau : X \times [0, 1] \to X$ of the identity map on X to κ so that κ and τ meet the requirements in this definition, we will say that (X', D') is a *retract of* (X, D) *relative to* κ. A metric space (X, D) is *contractible* if $(\{x\}, D|_{\{x\}})$ is a retract of (X, D) for every $x \in X$.

Metrics arise, in particular, from *weighted graphs* $G = (V, E, \omega)$, that is, graphs (V, E) together with an *edge-weight function* $\omega : E \to \mathbb{R}_{>0}$. The edge-weight function that maps every edge of a graph to 1 is denoted by $\mathbf{1}_E$. Clearly, any edge-weighting ω of a connected graph $G = (V, E)$ induces a metric $D_G = D_{(V,E,\omega)} : V \times V \to \mathbb{R}_{\geq 0}$ on V which is the (necessarily unique!) largest metric D that can be defined on V for which $D(u, v) \leq \omega(e)$ holds for every edge $e = \{u, v\} \in E$ with endpoints u and v. In case $G = (V, E)$ is just a connected graph for which no edge-weight function ω has been specified, we will sometimes denote by D_G the metric induced by G considered as a weighted graph with edge-weight function $\mathbf{1}_E$. A subgraph $G' = (V', E')$ of a connected graph $G = (V, E)$ is called an *isometric subgraph* of G if it is connected and the map $\iota : V' \to V : x \mapsto x$ is an isometric embedding of $(V', D_{G'})$ into (V, D_G). And the same terminology is used for the weighted subgraphs $G' = (V', E', \omega')$ of a connected weighted graph $G = (V, E, \omega)$, i.e., the weighted graphs $G' = (V', E', \omega')$ with $V' \subseteq V$, $E' \subseteq E$, and $\omega' = \omega|_{E'}$.

In the following chapters, we will also encounter *gated subsets* of metric spaces: A subset $Y \subseteq X$ of the point set X of a metric space (X, D) is dubbed a gated subset of X if there exists, for every x in X, some $x_Y \in Y$, called a *gate* of x in Y, such that $D(x, y) = D(x, x_Y) + D(x_Y, y)$ holds for all $y \in Y$. In the following lemma, we present three simple facts about gated subsets:

Lemma 1.4 *Let (X, D) be a proper metric space and Y a gated subset of X. Then the following holds.*

(i) *For every $x \in X$, there is a unique gate x_Y for x in Y.*
(ii) *Y is a* convex *subset of X, that is, $[y', y]_D \subseteq Y$ holds for all $y, y' \in Y$.*
(iii) *If Z is another gated subset of X, then $D\left(z_Y, z'_Y\right) = D(z, z')$ and $D(z, z_Y) = D\left(z', z'_Y\right)$ holds for all z, z' in Z that are of the form $z = y_Z$ and $z' = y'_Z$ for some $y, y' \in Y$.*

Proof (i) Suppose $g, g' \in Y$ are both gates for some $x \in X$. Then we have $D(x, g') = D(x, g) + D(g, g') = D(x, g') + 2D(g, g')$, which implies that

$D(g, g') = 0$ holds. Hence, since D is a proper metric, $g = g'$ must hold, as required.

(ii) Consider arbitrary elements $y, y' \in Y$ and $x \in X$ such that $D(y, x) + D(x, y') = D(y, y')$ holds and note that $D(y, y') = D(x, y) + D(x, y') = D(x, x_Y) + D(x_Y, y) + D(x, x_Y) + D(x_Y, y') \geq 2\, D(x, x_Y) + D(y, y') \geq D(y, y')$ and, therefore, $D(x, x_Y) = 0$ must hold. Hence, since D is a proper metric, $x = x_Y \in Y$ must hold, as required.

(iii) We leave the simple proof of this observation to the interested reader (see also [68] where a detailed proof is presented). □

Next, recall that medians can also be defined in metric spaces: Given an arbitrary metric space (X, D), an element m in X is called a *median* of some three elements $a, b, c \in X$ if m is contained in the intersection $Med_D(a, b, c) := [a, b] \cap [b, c] \cap [c, a]$. Clearly, one has

$$D(a, m) = \frac{1}{2}(D(a, b) + D(a, c) - D(b, c)) \tag{1.4}$$

whenever m is a median of three points $a, b, c \in X$.

Note also that, given any four points $a, b, c, d \in X$ such that there exists some point $m \in Med_D(a, b, c) \cap Med_D(a, b, d)$, one must have the following *4-point condition*

$$\begin{aligned} D(a, b) + D(c, d) &\leq D(a, m) + D(m, b) + D(c, m) + D(d, m) \\ &= D(a, c) + D(b, d) = D(a, d) + D(b, c) \end{aligned} \tag{1.5}$$

as well as (cf. (1.3)) $Med_D(m, c, d) \subseteq Med_D(a, c, d) \cap Med_D(b, c, d)$ and, hence, also

$$Med_D(m, c, d) = Med_D(a, c, d) = Med_D(b, c, d) \tag{1.6}$$

in case all of these sets have cardinality 1.

Medians do not need to exist, nor are they necessarily unique in case they do — that is, the set $Med_D(a, b, c)$ can have any cardinality. However, if medians exist and are unique for any three points $a, b, c \in X$, the space (X, D) is called a *median space*, and the unique element $m \in Med_D(a, b, c)$ is also denoted by $med(a, b, c) = med_D(a, b, c)$ and called *the* median of a, b, and c.

We conclude this section with the following observation that ensures us that medians in trees as introduced in Section 1.2 are also the medians in the associated metric spaces associated with a tree considered above:

Lemma 1.5 *Assume that $T = (V, E)$ is a tree, that $\omega : E \to \mathbb{R}_{>0}$ is an edge-weighting for T, and that $D_\omega := D_{(V,E,\omega)} : V \times V \to \mathbb{R}_{\geq 0}$ is the metric on V induced by the weighted graph (V, E, ω). Then, the following hold:*

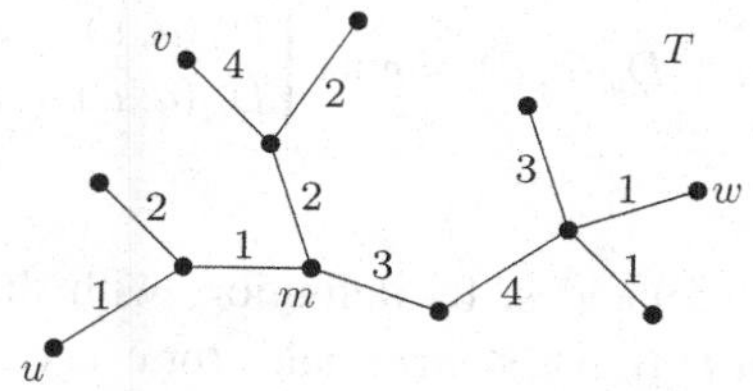

Figure 1.4 Vertex m is the median of vertices u, v, and w in the metric space induced by the weighted tree T.

(i) *One has* $D_\omega(u, v) = \sum_{e \in E_T[u,v]} \omega(e)$ *and, therefore, also* $[u, v]_{D_\omega} = V_T[u, v]$ *for any two vertices* $u, v \in V$.

(ii) *The metric space* (V, D_ω) *is a median space, and, for any three vertices* u, v, w *in* V*, the median* $med_{D_\omega}(u, v, w)$ *of* u, v, w *in* V *relative to* D_ω *coincides with the median* $med_T(u, v, w)$ *of* u, v, w *in* T.

(iii) *In particular, neither the set* $[u, v]_{D_\omega}$ *nor the vertex* $med_{D_\omega}(u, v, w)$ *depends on the specific edge-weighting* ω.

(iv) *Furthermore, a subset* U *of* V *is a gated subset of* V *relative to* D_ω *if and only if it is a convex subset of* V *relative to* D_ω *if and only if it is a connected subset of* V *(relative to* T*) in which case the gate* v_U *of any vertex* $v \in V$ *relative to* U *lies on every path of the form* $\mathbf{p}_T(v, u)$ *with* $u \in U$ *and is therefore, of course, the first vertex in* U *that lies on any such path, and the smallest connected subset of* V *containing* U *and* v *is the union of* U *and the set* $V_T[v, v_U]$.

(v) *In particular, given any three vertices* $a, b, c \in V$*, the set* $V_T[\{a, b, c\}] = V_T[a, b] \cup V_T[b, c] \cup V_T[c, a]$ *is, with* $m := med_T(a, b, c)$ *denoting their median, the union of the three sets* $V_T[a, m]$*,* $V_T[b, m]$*, and* $V_T[c, m]$ *(perhaps of cardinality* 1*) any two of which intersect in* m*, only (see Figure* 1.4*).*

(vi) *And given yet another vertex* $d \in V$*, one can relabel the vertices* a, b, c, d *in* V *so that* $m' := med_T(a, c, d)$ *coincides with* $med_T(b, c, d)$ *and, hence, also* $m = med_T(a, b, c)$ *with* $med_T(a, b, d)$ *(see* (1.6)*, cf. also Figure* 1.4*) in which case the set* $V_T[\{a, b, c, d\}]$ *is the union of the vertex sets* $V_T(a, m)$*,* $V_T(b, m)$*,* $V_T(c, m')$*,* $V_T(d, m')$*, and* $V_T(m, m')$ *(also perhaps of cardinality* 1*) of the five edge-disjoint paths* $\mathbf{p}_T(a, m)$*,* $\mathbf{p}_T(b, m)$*,* $\mathbf{p}_T(c, m')$*,* $\mathbf{p}_T(d, m')$*, and* $\mathbf{p}_T(m, m')$*, respectively. In particular, given any four vertices* $a, b, c, d \in V$*, the larger two of the three sums* $D_\omega(a, b) + D_\omega(c, d)$*,* $D_\omega(a, c) + D_\omega(b, d)$*, and* $D_\omega(a, d) + D_\omega(b, c)$ *must coincide in view of* (1.5)*, that is, the following "4-point condition"*

$$D_\omega(a,b) + D_\omega(c,d) \leq \max \begin{Bmatrix} D_\omega(a,c) + D_\omega(b,d) \\ D_\omega(a,d) + D_\omega(b,c) \end{Bmatrix} \tag{1.7}$$

must always hold.

Proof (i) Indeed, D_ω coincides, by definition, with the largest metric D' defined on V for which $D'(u,v) \leq \omega(e)$ holds for every edge $e = \{u,v\} \in E$ and hence, in view of the triangle inequality, with the largest metric D' defined on V for which $D'(u,v) \leq \sum_{e \in E_T[u,v]} \omega(e)$ holds for any two vertices $u, v \in V$. Thus, $D_\omega(u,v)$ must coincide with $\sum_{e \in E_T[u,v]} \omega(e)$ for all $u, v \in V$ as the map $V \times V \to \mathbb{R}_{\geq 0} : (u,v) \mapsto \sum_{e \in E_T[u,v]}$ is clearly a metric: By definition, it is symmetric and non-negative, and it satisfies the triangle inequality in view of (1.1). In particular, we have $D_\omega(u,v) + D_\omega(v,w) = D_\omega(u,w)$ for three vertices $u, v, w \in V$ if and only if $v \in V_T[u,w]$ holds.

(ii) It suffices to note that, given any three vertices u, v, w in V of a tree T, there exist a unique vertex $m = m_T(u,v,w) \in V$ in the intersection $V_T[u,v] \cap V_T[v,w] \cap V_T[w,u]$ (cf. Figure 1.4).

(iii) This is an obvious consequence of (i) and (ii).

(iv) It is obvious that, in a (weighted or unweighted) tree, a subset of the vertex set is connected if and only if it is convex (relative to the induced metric), and we have seen already above that any gated subset U of V must be convex and, hence, connected. So, $V_T[u,u'] \subseteq U$ must hold for any gated subset U of V and all $u, u' \in U$. Conversely, if this holds, if v is an arbitrary vertex in V, and if $u, u' \in U$ are two arbitrary vertices in U, the median $m := med_T(v,u,u')$ of v, u, u' must be — as a vertex in $V_T(u,u')$ — contained in U, implying that the first vertex in U that lies on the path $\mathbf{p}_T(v,u)$ from v to u as well as the first vertex in U that lies on the path $\mathbf{p}_T(v,u')$ from v to u' must lie on the path $\mathbf{p}_T(v,m)$ from v to m and that, therefore, these two vertices must coincide. So, this vertex must indeed be the gate v_U of v in U, and the smallest connected subset of V containing U and v must be the union of U and the vertex set $V_T[v,v_U]$ of the path from v to this vertex.

(v) This follows immediately from (iv) applied to, e.g., $v := c$ and $U := V_T[a,b]$.

(vi) And the last assertion follows from (v) and (iv) applied to, e.g., $v := d$ and $U := V_T[\{a,b,c\}]$: Indeed, putting $m := med_T(a,b,c)$ as above, we have $V_T[\{a,b,c\}] = V_T[a,m] \cup V_T[b,m] \cup V_T[c,m]$. So, without loss of generality, we may assume that the gate d_U is contained in $V_T[c,m] = V_T[c,a] \cap V_T[c,b]$ in which case this gate must be contained in

$$V_T[d,a] \cap V_T[d,b] \cap V_T[d,c] \cap V_T[c,a] \cap V_T[c,b]$$

and, hence, coincide with $m' = med_T(a,c,d)$ as well as with $med_T(b,c,d)$. The remaining assertions are immediate consequences. □

1.4 Computational complexity

Although not a central topic of this book, we will occasionally refer to concepts from the theory of algorithms and computational complexity. The reason for this is twofold. First, many conditions that characterize combinatorial objects of a certain type immediately give rise to algorithms that allow one to decide whether a given object is of the required type or not. Second, computational complexity theory offers ways to formalize the idea that it is "computationally hard" to decide whether a given object has a required property. This indicates, for example, that there are limits when it comes to combinatorially characterizing the class of objects satisfying this property. In the following, we aim at providing the reader with an intuitive understanding of the basic ideas relating to these concepts [79].

To measure the efficiency of an algorithm, we first need some way to quantify the *size of the input*. For specific problems it is usually quite clear what this size should be. For example, if the input to an algorithm is a graph $G = (V, E)$ then it is natural to view $|V| + |E|$ as a quantity that captures "how big" the graph is.

Now, for any function $g : \mathbb{N} \to \mathbb{R}_{>0}$, we say that an algorithm has *run time* $O(g(n))$ if the maximum number of steps in a run of the algorithm on an input of size n is in $O(g(n))$, that is, there exists some constant $b > 0$ such that, for all inputs of size n, the number of steps is at most $b\,g(n)$. In particular, we call a (deterministic) algorithm a *polynomial time* algorithm if its run time is in $O(n^k)$ for some constant $k > 0$. Note that many problems can be solved by polynomial time algorithms and these problems are usually thought of as those that can be solved efficiently on a computer. For example, there is a polynomial time algorithm that, given as input a graph G, outputs "yes" if G is connected and "no" if G is disconnected. Such problems where the output is either "yes" or "no" are called *decision problems* and the class of decision problems that can be solved by a polynomial time algorithm is denoted by P.

There are several decision problems for which so far no polynomial time algorithm has been found. A simple example of this is the CLIQUE problem. Here, the input consists of a graph $G = (V, E)$ together with an integer c, and the output is "yes" if there exists a *clique* of c vertices in G, that is, a subset $C \subseteq V$ of c vertices such that every pair of the form $\{u, v\}$ with $u, v \in C$ is an edge in G, and "no" if there is no such subset of V. However, even though we do not know whether or not CLIQUE belongs to the class P, we can at least check with a polynomial time algorithm whether a *given* subset $C \subseteq V$ of c vertices forms a clique in G or not. In particular, one could also view such a given subset C as a *certificate* that somebody hands in to show that the answer is "yes". Formalizing the concept of a certificate in a suitable way, the class NP consists of those decision problems which admit to check a proposed certificate using a polynomial time algorithm.

It is one of the big open problems in theoretical computer science whether the classes P and NP are the same or not. A very fruitful approach to better understand this problem has been to relate problems in the class NP by so-called *reductions*. That is, given two problems Π_1 and Π_2 in NP, we say that Π_1 *reduces* to Π_2 if there exists a polynomial time algorithm A that transforms every input I for Π_1 into an input $A(I)$ for Π_2 such that the answer for I is "yes" if and only if the answer for $A(I)$ is "yes".

The idea behind this is that the existence of such a reduction indicates that problem Π_2 is at least as hard as problem Π_1 or, more formally, if Π_1 does not belong to P, then also Π_2 cannot belong to P. A problem Π in NP is called NP-*complete* if every problem in NP reduces to Π. The problem CLIQUE mentioned above is known to be an example of an NP-complete problem. It is an intriguing situation that the existence of a polynomial time algorithm for *any* one of the very many problems that have been shown to be NP-complete to date would immediately show that $P = NP$ holds. Therefore, one value of showing that a particular decision problem is NP-complete lies in the fact that this shows that there may be fundamental obstacles to solving this problem efficiently, i.e., the fundamental obstacles that, so far, have prevented progress in dealing with any NP-complete problem. Nonetheless, there are many cases where slight variants of a given NP-complete problem are in P or where, even though a problem is known to be NP-complete, algorithms exist that can handle some practically relevant inputs for this problem quite efficiently. So, knowing that a problem is NP-complete does not at all mean that it does not make sense to try to develop good algorithms to tackle it anyway.

2
Encoding X-trees

In this chapter, we introduce the concept of an X-tree, one of the main objects of study in this book. In particular, after formally defining X-trees in the next section, we will show how they may be encoded using either one of three basic combinatorial entities: Splits, metrics, and quartets.

2.1 X-trees

We first formally define X-trees. To this end, consider a fixed finite non-empty set X of cardinality, say, n representing the set of species, genes, proteins, or whatever "Operational Taxonomic Units" (or "OTUs") one wants to subject to some phylogenetic analysis.

Definition 2.1 *An* X-tree $\mathcal{T} = (V, E, \varphi)$ *is a triple consisting of a finite set* V *called its vertex set, a set* $E \subseteq \binom{V}{2}$ *called its edge set, and a "labeling" map* $\varphi : X \to V$ *such that*

- *the graph* (V, E) *is a tree, called the* underlying tree *of* $\mathcal{T}$ *and denoted also by* $\underline{\mathcal{T}}$, *and*
- *the image* $\varphi(X)$ *contains* $V^{(1)} \cup V^{(2)}$, *the union of all vertices of* $\underline{\mathcal{T}}$ *of degree* 1 *and* 2.

An X*-tree* $\mathcal{T} = (V, E, \varphi)$ *will sometimes also be called just a* labeled tree *— or even just a tree when it is clear from the context that this refers to a labeled tree, and the labeling set* X *of a labeled tree* $\mathcal{T}$ *will also be called the* support *of* $\mathcal{T}$ *and denoted by* $supp(\mathcal{T})$.

An X*-tree* $\mathcal{T} = (V, E, \varphi)$ *will be called* simple *if* $X \subseteq V$ *holds and* φ *maps every* $x \in X$ *onto itself (so,* X *must contain all vertices of degree smaller than* 3 *in this case). It will be called a* phylogenetic X-tree *or, for short, just a*

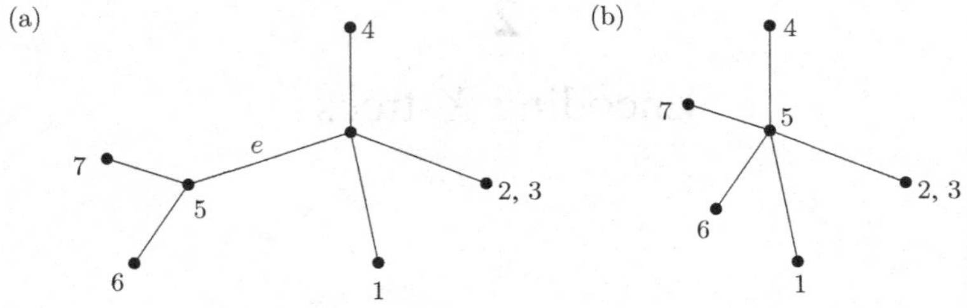

Figure 2.1 Two X-trees for $X := \langle 7 \rangle$.

phylogenetic tree, *if φ induces a bijection between X and the set $V^{(1)}$ of leaves of $\underline{T}$. Note that the support X of a simple phylogenetic X-tree is nothing but its set of leaves. As the map $\varphi = Id_X$ does not need to be specified in this case, we may as well omit it.*

Further, $\mathcal{T}$ will be called a binary X-tree *if it is a phylogenetic X-tree that, as a tree, is binary, i.e., every vertex in $V - V^{(1)}$ has degree 3. Clearly, it follows from Lemma 1.2 that $|E| \leq 2n - 3$ must hold for every X-tree $\mathcal{T} = (V, E, \varphi)$ and that equality holds if and only if $\mathcal{T}$ is binary.*

Two X-trees $\mathcal{T}_1 = (V_1, E_1, \varphi_1)$ and $\mathcal{T}_2 = (V_2, E_2, \varphi_2)$ are called isomorphic *if there exists a bijective map $\iota : V_1 \to V_2$, called an* (X-tree) isomorphism, *such that*

(i) $\{u, v\} \in E_1 \Leftrightarrow \{\iota(u), \iota(v)\} \in E_2$ *holds for all $u, v \in V_1$, and*
(ii) $\varphi_2(x) = \iota(\varphi_1(x))$ *holds for every $x \in X$.*

Below, we will also have to deal with "isomorphisms" between further structures that we encounter. We will do this then without first stating each time the obvious corresponding variant of the respectively pertinent definitions explicitly. It is easy to see (and we leave the simple, but instructive proof to the reader) that every vertex v in an X-tree $\mathcal{T} = (V, E, \varphi)$ is the median of three labeled vertices, i.e., three vertices in the image $\varphi(X)$ of X relative to the labeling map φ (as, in any finite tree T, any vertex of degree at least 3 is the median of three leaves) and that, if any two X-trees $\mathcal{T}_1$ and $\mathcal{T}_2$ are isomorphic, there exists exactly one X-tree isomorphism from $\mathcal{T}_1$ onto $\mathcal{T}_2$ which is then also called the *canonical* isomorphism from $\mathcal{T}_1$ onto $\mathcal{T}_2$. Some examples of X-trees are shown in Figure 2.1.

We will freely apply terminology developed in Chapter 1 for trees also to X-trees. For example, any vertex or edge of $\underline{T}$ is also called a vertex or an edge of $\mathcal{T}$, the path $\mathbf{p}_{\underline{\mathcal{T}}}(u, v)$ from a vertex u to a vertex v in $\underline{\mathcal{T}}$ will also be denoted by $\mathbf{p}_{\mathcal{T}}(u, v)$, the median $med_{\underline{\mathcal{T}}}(u, v, w)$ of any three vertices of $\mathcal{T}$ relative to $\underline{\mathcal{T}}$ will also be denoted by $med_{\mathcal{T}}(u, v, w)$, and so on. We will also refer to vertices of the form $\varphi(x)$ for some $x \in X$ by just referring to x if no misunderstanding can arise: For example, we will write $e_{\mathcal{T}}(x)$ for the unique edge e of $\underline{\mathcal{T}}$ that is incident to a leaf in $\underline{\mathcal{T}}$ of the form $\varphi(x)$, and $v_{\mathcal{T}}(x)$ for the unique vertex in $e_{\mathcal{T}}(x)$ distinct from x, we will write $\mathbf{p}_{\mathcal{T}}(x, v)$ for $\mathbf{p}_{\mathcal{T}}(\varphi(x), v)$ for all $x \in X$ and $v \in V$, we will write $med_{\mathcal{T}}(x, u, v)$ instead of $med_{\mathcal{T}}(\varphi(x), u, v)$ for all $x \in X$ and $u, v \in V$, two elements $x, y \in X$ will be said to form a cherry in $\mathcal{T}$ whenever the vertices $\varphi(x)$ and $\varphi(y)$ form a cherry in $\underline{\mathcal{T}}$, the unique edge $e \in E_v$ separating a vertex $v \in V$ from a vertex of the form $\varphi(x) \neq v$ for some $x \in X$ will be denoted by $e^{v \to x}$, and so on.

Also, a phylogenetic X-tree whose underlying tree is a star tree — i.e., a phylogenetic X-tree that has precisely one interior vertex (which then must necessarily be of degree n) — will be called a *phylogenetic star tree* (for X).

Further, given an X-tree $\mathcal{T} = (V, E, \varphi)$, contracting $T := \underline{\mathcal{T}}$ relative to a contracting map $\psi : V \to V'$ for T yields an X-tree $\mathcal{T}' := \psi \circ \mathcal{T} := (V', \psi E, \psi \circ \varphi)$. As for trees, we will say that $\mathcal{T}'$ is obtained from $\mathcal{T}$ by *contraction*. Conversely, we will also say that the X-tree $\mathcal{T}$ is obtained from an X-tree $\mathcal{T}'$ by *expansion* if $\mathcal{T}'$ is obtained from $\mathcal{T}$ by contraction. For example, consider the two $\langle 7 \rangle$-trees $\mathcal{T}$ and $\mathcal{T}'$ in Figure 2.1(a) and (b), respectively: $\mathcal{T}$ results from expanding $\mathcal{T}'$, and $\mathcal{T}'$ results from contracting the edge e in $\mathcal{T}$.

Similarly, given an X-tree $\mathcal{T} = (V, E, \varphi)$, we define the restriction $\mathcal{T}|_Y$ of $\mathcal{T}$ to some subset Y of X to be the Y-tree $\mathcal{T}' = (V', E', \varphi')$ whose underlying tree $\underline{\mathcal{T}'} = (V', E')$ is the restriction $\underline{\mathcal{T}}|_{\varphi(Y)}$ of the tree $\underline{\mathcal{T}}$ to $\varphi(Y)$ while its labeling map φ' is the restriction $\varphi|_Y$ of φ to Y. And we define a labeled tree $\mathcal{T}'$ to be displayed by $\mathcal{T}$ if its support $Y := supp(\mathcal{T}')$ is contained in X and the restriction $\mathcal{T}|_Y$ of $\mathcal{T}$ to Y is isomorphic to $\mathcal{T}'$. Clearly, the restriction $\mathcal{T}|_Y$ of a (binary) phylogenetic X-tree $\mathcal{T}$ to a subset Y of X must be a (binary) phylogenetic tree, too.

2.2 Encoding X-trees with splits

In this and the remaining sections of this chapter, we describe how to "encode" a given X-tree in terms of three basic combinatorial entities. The first entity that we now define is one of the most fundamental objects in phylogenetic combinatorics.

Definition 2.2 *As before, let X be a non-empty finite set of cardinality n. Every split of X will also be called an X-*split. *The set of all X-splits will be denoted by $\Sigma(X)$, the set of all trivial X-splits by $\Sigma_{triv}(X)$, and the set of all non-trivial X-splits by $\Sigma^*(X)$. Any subset Σ of $\Sigma(X)$ is called a* split system *(*over X*) or a* system of X-splits.

Remark 2.3 *Splits play such a fundamental role in phylogenetic combinatorics because, after all, the goal of phylogenetic analysis is to detect / identify the various phylogenetic "splits", the splits that separate, e.g., plants from animals, mammals from non-mammals, or Gram-positive from Gram-negative bacteria.*

We now describe how we can associate a specific split system $\Sigma_{\mathcal{T}}$ to any X-tree $\mathcal{T}$: Given an X-tree $\mathcal{T} = (V, E, \varphi)$ and an edge $e = \{u, v\} \in E$ of $\mathcal{T}$, we associate to e the split $S_e := \varphi^{-1}\big(\mathcal{T}^{(e)}(v)\big)|\varphi^{-1}\big(\mathcal{T}^{(e)}(u)\big)$. And to $\mathcal{T}$, we associate the split system $\Sigma_{\mathcal{T}} := \{S_e : e \in E\}$ consisting of all those splits. For example, the split associated with the edge e of the X-tree $\mathcal{T}$ in Figure 2.1(a) is $\{1, 2, 3, 4\}|\{5, 6, 7\}$, and the split system associated with that tree consists of the six splits

$$\{1\}|\{2, 3, 4, 5, 6, 7\}, \quad \{2, 3\}|\{1, 4, 5, 6, 7\}, \quad \{4\}|\{1, 2, 3, 5, 6, 7\},$$
$$\{6\}|\{1, 2, 3, 4, 5, 7\}, \quad \{7\}|\{1, 2, 3, 4, 5, 6\}, \quad \text{and} \quad \{1, 2, 3, 4\}|\{5, 6, 7\}.$$

Note that the requirement that every unlabeled vertex of an X-tree $\mathcal{T}$ has degree at least 3 implies that $S_e \neq S_{e'}$ holds for any two distinct edges e, e' of $\mathcal{T}$. In particular, $|E| = |\Sigma_{\mathcal{T}}|$ holds for every X-tree $\mathcal{T} = (V, E, \varphi)$.

Note also that e is an interior edge of $\mathcal{T}$ if and only if there exists, for every subset $A \in S_e$, a proper subset A' of A such that also $A'|(X - A')$ belongs to $\Sigma_{\mathcal{T}}$. That is, we can recognize whether e is a pendant or an interior edge of $\mathcal{T}$ by comparing the two sets in S_e with those in all the other splits in $\Sigma_{\mathcal{T}}$. And if e is a pendant edge in $\mathcal{T}$ and A is a subset in S_e such that no proper subset A' of A exists for which $A'|(X - A') \in \Sigma_{\mathcal{T}}$ holds, then φ must map every element $a \in A$ onto one and the same leaf of $\mathcal{T}$ — and actually the unique leaf in e in case V contains at least three elements.

In other words, the split system $\Sigma_{\mathcal{T}}$ "encodes" whether or not a given edge e is a pendant edge of $\mathcal{T}$. And it also encodes whether two elements $a, a' \in X$ will be mapped by φ onto one and the same leaf: One apparently has $\varphi(a) = \varphi(a') \in V^{(1)}$ if and only if there exists an inclusion-minimal subset A of X in the collection $\mathcal{C}(\Sigma) := \{A \subseteq X : A|(X - A) \in \Sigma\}$ of all parts of splits in Σ that contains both, a and a'. Clearly, considering each split $S = A|B$ as a 2-subset $S = \{A, B\}$ of the power set $\mathbb{P}(X)$ of X, we also have $\mathcal{C}(\Sigma) = \bigcup_{S \in \Sigma} S$.

More generally, the following result ensures that the split system associated to an X-tree $\mathcal{T}$ can indeed be viewed as a "loss-free encoding" of (the isomorphism class of) $\mathcal{T}$:

Theorem 2.4 *Two X-trees $\mathcal{T}_1$ and $\mathcal{T}_2$ are isomorphic if and only if the associated split systems $\Sigma_{\mathcal{T}_1}$ and $\Sigma_{\mathcal{T}_2}$ coincide.*

Proof Clearly, the split systems $\Sigma_{\mathcal{T}_1}$ and $\Sigma_{\mathcal{T}_2}$ coincide if the X-trees $\mathcal{T}_1$ and $\mathcal{T}_2$ are isomorphic. To prove the converse, we use induction with respect to the number of splits in the split system $\Sigma := \Sigma_{\mathcal{T}_1} = \Sigma_{\mathcal{T}_2}$: Indeed, our claim is obvious in case $\Sigma = \emptyset$. And if Σ contains only one split $A|B \in \Sigma(X)$, every X-tree $\mathcal{T} = (V, E, \varphi)$ with $\Sigma_{\mathcal{T}} = \Sigma = \{A|B\}$ must be isomorphic to the X-tree $(\{u, v\}, \{\{u, v\}\}, \varphi)$ that consists of the two vertices u and v, has the unique edge $e := \{u, v\}$, and a labeling map φ that maps all elements in A onto one, and all elements in B onto the other endpoint of e.

Now assume that $|\Sigma| > 1$ holds and consider two X-trees $\mathcal{T}_1 = (V_1, E_1, \varphi_1)$ and $\mathcal{T}_2 = (V_2, E_2, \varphi_2)$ for which $\Sigma = \Sigma_{\mathcal{T}_1} = \Sigma_{\mathcal{T}_2}$ holds. Note that this also implies that $|\Sigma| = |E_1| = |E_2| > 1$ and $|V_1| = |V_2| = |\Sigma| + 1 > 2$ must hold. Select a leaf a_1 in V_1 and an element $x \in X$ with $\varphi_1(x) = a_1$, let $e_1 = \{a_1, v_1\} := e_{\mathcal{T}_1}(a_1)$ denote the unique pendant edge in E_1 with $a_1 \in e_1$, let $S = A|B := S_{e_1}$ denote the split associated to e_1 in $\mathcal{T}_1$, and assume that, say, $x \in A$ and, therefore, $A = \varphi_1^{-1}(a_1)$ holds.

Clearly, as e_1 is a pendant edge of $\mathcal{T}_1$, the edge $e_2 = \{a_2, v_2\} \in E_2$ with $S_{e_2} = S$ must be a pendant edge of $\mathcal{T}_2$. So, we may assume, without loss of generality, that a_2 is that element in e_2 that is a leaf of $\mathcal{T}_2$ and that, in consequence, φ_2 maps all elements in A onto a_2.

Now, let

$$\mathcal{T}_i' = (V_i - a_i, E_i - e_i, \psi^{e_i} \circ \varphi_i)$$

denote, for $i = 1, 2$, the X-tree obtained from $\mathcal{T}_i$ by contracting the edge e_i relative to the contracting map $\psi^{e_i} : V_i \to V_i - a_i$ introduced in Chapter 1.

Then, putting $\varphi_i' := \psi^{e_i} \circ \varphi_i$ for $i = 1, 2$, we have $\varphi_i'(x) = \varphi_i(x)$ for all $x \in X - A$, and $\varphi_i'(a) = \psi^{e_i}(a_i) = v_i$ for all $a \in A$. Further, we have $\Sigma_{\mathcal{T}_1'} = \Sigma_{\mathcal{T}_2'} = \Sigma - S$ by construction. So, by induction, the X-trees $\mathcal{T}_1'$ and $\mathcal{T}_2'$ are isomorphic.

Let ι' denote the corresponding unique X-tree isomorphism from $\mathcal{T}_1'$ to $\mathcal{T}_2'$. Clearly, the fact that $v_2 = \varphi_2'(a) = \iota'(\varphi_1'(a)) = \iota'(v_1)$ holds for any $a \in A$ implies that $v_2 = \iota'(v_1)$ must hold. Thus, defining a map $\iota : V_1 \to V_2$ by putting $\iota(a_1) := a_2$ and $\iota(v) := \iota'(v)$ for all $v \in V_1 - a_1$, it can be checked easily that ι is indeed the required X-tree isomorphism from $\mathcal{T}_1$ to $\mathcal{T}_2$. We leave the details to the reader. □

2.3 Encoding X-trees with metrics

In this section, we will see that an alternative way of encoding an X-tree $\mathcal{T} = (V, E, \varphi)$ is given by the metric $D_{\mathcal{T}} : X \times X$ that is defined by $D_{\mathcal{T}}(x, y) = D_{\underline{\mathcal{T}}}(\varphi(x), \varphi(y))$ for all $x, y \in X$, that is, (x, y) is mapped to the length of the path $\mathbf{p}_{\mathcal{T}}(x, y)$ from $\varphi(x)$ to $\varphi(y)$ in $\mathcal{T}$. Note that, given any X-tree $\mathcal{T} = (V, E, \varphi)$, there are two metrics associated with $\mathcal{T}$, the metric $D_{\underline{\mathcal{T}}}$ defined on V and the metric $D_{\mathcal{T}}$ defined on X which, though closely related, should always be clearly distinguished from each other and that, while the metric $D_{\underline{\mathcal{T}}}$ is always proper, the metric $D_{\mathcal{T}}$ need not be a proper metric on X. For example, we have $D_{\mathcal{T}}(4, 6) = 3$ and $D_{\mathcal{T}}(2, 3) = 0$ for the metric $D_{\mathcal{T}}$ defined on $X := \langle 7 \rangle$ by the X-tree $\mathcal{T}$ displayed in Figure 2.1(a).

It is easily seen that "$\varphi(x) = \varphi(y) \Leftrightarrow D_{\mathcal{T}}(x, y) = 0$" and "$\{\varphi(x), \varphi(y)\} \in E \Leftrightarrow D_{\mathcal{T}}(x, y) = 1$" holds, for every X-tree $\mathcal{T} = (V, E, \varphi)$, for all $x, y \in X$. Further, an element $x \in X$ is mapped by φ onto a leaf of $\mathcal{T}$ if and only if $D_{\mathcal{T}}(a, b) = D_{\mathcal{T}}(a, x) + D_{\mathcal{T}}(x, b)$ implies $D_{\mathcal{T}}(a, x) = 0$ or $D_{\mathcal{T}}(x, b) = 0$ for all $a, b \in X$.

The following theorem shows that $\mathcal{T}$ is completely determined, up to isomorphism, by $D_{\mathcal{T}}$ implying that metrics, too, provide a way to encode X-trees.

Theorem 2.5 *Two X-trees $\mathcal{T}_1$ and $\mathcal{T}_2$ are isomorphic if and only if the associated metrics $D_{\mathcal{T}_1}$ and $D_{\mathcal{T}_2}$ coincide.*

Proof Clearly, if two X-trees are isomorphic, the associated metrics must coincide. Conversely, suppose that $\mathcal{T}_1 = (V_1, E_1, \varphi_1)$ and $\mathcal{T}_2 = (V_2, E_2, \varphi_2)$ are two X-trees such that $D_{\mathcal{T}_1} = D_{\mathcal{T}_2} =: D$ holds. We want to show that $\mathcal{T}_1$ must then be isomorphic to $\mathcal{T}_2$.

If $D(x, y) = 0$ held for all $x, y \in X$, then $\mathcal{T}_1$ and $\mathcal{T}_2$ would both be isomorphic to the X-tree $\mathcal{T} = (\{v\}, \emptyset, \varphi)$ that consists of a single vertex v, contains no edge, and has a labeling map that maps every $x \in X$ onto this single element v. So, we may assume from now on that $D(x, y) \neq 0$ holds for some $x, y \in X$ and that, therefore, also $|E_1|, |E_2| > 0$ holds.

Next, choose an arbitrary leaf $a_1 \in V_1$ of $\mathcal{T}_1$ and some $x \in X$ with $a_1 := \varphi_1(x)$, put $a_2 := \varphi_2(x)$, and note that, in view of the above remarks, a_2 must be a leaf of $\mathcal{T}_2$.

Now, for $i = 1, 2$, let $e_i := e_{\mathcal{T}_i}(a_i)$ denote the unique edge in E_i that is incident with a_i, and consider, just as in Theorem 2.4, the contraction

$$\mathcal{T}_i' = (T_i', \varphi_i') := (V_i - a_i, E_i - e_i, \psi^{e_i} \circ \varphi_i)$$

obtained from $\mathcal{T}_i$ by contracting the edge e_i relative to the contracting map $\psi^{e_i} : V_i \to V_i - a_i$. Clearly, denoting by D_i' the metric $D_{\mathcal{T}_i'}$ induced by $\mathcal{T}_i'$, we see that $D_i'(y,z)$ must coincide with $D_i(y,z)$ for all $y, z \in X - \varphi_i^{-1}(a_i)$, $D_i(y,z) = 0$ must hold for all $y, z \in \varphi^{-1}(a_i)$, and $D_i'(y,z) = D_i(y,z) - 1$ for all $y \in \varphi_i^{-1}(a_i)$ and $z \in X - \varphi_i^{-1}(a_i)$. So, we must have $D_1' = D_2'$ implying, by induction relative to, say, the total number $|E_1| + |E_2|$ of edges in $\mathcal{T}_1$ and $\mathcal{T}_2$, that a (necessarily unique) X-tree isomorphism ι' from $\mathcal{T}_1'$ to $\mathcal{T}_2'$ must exist.

Clearly, we have $v_2 = \iota'(v_1)$ in view of $v_2 = \varphi_2'(x) = \iota'\left(\varphi_1'(x)\right) = \iota'(v_1)$. Thus, defining — as in Theorem 2.4 — a map $\iota : V_1 \to V_2$ by $\iota(a_1) := a_2$ and $\iota(v) := \iota'(v)$ for all $v \in V_1 - a_1$, it can be checked easily that ι provides the required X-tree isomorphism from $\mathcal{T}_1$ to $\mathcal{T}_2$. We again leave the (rather instructive) details to the reader. □

2.4 Encoding X-trees with quartets

A third way to encode an X-tree $\mathcal{T}$ is by recording the shape of the subtrees that $\mathcal{T}$ induces on the 4-subsets of X. To make this more precise, we define the next key concept in phylogenetic combinatorics:

Definition 2.6 *Given some elements $a_1, a_2, b_1, b_2 \in X$, we will use the shorthand "$a_1a_2|b_1b_2$" to denote the unordered pair $\{\{a_1,a_2\},\{b_1,b_2\}\}$ of the two subsets $\{a_1,a_2\}$ and $\{b_1,b_2\}$. In case the four elements a_1, a_2, b_1, b_2 are all distinct, this "pair of pairs" will also be called a* quartet *or, more specifically, a quartet on X. In other words, a quartet q on X is a partition of a 4-subset Y of X — also dubbed the* support *of q and denoted by* supp(q) *— into two disjoint 2-subsets. Clearly, one has $a_1a_2|b_1b_2 = a_2a_1|b_1b_2 = b_1b_2|a_1a_2$ and so on, while there are exactly three distinct quartets with support Y: $a_1a_2|b_1b_2$, $a_1b_1|a_2b_2$, and $a_1b_2|a_2b_1$.*

We denote the set of all quartets on X by $\mathcal{Q}(X)$. Note that $|\mathcal{Q}(X)| = 3\binom{n}{4} = \frac{n(n-1)(n-2)(n-3)}{8}$ holds for every n-set X. Any subset $\mathcal{Q}$ of $\mathcal{Q}(X)$ is called a *quartet system* over X.

To see how to encode an X-tree in terms of quartet systems, let $\mathcal{T} = (V, E, \varphi)$ be an X-tree. We define a quartet $a_1a_2|b_1b_2 \in \mathcal{Q}(X)$ to be *displayed by $\mathcal{T}$* if the paths $\mathbf{p}_{\mathcal{T}}(a_1,a_2)$ and $\mathbf{p}_{\mathcal{T}}(b_1,b_2)$ have no vertex in common — or, equivalently, if some edge $e \in E$ exists that separates $\varphi(a_1)$ and $\varphi(a_2)$ from $\varphi(b_1)$ and $\varphi(b_2)$. And we denote the collection of all quartets in $\mathcal{Q}(X)$ that are displayed by $\mathcal{T}$ by $\mathcal{Q}_{\mathcal{T}}$.

Clearly, a quartet $q \in \mathcal{Q}(X)$ with support Y is displayed by $\mathcal{T}$ if and only if q is contained in the split system $\Sigma_{\mathcal{T}'}$ of the tree $\mathcal{T}' := \mathcal{T}|_Y$ obtained by restricting $\mathcal{T}$ to the support Y of q, in which case q must actually coincide with the only non-trivial split in $\Sigma_{\mathcal{T}'}$ or, equivalently, if and only if $D_{\mathcal{T}}$

$(a_1, a_2) + D_{\mathcal{T}}(b_1, b_2) < D_{\mathcal{T}}(a_1, b_1) + D_{\mathcal{T}}(a_2, b_2)$ and, therefore, also $D_{\mathcal{T}}(a_1, a_2) + D_{\mathcal{T}}(b_1, b_2) < D_{\mathcal{T}}(a_1, b_1) + D_{\mathcal{T}}(a_2, b_2) = D_{\mathcal{T}}(a_1, b_2) + D_{\mathcal{T}}(a_2, b_1)$ (cf. Assertion (vi) in Lemma 1.5) holds for the metric $D_{\mathcal{T}}$. In particular, given a 4-subset Y of X and a quartet $q \in \mathcal{Q}(Y)$ that is displayed by $\mathcal{T}$, then it is the only quartet in $\mathcal{Q}_{\mathcal{T}}$ with support Y — that is, at most one of the three quartets in $\mathcal{Q}(Y)$ can be contained in $\mathcal{Q}_{\mathcal{T}}$.

For example, the quartet $12|67$ is displayed by the X-tree $\mathcal{T}$ in Figure 2.1(a), but neither $16|27$ nor $17|26$.

Clearly, given any 4-subset Y of X and a quartet q with support Y, there can be various non-isomorphic Y-trees that display q. Yet, there is, up to isomorphism, only one phylogenetic Y-tree that displays q, henceforth denoted by $\mathcal{T}_q$. So, if $\mathcal{T}$ is a phylogenetic X-tree, q is displayed by $\mathcal{T}$ if and only if $\mathcal{T}|_Y$ is isomorphic to $\mathcal{T}_q$.

Next, we denote by $\overline{\mathcal{Q}}_{\mathcal{T}}$ the collection of all 4-subsets Y of X with $\mathcal{Q}(Y) \cap \mathcal{Q}_{\mathcal{T}} = \emptyset$. Clearly, given any subset Z of X, one has $\mathcal{Q}_{\mathcal{T}|_Z} = \mathcal{Q}(Z) \cap \mathcal{Q}_{\mathcal{T}}$, i.e., the quartet system associated to the Z-tree $\mathcal{T}|_Z$ obtained by restricting $\mathcal{T}$ to Z consists of all quartets in $\mathcal{Q}_{\mathcal{T}}$ whose support is contained in Z: Indeed, this follows immediately from the fact that $\mathcal{T}|_Y = (\mathcal{T}|_Z)|_Y$ holds for every 4-subset Y of Z.

As noted above, given any 4-subset Y of X, $|\mathcal{Q}(Y) \cap \mathcal{Q}_{\mathcal{T}}| \leq 1$ must always hold, and it is obvious that $\mathcal{T}$ is binary if and only if $|\mathcal{Q}(Y) \cap \mathcal{Q}_{\mathcal{T}}| = 1$ holds for every 4-subset Y of X and, therefore, if and only if $|\mathcal{Q}_{\mathcal{T}}| = \binom{n}{4}$ holds. Furthermore, $|\mathcal{Q}(Y) \cap \mathcal{Q}_{\mathcal{T}}| = 0$ holds for some 4-subset $Y = \{a, b, c, d\}$ of X if and only if there exists a (necessarily unique) vertex $v = v_{\mathcal{T}}(Y)$ that is the median of any three of the four vertices $\varphi(a), \varphi(b), \varphi(c), \varphi(d)$. Note also that, given a 4-subset $Y = \{a, b, c, d\}$ of X and an interior vertex v of a phylogenetic X-tree $\mathcal{T}$, one has $Y \in \overline{\mathcal{Q}}_{\mathcal{T}}$ and $v = v_{\mathcal{T}}(Y)$ if and only if the four edges $e^{v \to a}, e^{v \to b}, e^{v \to c}, e^{v \to d} \in E_v$ that separate v from those four vertices $\varphi(a), \varphi(b), \varphi(c)$, and $\varphi(d)$, respectively, are all distinct.

The following theorem states that quartets can also be used to encode phylogenetic X-trees:

Theorem 2.7 *Two phylogenetic X-trees $\mathcal{T}_1$ and $\mathcal{T}_2$ are isomorphic if and only if the associated quartet systems $\mathcal{Q}_{\mathcal{T}_1}$ and $\mathcal{Q}_{\mathcal{T}_2}$ coincide.*

Proof Clearly, if any two X-trees — whether phylogenetic or not — are isomorphic, then the associated quartet systems must be equal.

To show that the converse holds for phylogenetic X-trees, assume — without loss of generality — that $n \geq 4$ holds. Note also that, by induction, it suffices to show that two simple phylogenetic X-trees $\mathcal{T}_1$ and $\mathcal{T}_2$ must be isomorphic if we can find some element $a \in X$ such that the quartet systems

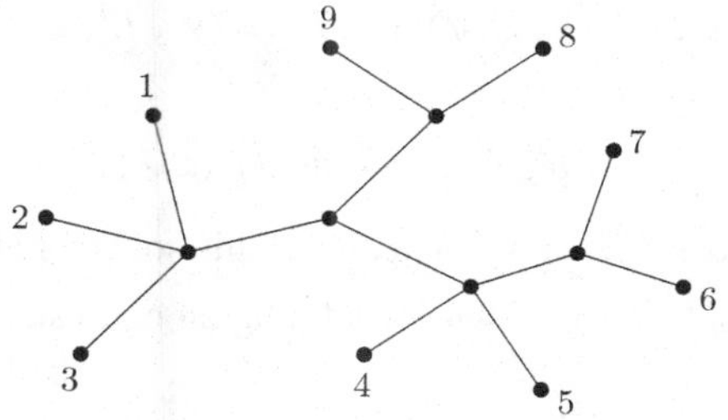

Figure 2.2 An example of a phylogenetic $\langle 9\rangle$-tree $\mathcal{T}$ used in the proof of Theorem 2.7.

$\mathcal{Q}_1(a) := \{q \in \mathcal{Q}_{\mathcal{T}_1} : a \in supp(q)\}$ and $\mathcal{Q}_2(a) := \{q \in \mathcal{Q}_{\mathcal{T}_2} : a \in supp(q)\}$ and the restrictions $\mathcal{T}_1|_{X-a}$ and $\mathcal{T}_2|_{X-a}$ coincide.

Actually, while it follows from Theorem 2.7 that this holds for any element $a \in X$, the proof is particularly simple in case a is part of a cherry in $\mathcal{T}_1$ or, equivalently, in $\mathcal{T}_2$. Indeed, given any phylogenetic X-tree $\mathcal{T} = (V, E, \varphi)$, two elements $a, a' \in X$ form a cherry in $\mathcal{T}$ if and only if there is no quartet in $\mathcal{Q}_{\mathcal{T}}$ of the form $ab|a'b'$, and they form a cherry that is attached to a vertex of degree 3 if and only if $aa'|bb' \in \mathcal{Q}_{\mathcal{T}}$ holds for any two distinct elements $b, b' \in X - \{a, a'\}$. For example, the pair 8, 9 in the phylogenetic X-tree in Figure 2.2 forms such a cherry as $89|bb' \in \mathcal{Q}_{\mathcal{T}}$ holds for any two distinct elements $b, b' \in \{1, 2, \ldots, 7\}$ while the pair 4, 5 does not form such a cherry in view of, e.g., $45|16 \notin \mathcal{Q}_{\mathcal{T}}$.

Thus, given an element $a \in X$ that is part of a cherry $a, a' \in X$ in $\mathcal{T}$, the interior vertex $v_{\mathcal{T}}(a)$ of $\mathcal{T}$ to which the pendant edge $e_{\mathcal{T}}(a)$ in $\mathcal{T}$ containing a is attached coincides with the interior vertex $v_{\mathcal{T}'}(a')$ of the $(X - a)$-tree $\mathcal{T}' := \mathcal{T}|_{X-a}$ to which the pendant edge $e_{\mathcal{T}'}(a')$ in $\mathcal{T}'$ containing a' is attached if and only if there exist two distinct elements $b, b' \in X$ with $aa'|bb' \notin \mathcal{Q}_{\mathcal{T}}$ in which case $\mathcal{T}$ can be obtained from the tree $\mathcal{T}'$ by just attaching a by a single pendant edge to the vertex $v_{\mathcal{T}'}(a')$.

Otherwise, $\mathcal{T}$ can be obtained from the tree $\mathcal{T}'$ by adding the element $a \in X$ and an additional vertex u to the vertex set V' of $\mathcal{T}'$, replacing the edge $\{a', v_{\mathcal{T}'}(a')\}$ of $\mathcal{T}'$ by three edges $\{a, u\}$, $\{a', u\}$, and $\{u, v_{\mathcal{T}'}(a')\}$.

So, whether or not such an element $a \in X$ that is part of a cherry $a, a' \in X$ in $\mathcal{T}$ is attached to a vertex of degree 3 or larger than 3, we can reconstruct $\mathcal{T}$ — up to isomorphism — from $\mathcal{T}'$ and the set of quartets in $\mathcal{Q}_{\mathcal{T}}$ whose support contains a. □

Of course, there are still further ways of encoding X-trees in terms of other data. For example, it is a simple (and worthwhile) exercise to show that two phylogenetic X-trees $\mathcal{T} = (V, E, \varphi)$ and $\mathcal{T}' = (V', E', \varphi')$ are isomorphic if and only if

$$med_{\mathcal{T}}(a, b, c) = med_{\mathcal{T}}(a', b', c') \iff med_{\mathcal{T}'}(a, b, c) = med_{\mathcal{T}'}(a', b', c')$$

holds for all a, b, c, a', b', c' in X if and only if this holds for all a, b, c, a', b', c' in X with $a = a'$ and $b = b'$. That is, defining an equivalence relation $\stackrel{\mathcal{T}}{\equiv}$ on $\binom{X}{3}$ by putting

$$\{a, b, c\} \stackrel{\mathcal{T}}{\equiv} \{a', b', c'\} \iff med_{\mathcal{T}}(a, b, c) = med_{\mathcal{T}}(a', b', c'),$$

two phylogenetic X-trees $\mathcal{T}$ and $\mathcal{T}'$ are isomorphic if and only if the two equivalence relations $\stackrel{\mathcal{T}}{\equiv}$ and $\stackrel{\mathcal{T}'}{\equiv}$ coincide.

And the same holds if and only if the collection $\mathbf{M}(\mathcal{T})$ of subsets $\mathcal{L}$ of $\binom{X}{2}$ for which there is no non-zero-map $\omega \in \mathbb{R}^E$ such that $D_{(\mathcal{T},\omega)}(a, b) := \sum_{e \in E_{\mathcal{T}}[a,b]} \omega(e)$ vanishes for all $\{a, b\} \in \mathcal{L}$ coincides with the corresponding collection $\mathbf{M}(\mathcal{T}')$.

In the next chapter, however, rather than dwelling on more such facts (interesting or even intriguing as they may be), we will investigate the more fundamental problem of specifying those combinatorial properties that characterize the split or quartet systems or metrics that actually encode X-trees.

3

Consistency of X-tree encodings

As before, we will always denote by X a fixed finite non-empty set of cardinality n representing the set of items one may want to subject to phylogenetic analysis and to which, therefore, our concepts, constructions, and conclusions may be applied. In the previous chapter, we have seen that X-trees can be encoded by split systems, metrics, or quartet systems. However, not all split systems, metrics, or quartet systems encode an X-tree. Indeed, it is not hard to find examples of such objects that are not encodings of some X-tree. For example, there is no X-tree $\mathcal{T}$ for $X := \{a, b, c, d, e\}$ with $\Sigma_{\mathcal{T}} = \{ab|cde, ac|bde\}$. In this chapter, we explore how those split systems, metrics, or quartet systems that are encodings of X-trees can be characterized.

3.1 The 4-point condition

As we have seen in the introduction, the edges of phylogenetic trees in general come with a *weight* assigned to them (for example, a number that is proportional to genetic distance). It is easy to formally take account of this fact: We just define a *weighted* X-tree $\mathcal{T}$ to be a quadruple (V, E, ω, φ) such that (V, E, φ) is an X-tree and $\omega : E \to \mathbb{R}_{>0}$ is an edge-weighting — implying that the triple $\underline{\mathcal{T}} := (V, E, \omega)$ is a weighted graph called the underlying weighted tree of $\mathcal{T}$. An example of such a tree is given in Figure 3.1.

Generalizing the corresponding construction for unweighted X-trees, any weighted X-tree $\mathcal{T} = (V, E, \omega, \varphi)$ induces a metric $D_{\mathcal{T}}$ on X defined by putting $D_{\mathcal{T}}(x, y) := D_{\underline{\mathcal{T}}}\big(\varphi(x), \varphi(y)\big)$ for all $x, y \in X$. Of course, if ω coincides with the map $\mathbf{1}_E$ that assigns the weight 1 to every edge of $\mathcal{T}$, then we

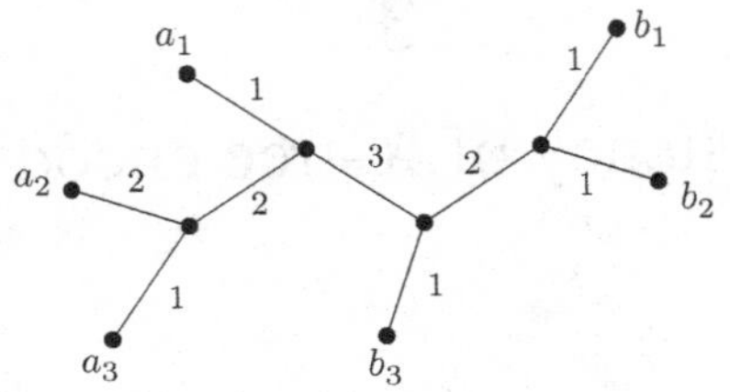

Figure 3.1 A weighted X-tree for $X = \{a_1, a_2, a_3, b_1, b_2, b_3\}$.

obtain the metric $D_{\mathcal{T}}$ associated to the unweighted X-tree (V, E, φ), as defined in the previous chapter.[1]

Before we present a characterization of those metrics on X that are induced by weighted X-trees, we first want to show that any weighted X-tree $\mathcal{T} = (V, E, \omega, \varphi)$ is also uniquely determined, up to canonical isomorphism, by the metric $D_{\mathcal{T}}$ associated to it. To this end, note that, given a weighted X-tree $\mathcal{T} = (V, E, \omega, \varphi)$, the following holds:

($\mathcal{T}$-i) One has "$\varphi(x) = \varphi(y) \iff D_{\mathcal{T}}(x, y) = 0$" for any two elements $x, y \in X$.

($\mathcal{T}$-ii) An element $x \in X$ is mapped by φ onto a leaf of $\mathcal{T}$ if and only if the *eccentricity*

$$exc_{\mathcal{T}}(x|X) := \min\{D_{\mathcal{T}}(x, a) + D_{\mathcal{T}}(x, b) - D_{\mathcal{T}}(a, b) : \\ a, b \in X,\ \varphi(a), \varphi(b) \neq \varphi(x)\}$$

of x relative to X and $\mathcal{T}$ is positive.

($\mathcal{T}$-iii) If this is the case, $exc_{\mathcal{T}}(x|X)$ coincides with twice the ω-value $\omega(e)$ of the unique edge $e = e_x \in E$ containing $\varphi(x)$, and the unique vertex $v \in V$ contained in e with $v \neq \varphi(x)$ coincides, for all $a, b \in X$ with $exc_{\mathcal{T}}(x|X) = D_{\mathcal{T}}(x, a) + D_{\mathcal{T}}(x, b) - D_{\mathcal{T}}(a, b)$, with the median $med_{(V,E,\varphi)}(x, a, b)$ of x, a, b relative to the X-tree (V, E, φ).

So, to show that the weighted X-tree $\mathcal{T} = (V, E, \omega, \varphi)$ is uniquely determined, up to canonical isomorphism, by the associated metric $D_{\mathcal{T}}$, we can now proceed exactly as in the proof of Theorem 2.5 (where we addressed only the special case that the edge-weight function is $\mathbf{1}_E$): That is, supposing that

[1] Note that, as in the case of unweighted X-trees, there are two closely related, yet distinct metrics associated with any given weighted X-tree $\mathcal{T} = (V, E, \omega, \varphi)$ that need to be clearly distinguished from each other, the metric $D_{\underline{\mathcal{T}}}$ defined on V by the underlying weighted tree $\underline{\mathcal{T}}$ and the metric $D_{\mathcal{T}}$ defined on X by $\mathcal{T}$.

$\mathcal{T}_1 = (V_1, E_1, \omega_1, \varphi_1)$ and $\mathcal{T}_2 = (V_2, E_2, \omega_2, \varphi_2)$ are two X-trees such that $D_{\mathcal{T}_1} = D_{\mathcal{T}_2} =: D$ holds, we show that $\mathcal{T}_1$ must then be isomorphic to $\mathcal{T}_2$ by

(i) choosing some leaf $v_1 \in V_1$ of $\mathcal{T}_1$ and some $x \in X$ with $v_1 := \varphi_1(x)$,
(ii) putting $v_2 := \varphi_2(x)$,
(iii) noting that, in view of $exc_{\mathcal{T}_1}(x|X) = \min_{a,b \in X; \varphi(a), \varphi(b) \neq \varphi(x)} \big(D(x,a) + D(x,b) - D(a,b)\big) = exc_{\mathcal{T}_2}(x|X)$, v_2 must also be a leaf (of $\mathcal{T}_2$),
(iv) and that, denoting the unique edge in E_i that is incident with v_i by $e_i = \{u_i, v_i\}$, one must have $\omega_1(e_1) = \omega_2(e_2)$ in view of ($\mathcal{T}$-iii).

So, just as before, one can consider the two trees $\mathcal{T}_i' = \big(V_i', E_i', \omega_i', \varphi_i'\big)$ defined by putting $V_i' := V_i - v_i$, $E_i' := E_i - e_i$, $\varphi_i' := \psi^{e_i} \circ \varphi_i$, and $\omega_i' := \omega_i|_{E_i - e_i}$, and note that both induce the same metric D' and must therefore, by induction, be canonically isomorphic which, again just as before, implies that also $\mathcal{T}_1$ and $\mathcal{T}_2$ must be canonically isomorphic.

We now turn to the characterization of those metrics that are induced by weighted X-trees. We define a metric D on X to be *treelike* if there exists a weighted X-tree $\mathcal{T} = (V, E, \omega, \varphi)$ with $D = D_{\mathcal{T}}$. It follows immediately from the 4-point condition stated in Equation (1.7) that

$$D_{\mathcal{T}}(a,b) + D_{\mathcal{T}}(c,d) \leq \max\{D_{\mathcal{T}}(a,c) + D_{\mathcal{T}}(b,d), D_{\mathcal{T}}(a,d) + D_{\mathcal{T}}(b,c)\}$$

must also hold, for the metric $D_{\mathcal{T}}$ induced by a weighted X-tree $\mathcal{T}$, for every four elements $a, b, c, d \in X$. That is, the two larger ones of the three sums

$$D_{\mathcal{T}}(a,b) + D_{\mathcal{T}}(c,d), \ D_{\mathcal{T}}(a,c) + D_{\mathcal{T}}(b,d), \text{ and } D_{\mathcal{T}}(a,d) + D_{\mathcal{T}}(b,c)$$

must be equal — for example, we have

$$D_{\mathcal{T}}(a_1,b_1) + D_{\mathcal{T}}(a_3,b_3) = (1+3+2+1) + (1+2+3+1) = 14$$
$$D_{\mathcal{T}}(a_1,b_3) + D_{\mathcal{T}}(a_3,b_1) = (1+3+1) + (1+2+3+2+1) = 14$$

and

$$D_{\mathcal{T}}(a_1,a_3) + D_{\mathcal{T}}(b_1,b_3) = (1+2+1) + (1+2+1) = 8$$

for the distance sums between the points $a_1, a_3, b_1, b_3 \in X$ relative to the weighted $\{a_1, a_2, a_3, b_1, b_2, b_3\}$-tree $\mathcal{T}$ in Figure 3.1. The following theorem that was established already in the 1960s in [136] and independently rediscovered by at least three more authors [37, 50, 121], states that this property characterizes treelike metrics. The proof presented here is adapted from [37] — later, when we deal with the *tight-span construction* in Chapter 5, we will review a more structural and non-inductive proof working for sets X of any cardinality that was given in [50].

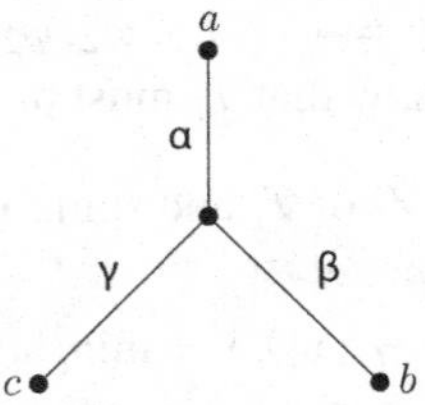

Figure 3.2 Every metric D on a set $X = \{a, b, c\}$ with three elements is treelike: put $\alpha = \frac{1}{2}\big(D(a,b)+D(a,c)-D(b,c)\big)$, $\beta = \frac{1}{2}\big(D(a,b)+D(b,c)-D(a,c)\big)$ and $\gamma = \frac{1}{2}\big(D(a,c)+D(b,c)-D(a,b)\big)$ so that $\alpha+\beta = D(a,b)$, etc., holds.

Theorem 3.1 *A metric D on X is treelike if and only if it satisfies the 4-point condition.*

Proof In view of the remarks above, it suffices to show that, if D satisfies the 4-point condition, then D is treelike. We will use induction on n. It is not hard to check that for $n \leq 3$, every metric D on X is treelike. In particular, an X-tree corresponding to a metric D defined on a 3-set together with its edge-weights is presented in Figure 3.2 (the reader should check that they sum up as required!). Note that, since D satisfies the triangle inequality, the weights α, β, and γ are non-negative (edges with weight 0 are contracted in the resulting weighted X-tree).

By induction, we may also assume that D is a proper metric, i.e., that $D(a,b) \neq 0$ holds for any two distinct elements $a, b \in X$.

Now assume that $n \geq 4$ holds. Select three distinct elements $a, b, c \in X$ so that $D(a,c) + D(b,c) - D(a,b)$ is as large as possible and, switching a and b if necessary, so that also the real number

$$\alpha := \frac{1}{2}\big(D(a,b) + D(a,c) - D(b,c)\big)$$

which — in view of $D(a,c) \leq D(a,b) + D(b,c)$, is necessarily contained in $[0, D(a,b)]$ — actually is a positive number (as $D(a,b)+D(a,c)-D(b,c) = D(a,b) + D(b,c) - D(a,c) = 0$ would imply $0 = \big(D(a,b) + D(a,c) - D(b,c)\big) + \big(D(a,b) + D(b,c) - D(a,c)\big) = 2D(a,b)$ in contradiction to $a \neq b$).

Now, let d be an arbitrary element of $X - \{a, b\}$. Then, by our choice of a, b, and c, we have

$$D(a,b) + D(c,d) \leq D(a,c) + D(b,d)$$

as well as

$$D(a,b)+D(c,d)\leq D(a,d)+D(b,c)$$

as, otherwise, e.g., $D(a,b)+D(c,d)>D(a,c)+D(b,d)$ or, equivalently, $D(d,c)-D(b,d)>D(a,c)-D(a,b)$ would imply $D(b,c)+D(d,c)-D(b,d)>D(a,c)+D(b,c)-D(a,b)$ in contradiction to our choice of a, b, and c. So, the 4-point condition implies that

$$D(a,c)+D(b,d)=D(a,d)+D(b,c)\geq D(a,b)+D(c,d) \tag{3.1}$$

and, therefore, $D(a,b)-D(b,c)\leq D(a,d)-D(c,d)$ and — adding $D(a,b)-D(b,c)-D(b,d)$ to both sides of the first equation — also

$$\frac{D(a,b)+D(a,d)-D(b,d)}{2}=\alpha$$

must hold for all $d\in X-\{a,b\}$ and that, in consequence, also

$$D(a,d)+D(b,d')=D(a,d')+D(b,d)\geq D(a,b)+D(d,d') \tag{3.2}$$

must hold for all $d,d'\in X-\{a,b\}$.

Now, put $X':=X-b$ and let $\mathcal{T}'=(V',E',\omega',\varphi')$ denote a weighted X'-tree for which $D_{\mathcal{T}'}$ coincides with $D':=D|_{X'}$. Note that we may also assume, without loss of generality, that b is not an element of the vertex set V' of $\mathcal{T}'$ (as we are free to choose the elements in V' from an arbitrary set that we may assume to be disjoint from any set we may want to specify). Note first that $\varphi'(a)$ must be a leaf in $\mathcal{T}'$ and that the unique edge $e'\in E'$ with $\varphi'(a)\in e'$ must have a weight $\omega'(e')$ at least as large as α: Indeed, if $\varphi'(a)$ were not a leaf in $\mathcal{T}'$, there would exist elements $d,d'\in X-\{a,b\}$ with $D(d,d')=D(d,a)+D(a,d')$. This, however, would — in view of Equation (3.2) — imply

$$D(b,d)\geq D(a,b)+D(d,d')-D(a,d')=D(a,b)+D(a,d)$$

in contradiction to $2\alpha=D(a,b)+D(a,d)-D(b,d)>0$.

Next, to show that $\omega'(e')\geq\alpha$ holds, let u' denote the unique interior vertex of V' with $e'=\{u',\varphi'(a)\}$ and note that there must exist an element $d\in X'-\{a,c\}$ with $u'=med_{\mathcal{T}'}(a,c,d)$. Indeed, this is obvious in case $\deg(u')\geq 3$, and it also holds in case $\deg(u')<3$ as this implies $u'\in\varphi'(X')$ in which case we may just choose any $d\in X'$ with $u'=\varphi'(d)$. In view of Lemma 1.5 and Equation (1.4), we then have

$$\begin{aligned}\omega'(e')=D_{\mathcal{T}'}\big(u',\varphi'(a)\big)&=D_{\mathcal{T}'}\big(med_{\mathcal{T}'}(a,c,d),\varphi'(a)\big)\\&=\frac{D(c,a)+D(d,a)-D(c,d)}{2}\end{aligned}$$

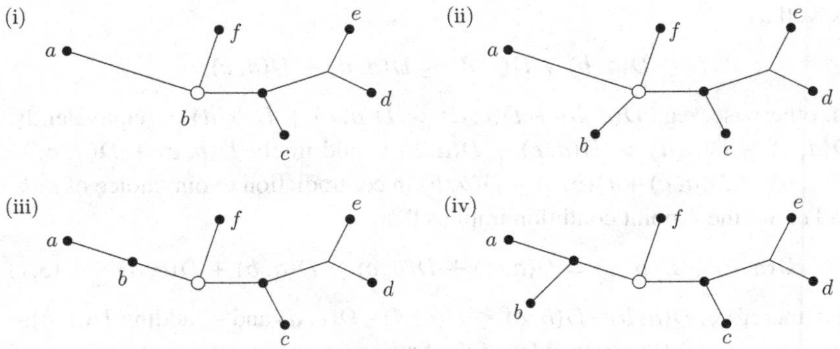

Figure 3.3 Examples illustrating the four cases (i)–(iv) in the proof of Theorem 3.1. Vertex u', as defined in the text, is indicated by an empty circle.

and, therefore, in view of Equation (3.1), also

$$\omega'(e') \geq \frac{D(c,a) + D(a,b) - D(b,c)}{2} = \alpha,$$

as claimed.

Now, we distinguish the four cases (i) $\omega'(e') = \alpha = D(a,b)$, (ii) $\omega'(e') = \alpha < D(a,b)$, (iii) $\omega'(e') > \alpha = D(a,b)$, and (iv) $\omega'(e') > \alpha < D(a,b)$ that are possible in view of $\omega'(e') \geq \alpha$ and $D(a,b) \geq \alpha$ (see Figure 3.3).

In case (i), we may put $\mathcal{T} = (V', E', \omega', \varphi)$ with the single change relative to $\mathcal{T}'$ that $\varphi : X \to V'$ is defined by extending $\varphi' : X' \to V'$ to X by putting $\varphi(b) := u'$.

In case (ii), we put $\mathcal{T} = (V, E, \omega, \varphi)$ with $V := V' + b$, $E := E' + e$ where the additional edge e is the 2-subset $\{b, u'\}$, $\omega : E \to \mathbb{R}_{>0}$ is defined by extending $\omega' : E' \to \mathbb{R}_{>0}$ by putting $\omega(e) := D(a,b) - \alpha$, and φ is defined by extending the map $\varphi' : X' \to V'$ to a map $\varphi : X \to V$ by putting $\varphi(b) := b$.

In case (iii), we put $\mathcal{T} = (V, E, \omega, \varphi)$ with $V := V' + b$, $E := (E' - e') \dot{\cup} \{e_1, e_2\}$ where the two additional edges e_1 and e_2 are the 2-subsets $e_1 := \{b, \varphi'(a)\}$ and $e_2 := \{b, u'\}$, $\omega : E \to \mathbb{R}_{>0}$ is defined by first restricting $\omega' : E' \to \mathbb{R}_{>0}$ to $E' - e'$ and then extending this restriction to a map from E into $\mathbb{R}_{>0}$ by putting $\omega(e_1) := D(a,b)$ and $\omega(e_2) := \omega'(e') - \alpha$, and φ is defined by extending the map $\varphi' : X' \to V'$ to a map $\varphi : X \to V$ by putting $\varphi(b) := b$.

And finally, in the "generic" fourth case, we put $\mathcal{T} = (V, E, \omega, \varphi)$ with $V := V' \dot{\cup} \{u, b\}$ where u is a new element that has not been used before, $E := (E' - e') \dot{\cup} \{e_1, e_2, e_3\}$ where the three additional edges e_1, e_2, and e_3 are the 2-subsets $e_1 := \{u, \varphi'(a)\}$, $e_2 := \{b, u\}$, and $e_3 := \{u, u'\}$, $\omega : E \to \mathbb{R}_{>0}$

is defined by first restricting $\omega' : E' \to \mathbb{R}_{>0}$ to $E' - e'$ and then extending this restriction to a map from E into $\mathbb{R}_{>0}$ by putting $\omega(e_1) := \alpha$, $\omega(e_2) := D(a,b) - \alpha$, and $\omega(e_3) := \omega'(e') - \alpha$, and φ again is defined by extending the map $\varphi' : X' \to V'$ to a map $\varphi : X \to V$ by putting $\varphi(b) := b$.

Using the discussion above that allowed us to conclude that $\omega'(e') \geq \alpha$ must always hold, it is not difficult to verify in a step-by-step manner that these constructions indeed yield a weighted X-tree $\mathcal{T} = (V, E, \omega, \varphi)$ with $D = D_{\mathcal{T}}$. We leave the straightforward (though, of course, a bit cumbersome) verifications to the reader. □

Using Theorem 3.1, we can now also characterize those metrics that correspond to unweighted X-trees:

Corollary 3.2 *Let D be a metric on X. There exists an unweighted X-tree $\mathcal{T} = (V, E, \varphi)$ with $D = D_{\mathcal{T}}$ if and only if D satisfies the* 4-point condition *and the following additional property* (UT) *(for* "Unweighted Tree"*):*

(UT) *For all $a, a', b \in X$ with $D(a,b) + D(a',b) > D(a,a')$, there exists some $b' \in X$ such that $D(a,b) + D(a',b') = 2 + D(a,a') + D(b,b')$.*

Proof First assume that $D = D_{\mathcal{T}}$ holds for some unweighted X-tree $\mathcal{T} = (V, E, \varphi)$. Then, D must satisfy the 4-point condition by Theorem 3.1. To show that D also satisfies Condition (UT), consider any three elements a, a', b in X with $D(a,b) + D(a',b) > D(a,a')$, and let $u \in V$ denote the median $med_{\mathcal{T}}(a, a', b)$ as illustrated in Figure 3.4(a). Note that our assumption $D(a,b) + D(a',b) > D(a,a')$ implies that $D_{\underline{\mathcal{T}}}(\varphi(b), u) > 0$ must hold. Next, let v be the first vertex on the path $\mathbf{p}_{\mathcal{T}}(u,b)$ from u to $\varphi(b)$, i.e., that vertex on that path that is adjacent to u, implying that $D_{\underline{\mathcal{T}}}(\varphi(b), u) = 1 + D_{\underline{\mathcal{T}}}(\varphi(b), v)$ must hold. By the definition of X-trees, there must exist some $b' \in X$ such that the paths $\mathbf{p}_{\mathcal{T}}(u,b)$ and $\mathbf{p}_{\mathcal{T}}(u,b')$ have only the edge $\{u, v\}$ in common (possibly $b = b'$ in case v is a leaf and coincides with $\varphi(b)$) implying that also

$$D_{\underline{\mathcal{T}}}(\varphi(b'), u) = 1 + D_{\underline{\mathcal{T}}}(\varphi(b'), v), \;\; D(b,b') = D_{\underline{\mathcal{T}}}(\varphi(b), v) + D_{\underline{\mathcal{T}}}(\varphi(b'), v),$$

and

$$\begin{aligned} D(a', b') &= D_{\underline{\mathcal{T}}}(\varphi(a'), u) + D_{\underline{\mathcal{T}}}(\varphi(b'), u) \\ &= D_{\underline{\mathcal{T}}}(\varphi(a'), u) + 1 + D_{\underline{\mathcal{T}}}(\varphi(b'), v) \end{aligned}$$

and, therefore, also $D(a,b) + D(a',b') = 2 + D(a,a') + D(b,b')$ holds, as claimed.

To prove the converse, assume that D satisfies the 4-point condition and Condition (UT). In view of Theorem 3.1, there exists a weighted X-tree

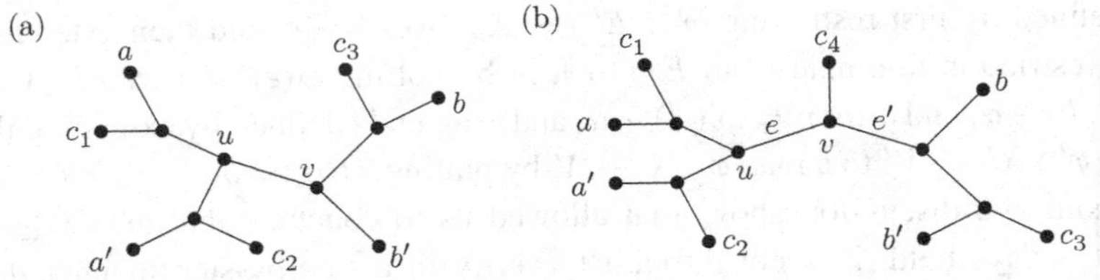

Figure 3.4 Examples illustrating the construction of the proof of Corollary 3.2, in (a) the "if"- and in (b) the "only if"-direction. The labels c_i, $1 \leq i \leq 4$, indicate elements in X distinct from the elements a, a', b, and b' mentioned in the text.

$\mathcal{T} = (V, E, \omega, \varphi)$ with $D = D_{\mathcal{T}}$. It remains to show that ω assigns the weight 1 to every edge of $\mathcal{T}$. If not, there would exist an $e = \{u, v\} \in E$ and an element $b \in X$ such that e is the first edge on the path from u to $\varphi(b)$ and, simultaneously, the last edge on that path with $\omega(e) \neq 1$. It follows from the definition of an X-tree that there exist elements $a, a' \in X$ such that the path $\mathbf{p}_{\mathcal{T}}(a, a')$ passes through u but not through v (possibly $a = a'$ in case u is a leaf). An example is in Figure 3.4(b). By the choice of a, a', and b, we clearly have $D(a, b) + D(a', b) > D(a, a')$. Hence, by (UT), there exists some $b' \in X$ such that $D(a, b) + D(a', b') = 2 + D(a, a') + D(b, b')$. Note that this implies that the path $\mathbf{p}_{\mathcal{T}}(b, b')$ does not pass through u since, otherwise, $D(a, a') + D(b, b') = D_{\mathcal{T}}(\varphi(a), u) + D_{\mathcal{T}}(u, \varphi(a')) + D_{\mathcal{T}}(\varphi(b), u) + D_{\mathcal{T}}(u, \varphi(b')) \geq D(a, b) + D(a', b') = 2 + D(a, a') + D(b, b')$ would hold, which is clearly impossible. Hence, e is an edge on both, the path $\mathbf{p}_{\mathcal{T}}(a, b)$ and the path $\mathbf{p}_{\mathcal{T}}(a', b')$, but not on either one of the two paths $\mathbf{p}_{\mathcal{T}}(a, a')$ or $\mathbf{p}_{\mathcal{T}}(b, b')$, implying that

$$2\,\omega(e) \leq D(a, b) + D(a', b') - D(a, a') - D(b, b') = 2$$

and, therefore (in view of $\omega(e) \neq 1$), that $\omega(e) < 1$ and so $2\,\omega(e) < D(a, b) + D(a', b') - D(a, a') - D(b, b')$ must hold. Hence, there must be an edge e' following the edge e along the path from u to $\varphi(b)$ as well as the path from u to $\varphi(b')$ and we must have $2\,\omega(e) + 2\,\omega(e') \leq D(a, b) + D(a', b') - D(a, a') - D(b, b') = 2$ for that edge e' in contradiction to our choice of e as this implies that $\omega(e') = 1$ must hold. □

3.2 Compatibility

In the last section, we saw that treelike metrics can be characterized by the 4-point condition. We now turn our attention to the problem of deciding which split systems correspond to X-trees. Consider the $\{a, b, c, d, e, f, g, h\}$-tree in

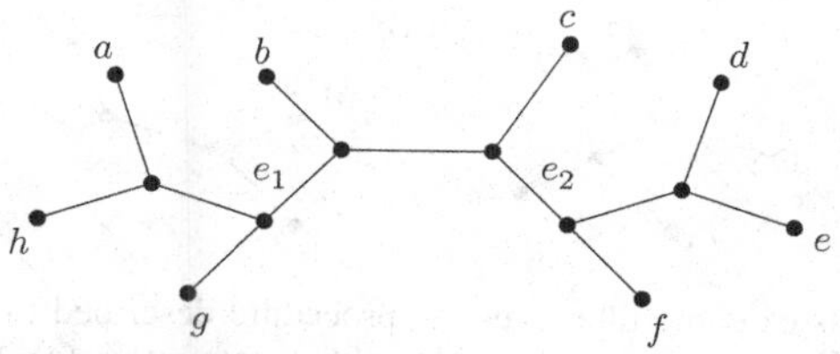

Figure 3.5 The splits S_{e_1} and S_{e_2} associated to the edges e_1 and e_2, respectively, are compatible.

Figure 3.5. Clearly, the splits associated to the edges e_1 and e_2 are, respectively, the splits $\{a, g, h\}|\{b, c, d, e, f\}$ and $\{d, e, f\}|\{a, b, c, g, h\}$. As we can see, the intersection of the sets $\{a, g, h\}$ and $\{d, e, f\}$ is empty while, correspondingly, the union of their complements $\{b, c, d, e, f\}$ and $\{a, b, c, g, h\}$ is all of X.

In general, given any two distinct edges $e_1, e_2 \in E$ of some X-tree $\mathcal{T} = (V, E, \varphi)$, the edge e_2 must be contained in exactly one of the two connected components of the graph $(V, E - e_1)$. So, its elimination will only further split one of the two subsets in the split S_{e_1} associated with e_1 implying that the graph $(V, E - \{e_1, e_2\})$ must have exactly three — and not four — connected components. Thus, if $A_1|B_1$ and $A_2|B_2$ are the two splits associated to two edges e_1 and e_2 of some X-tree $\mathcal{T} = (V, E, \varphi)$, then

(CP) at least one of the four intersections $A_1 \cap A_2$, $A_1 \cap B_2$, $B_1 \cap A_2$, and $B_1 \cap B_2$ is empty

or, equivalently,

(CP′) at least one of the four unions $A_1 \cup A_2$, $A_1 \cup B_2$, $B_1 \cup A_2$, and $B_1 \cup B_2$ coincides with X.

We call two splits $S_1 = A_1|B_1$ and $S_2 = A_2|B_2$ *compatible* if (CP) or, equivalently, (CP′) holds and, otherwise, we call them *incompatible*. Clearly, if the two splits $S_1 = A_1|B_1$ and $S_2 = A_2|B_2$ are compatible, then $A_1 \cap A_2 \neq \emptyset$ implies that either $A_1 \subseteq A_2$ and $B_2 \subseteq B_1$, or $A_2 \subseteq A_1$ and $B_1 \subseteq B_2$, or $B_2 \subseteq A_1$ and $B_1 \subseteq A_2$ must hold. In particular, if A is an inclusion-minimal set in $\{A_1, B_1, A_2, B_2\}$, then "$A \cap A' \neq \emptyset \Rightarrow A \subseteq A'$" must hold for any subset $A' \in \{A_1, B_1, A_2, B_2\}$.

Now suppose $\Sigma \subseteq \Sigma(X)$ is a split system such that any two splits in Σ are compatible. For brevity, we call such a split system *compatible*. Clearly, the above remark implies that there exists, for every inclusion-minimal set A in

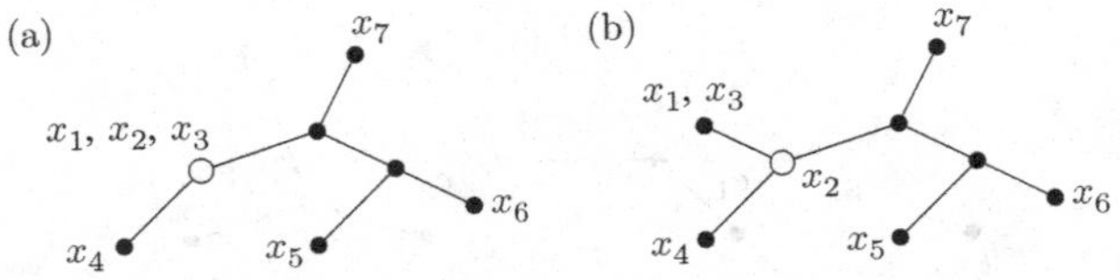

Figure 3.6 This example illustrates the procedure described in the text that constructs an X-tree from a compatible split system on X. (a) An X-tree $\mathcal{T}'$ with $X = \{x_1, \ldots, x_7\}$. Consider the split $S = \{x_1, x_3\}|X - \{x_1, x_3\}$ and note that $\{x_1, x_3\}$ is inclusion-minimal in $\mathcal{C}(\Sigma_{\mathcal{T}'} + S)$. Vertex v', as defined in the text, is indicated by an empty circle. (b) The resulting X-tree $\mathcal{T}$.

the collection $\mathcal{C}(\Sigma) = \bigcup \Sigma$ of all subsets of X that are part of a split in Σ and every $S \in \Sigma$, a unique subset $A' = A'(S, A) \in S$ with $A \subseteq A'$.

Remarkably, this simple observation implies easily that, conversely, if Σ is a compatible split system, then there exists an X-tree $\mathcal{T} = (V, E, \varphi)$ such that $\Sigma = \Sigma_{\mathcal{T}}$ holds: Indeed, assume that A is an inclusion-minimal set in $\mathcal{C}(\Sigma)$ and that, using induction relative to $|\Sigma|$, there exists an X-tree $\mathcal{T}' = (V', E', \varphi')$ such that $\Sigma' = \Sigma_{\mathcal{T}'}$ holds for the set $\Sigma' := \Sigma - S$ where S denotes the split $S_A := \{A, X - A\} \in \Sigma$. Next, note that $\varphi'(a) = \varphi'(a')$ must hold for all $a, a' \in A$ as Σ' does not contain any split S' that separates a and a'. Consequently, denoting the unique vertex in V' that coincides with $\varphi'(a)$ for all $a \in A$ by v', and denoting by $\mathcal{T} = (V, E, \varphi)$ the X-tree with

- $V := V' + v$ where v is an arbitrary new element not related to any of the elements involved in $\mathcal{T}'$,
- $E := E' + e$ where e is the 2-subset $\{v, v'\}$, and
- $\varphi : X \to V$ is defined by putting $\varphi(x) := \begin{cases} \varphi'(x) \text{ if } x \in X - A, \\ v \quad \text{otherwise,} \end{cases}$

it is easily checked that $\Sigma_{\mathcal{T}} = \Sigma_{\mathcal{T}'} + S = \Sigma$ holds (see Figure 3.6 for an example). In other words, we have shown that the following holds:

Theorem 3.3 *Let $\Sigma \subseteq \Sigma(X)$ be a split system. There exists an X-tree $\mathcal{T}$ with $\Sigma = \Sigma_{\mathcal{T}}$ if and only if Σ is compatible.*

In particular, $|\Sigma| \leq 2n - 3$ must hold for every compatible split system Σ on X.

In disguise, this also has been known for a long time and probably was considered to be just obvious in all those discussions taxonomists had for many years on character compatibility, cf. e.g., [71] — see also Chapter 9 where it is indicated how the required X-tree could be obtained in a direct non-inductive fashion by just forming the "Hasse graph" for the hierarchy that the "Combinatorial Farris Transform" would — according to Theorem 9.5 — associate to a compatible split system.

As stated above, the theorem appears in P. Buneman's celebrated paper [36]. An algorithm called "tree popping" that is similar in spirit to the construction of an X-tree from a compatible split system given above was described by C. Meacham [113].

We conclude this section with some remarks concerning the encoding of weighted X-trees in terms of splits. Clearly, we have to record the weight $\omega(e)$ of each edge e in any such tree $\mathcal{T} = (V, E, \omega, \varphi)$. But this can be done in a very natural way by introducing *weighted split systems* (for X), that is, maps ν from $\Sigma(X)$ into $\mathbb{R}_{\geq 0}$, and assigning, to any weighted X-tree $\mathcal{T} = (V, E, \omega, \varphi)$, the weighted split system $\nu_{\mathcal{T}}$ defined by

$$\nu_{\mathcal{T}}(S) := \begin{cases} \omega(e) & \text{if } S = S_e \text{ for some edge } e \text{ of } \mathcal{T}, \\ 0 & \text{else.} \end{cases}$$

Defining, as usual, the *support* of any map $\nu : \Sigma(X) \to \mathbb{R}$ by

$$supp(\nu) := \{S \in \Sigma(X) : \nu(S) \neq 0\},$$

the following theorem immediately follows from Theorem 3.3:

Theorem 3.4 *Let $\nu : \Sigma(X) \to \mathbb{R}_{\geq 0}$ be a weighted split system for X. Then, there exists a weighted X-tree $\mathcal{T} = (V, E, \omega, \varphi)$ with $\nu = \nu_{\mathcal{T}}$ if and only if any two splits in $supp(\nu)$ are compatible.*

More specifically, there is a canonical one-to-one correspondence between

(i) *the — necessarily closed — subset $\mathbb{T}(X)$ of $\mathbb{R}^{X \times X}$ consisting of all metrics in $\mathbb{R}^{X \times X}$ that satisfy the 4-point condition, and*
(ii) *the — also necessarily closed — subset $\mathcal{S}_X(\mathtt{tree})$ of $\mathbb{R}_{\geq 0}^{\Sigma(X)}$ consisting of all weighted split systems in $\mathbb{R}_{\geq 0}^{\Sigma(X)}$ with compatible support,*

given by restricting the map

$$\mathbb{R}^{\Sigma(X)} \to \mathbb{R}^{X \times X} : \nu \mapsto \Big(D_\nu : X \times X \to \mathbb{R} : (x, y) \mapsto \sum_{S \in \Sigma(X), S(x) \neq S(y)} \nu(S)\Big)$$

to $\mathcal{S}_X(\mathtt{tree})$ (see Chapter 7 for more details regarding this map). More specifically, this map induces an isomorphism between $\mathcal{S}_X(\mathtt{tree})$ and $\mathbb{T}(X)$ considered as "topologically stratified spaces" relative to their natural stratifications given in terms of their respective subspaces

$$\mathcal{S}_X^{(k)}(\mathtt{tree}) := \{\nu \in \mathcal{S}_X(\mathtt{tree}) : |supp(\nu)| \leq k\}$$

and

$$\mathbb{T}(X)^{(k)} := \{D_{\mathcal{T}} : \mathcal{T} \text{ is a weighted } X\text{-tree with at most } k \text{ edges}\}.$$

Here, k runs from $-1, 0, 1, \ldots$ up to $2n-3$, and the single strata, i.e., the connected components of the sets $\mathcal{S}_X^{(k)}(\texttt{tree}) - \mathcal{S}_X^{(k-1)}(\texttt{tree})$, $k = 0, 1, \ldots, 2n-3$ (or, respectively, of $\mathbb{T}(X)^{(k)} - \mathbb{T}(X)^{(k-1)}$), correspond exactly to the isomorphism classes of unweighted X-trees with exactly k edges and consist, given such an unweighted X-tree $\mathcal{T}$, exactly of all weighted split systems or tree metrics corresponding to weighted X-trees $\mathcal{T}'$ whose "underlying" unweighted X-tree is $\mathcal{T}$.

Remark 3.5 *The sets $\mathbb{T}(X)$ and $\mathcal{S}_X(\texttt{tree})$ have also been dubbed the* tree space *associated to X [23]. It follows immediately from the fact that, by definition, a collection Σ of splits is compatible if and only if any two splits in Σ are compatible that a map $\nu \in \mathbb{R}^{\Sigma(X)}$ is contained in $\mathcal{S}_X(\texttt{tree})$ if and only if all of its "two-dimensional components"*

$$\nu_{S_1,S_2} : \Sigma(X) \to \mathbb{R} : S \mapsto \begin{cases} \nu(S) & \text{in case } S \in \{S_1, S_2\}, \\ 0 & \text{otherwise,} \end{cases}$$

$\big(S_1, S_2 \in \Sigma(X)\big)$ are contained in $\mathcal{S}_X(\texttt{tree})$. Thus, according to a well-known result of Gromov's, $\mathcal{S}_X(\texttt{tree})$ is a CAT(0)-space *relative to the restriction of the canonical Euclidean L_2-metric on $\mathbb{R}^{\Sigma(X)}$ to its subset $\mathcal{S}_X(\texttt{tree})$ (see [82] for Gromov's result and, e.g., [30, 83] for more on CAT(κ)-spaces, $\kappa \in \mathbb{R}$). This fact has some nice consequences that have been analyzed extensively in [23], see also [118].*

3.3 Quartet systems

In light of our previous results, it is not surprising that encodings of X-trees in terms of quartets can also be characterized by certain local conditions. To identify such conditions, we will first discuss the structure of the quartet systems that arise from such trees.

Consider an arbitrary phylogenetic X-tree $\mathcal{T} = (V, E, \varphi)$, and the associated quartet system $\mathcal{Q} = \mathcal{Q}_{\mathcal{T}}$. Recalling that, for every 4-subset $\{a, b, c, d\} \subseteq X$, at most one of the three quartets $ab|cd$, $ac|bd$, and $ad|bc$ can be contained in $\mathcal{Q}$, we define a quartet system with this property *thin*. Clearly, there are thin quartet systems $\mathcal{Q}$ for which no phylogenetic X-tree $\mathcal{T}$ with $\mathcal{Q} = \mathcal{Q}_{\mathcal{T}}$ exists — consider, for example, the quartet system $\{a_1a_2|b_1x, a_1a_2|b_2x\}$ on $\{a_1, a_2, b_1, b_2, x\}$.

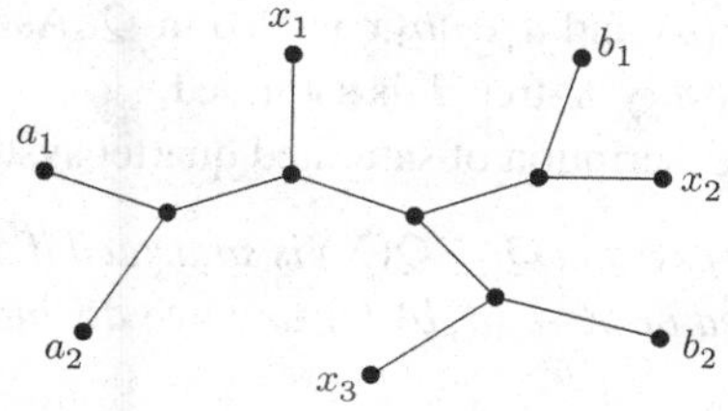

Figure 3.7 An example of an X-tree that is used in the text to illustrate the concept of a transitive quartet system.

The problem is that the presence of certain quartets in $\mathcal{Q}_\mathcal{T}$ implies that also certain other quartets must be contained in $\mathcal{Q}_\mathcal{T}$. To illustrate this, consider the phylogenetic $\{a_1, a_2, b_1, b_2, x_1, x_2, x_3\}$-tree $\mathcal{T}$ in Figure 3.7. In this example, we have both $a_1a_2|b_1x_1 \in \mathcal{Q}_\mathcal{T}$ and $a_1a_2|b_2x_1 \in \mathcal{Q}_\mathcal{T}$, but also $a_1a_2|b_1b_2 \in \mathcal{Q}_\mathcal{T}$. The same holds if we replace x_1 by x_2 or x_3.

Motivated by this example, we define a quartet system $\mathcal{Q} \subseteq \mathcal{Q}(X)$ to be *transitive* if, for any five distinct elements $a_1, a_2, b_1, b_2, x \in X$, the quartet $a_1a_2|b_1b_2$ is in $\mathcal{Q}$ whenever both of the quartets $a_1a_2|b_1x$ and $a_1a_2|xb_2$ are contained in $\mathcal{Q}$. To see that, for every X-tree $\mathcal{T} = (V, E, \varphi)$ (whether phylogenetic or not), the quartet system $\mathcal{Q}_\mathcal{T}$ is transitive, suppose that a_1, a_2, b_1, b_2, x are five distinct elements in X such that $a_1a_2|b_1x \in \mathcal{Q}_\mathcal{T}$ and $a_1a_2|b_2x \in \mathcal{Q}_\mathcal{T}$ hold. Then, by definition, the path $\mathbf{p}_\mathcal{T}(a_1, a_2)$ has no vertex in common with either the path $\mathbf{p}_\mathcal{T}(b_1, x)$ or the path $\mathbf{p}_\mathcal{T}(b_2, x)$ and, hence, also not with the path $\mathbf{p}_T(b_1, b_2)$ as the vertices of this path form a subset of the union of the vertex sets of the paths $\mathbf{p}_\mathcal{T}(b_1, x)$ and $\mathbf{p}_\mathcal{T}(b_2, x)$ — actually, the set of edges of $\mathbf{p}_\mathcal{T}(b_1, b_2)$ coincides with the symmetric difference of the set of edges of $\mathbf{p}_\mathcal{T}(b_1, x)$ and $\mathbf{p}_\mathcal{T}(b_2, x)$. Thus, we must also have $a_1a_2|b_1b_2 \in \mathcal{Q}_\mathcal{T}$, as claimed.

But even being thin and transitive does not ensure that a quartet system arises from an X-tree. To see this, assume that $a_1, a_2, b_1, b_2 \in X$ and $a_1a_2|b_1b_2 \in \mathcal{Q}_\mathcal{T}$ holds for some X-tree $\mathcal{T}$. Clearly, the quartet system $\mathcal{Q}$ containing just this single quartet is thin and transitive. However, if x is a further element in X, at least either one of the two quartets $a_1x|b_1b_2$ or $a_1a_2|b_1x$ must also be contained in $\mathcal{Q}_\mathcal{T}$ as $\varphi(x)$ must be contained in either one of the two connected components of the disconnected graph derived from $\mathcal{T}$ by eliminating any one of the edges separating a_1 and a_2 from b_1 and b_2 that must exist in view of $a_1a_2|b_1b_2 \in \mathcal{Q}_\mathcal{T}$.

In view of this, we call a quartet system $\mathcal{Q} \subseteq \mathcal{Q}(X)$ *saturated* if, for any five distinct elements $a_1, a_2, b_1, b_2, x \in X$ with $a_1a_2|b_1b_2 \in \mathcal{Q}$, at least one of the two quartets $a_1x|b_1b_2$ and $a_1a_2|b_1x$ is also in $\mathcal{Q}$. As seen just above, the quartet system $\mathcal{Q}_{\mathcal{T}}$ of every X-tree $\mathcal{T}$ is saturated.

A more convenient description of saturated quartet systems is given in

Lemma 3.6 *A quartet system $\mathcal{Q} \subseteq \mathcal{Q}(X)$ is saturated if and only if the binary relation "$\preceq_{a,b}$" defined on $X - \{a, b\}$ for any two distinct elements $a, b \in X$ by putting*

$$c \preceq_{a,b} d \iff ad|bc \notin \mathcal{Q}$$

for all $c, d \in X - \{a, b\}$ is transitive.

In particular, the induced binary relation "$\sim_{a,b}$" defined on $X - \{a, b\}$ by putting

$$c \sim_{a,b} d \iff c \preceq_{a,b} d \text{ and } d \preceq_{a,b} c \quad (\iff ad|bc, ac|bd \notin \mathcal{Q})$$

for all $c, d \in X - \{a, b\}$, is an equivalence relation on $X - \{a, b\}$ in this case and, putting

$$c \prec_{a,b} d \iff c \preceq_{a,b} d \text{ and not } d \preceq_{a,b} c \quad (\iff c \preceq_{a,b} d \text{ and not } c \sim_{a,b} d)$$

one has

$$ac|db \in \mathcal{Q} \iff c \prec_{a,b} d.$$

Furthermore, if $\mathcal{Q}$ is also thin, then "$\preceq_{a,b}$" induces a linear order "$\leq_{a,b}$" on the set $X/\!\sim_{a,b}$ of "$\sim_{a,b}$"-equivalence classes, i.e., given any two elements $c, d \in X - \{a, b\}$, we must have either $c \preceq_{a,b} d$ or $c \preceq_{a,b} d$ (as $ad|bc \in \mathcal{Q}$ and $ac|bd \in \mathcal{Q}$ cannot hold simultaneously in this case), and one has "$ab|cd \in \mathcal{Q} \Rightarrow c \sim_{a,b} d$" for any four distinct elements a, b, c, d in X (as neither $ad|bc \in \mathcal{Q}$ nor $ac|bd \in \mathcal{Q}$ can hold if $\mathcal{Q}$ is thin and $ab|cd \in \mathcal{Q}$ holds).

Proof If $c \preceq_{a,b} d$ and $d \preceq_{a,b} e$ or, equivalently, $ad|bc \notin \mathcal{Q}$ and $ae|bd \notin \mathcal{Q}$ holds for some $c, d, e \in X - \{a, b\}$, we cannot have $ae|bc \in \mathcal{Q}$ as, otherwise, either $ad|bc \in \mathcal{Q}$ or $ae|bd \in \mathcal{Q}$ must hold in case $\mathcal{Q}$ is saturated. So, we must have $c \preceq_{a,b} e$, as claimed.

Conversely, if "$\preceq_{a,b}$" is transitive for any two distinct elements $a, b \in X$, if $a_1a_2|b_1b_2 \in \mathcal{Q}$ holds for some $a_1, a_2, b_1, b_2 \in X$, and if x is an additional element in X, we must have either $a_1x|b_1b_2 \in \mathcal{Q}$ or $a_1a_2|b_1x \in \mathcal{Q}$. Indeed, otherwise, $a_1x|b_1b_2 \notin \mathcal{Q}$ and $a_1a_2|b_1x \notin \mathcal{Q}$ would imply $b_2 \preceq_{a_1,b_1} x$ and $x \preceq_{a_1,b_1} a_2$. Thus, if "$\preceq_{a_1,b_1}$" is transitive, it would also imply $b_2 \preceq_{a_1,b_1} a_2$ in contradiction to $a_1a_2|b_1b_2 \in \mathcal{Q}$. □

As was first observed by H. Colonius and H. Schultze in [44, 45], the properties thin, transitive, and saturated suffice to characterize quartet systems that arise from phylogenetic X-trees.

Theorem 3.7 *A quartet system $\mathcal{Q} \subseteq \mathcal{Q}(X)$ is of the form $\mathcal{Q} = \mathcal{Q}_{\mathcal{T}}$ for some phylogenetic X-tree $\mathcal{T}$ if and only if $\mathcal{Q}$ is thin, transitive, and saturated.*

Proof It follows from the discussion preceding Lemma 3.6 that every quartet system of this form is thin, transitive, and saturated. To show the converse, assume $\mathcal{Q} \subseteq \mathcal{Q}(X)$ is a thin, transitive, and saturated quartet system.

We proceed by induction on the size of $|\mathcal{Q}| + n$. Clearly if $\mathcal{Q}$ is empty, then the star tree for X is a phylogenetic X-tree $\mathcal{T} = (V, E, \varphi)$ with $\mathcal{Q} = \mathcal{Q}_{\mathcal{T}}$.

Now assume that $\mathcal{Q}$ is non-empty and that, hence, also $n \geq 4$ holds. Crucial to our proof is that, given any subset $Y \subseteq X$, the restriction

$$\mathcal{Q}|_Y := \{uv|xy \in \mathcal{Q} : u, v, x, y \in Y\}$$

must also be thin, transitive, and saturated. So, we may assume that there exists, for every proper subset Y of X, a phylogenetic Y-tree $\mathcal{T}(Y)$ for which $\mathcal{Q}_{\mathcal{T}(Y)} = \mathcal{Q}|_Y$ holds.

To construct a phylogenetic X-tree $\mathcal{T} = (V, E, \varphi)$ with $\mathcal{Q} = \mathcal{Q}_{\mathcal{T}}$, we choose two arbitrary distinct elements $a, b \in X$ and consider the phylogenetic $(X-b)$-tree $\mathcal{T}' := \mathcal{T}(X - b) = (V', E', \varphi')$ with $\mathcal{Q}_{\mathcal{T}'} = \mathcal{Q}|_{X-b}$. As $\mathcal{Q}$ is thin and saturated, the binary relation "$\sim_{a,b}$" must be an equivalence relation on $X - \{a, b\}$ and, in view of Lemma 3.7, the binary relation "$\preceq_{a,b}$" must be a linear ordering of $X/\sim_{a,b}$. So, let $A_1, \ldots, A_k$ denote the "$\sim_{a,b}$"-equivalence classes in $X - \{a, b\}$ indexed such that $A_i \preceq_{a,b} A_j$ holds for all $i, j \in \{1, \ldots, k\}$ with $i \leq j$.

Now, we choose an arbitrary element $c \in A_k$ and consider the path $u_0 := \varphi(a), u_1, \ldots, u_l := \varphi(c)$ from $\varphi(a)$ to $\varphi(c)$ in $\mathcal{T}'$. Let $j \in \{0, \ldots, l\}$ be the largest index in $\{0, \ldots, l\}$ such that $u_0, u_1, \ldots, u_j \in V(\mathbf{p}(a, c'))$ holds for every element $c' \in A_k$. Note that, since $\mathcal{T}'$ is a phylogenetic tree, every path from $\varphi(a)$ to any other vertex in $\mathcal{T}'$ must pass through u_1. So, $j \geq 1$ must hold. To finish our construction of $\mathcal{T}$, we distinguish two cases.

First consider the case that $ab|cc' \in \mathcal{Q}$ holds for any two distinct elements $c, c' \in A_k$. Then, we choose two distinct new vertices v and w not contained in V' and define $\mathcal{T}$ by

$$V := V \cup \{v, w\}, \quad E := (E' - \{u_{j-1}, u_j\}) \cup \{\{u_{j-1}, v\}, \{v, u_j\}, \{v, w\}\},$$

$\varphi(x) := \varphi'(x)$ for every $x \in X - b$, and $\varphi(b) := w$.

Otherwise, we choose just one new vertex w not contained in V' and define $\mathcal{T}$ by

$$V := V + w, \quad E := E' \cup \{\{u_j, w\}\},$$

and, as above, $\varphi(x) := \varphi'(x)$ for every $x \in X - b$, and $\varphi(b) := w$.

We leave it as an exercise to the reader to verify that, in both cases, $\mathcal{Q}_{\mathcal{T}}$ coincides indeed with $\mathcal{Q}$, as desired. □

Corollary 3.8 (i) *Given a phylogenetic X-tree $\mathcal{T}$ and elements $a, b, c, d, e \in X$ such that $ab|cd, bc|de \in \mathcal{Q}_{\mathcal{T}}$ holds. Then, $|\{a, b, c, d, e\}| = 5$ and $ab|de \in \mathcal{Q}_{\mathcal{T}}$ must also hold.*

(ii) *More generally, given any sequence $x_0, x_1, \dots, x_\ell$ of elements from X for which $x_0x_i|x_{i+1}x_{i+2} \in \mathcal{Q}_{\mathcal{T}}$ holds for all $i = 1, 2, \dots, \ell-2$, one must have $|\{x_0, x_1, \dots, x_\ell\}| = \ell + 1$ and $x_ix_j|x_kx_l \in \mathcal{Q}_{\mathcal{T}}$ for all $i, j, k, l \in \{1, 2, \dots, \ell\}$ with $i < j < k < l$.*

Proof (i) Indeed, we must have $a \neq e$ as $ab|cd \in \mathcal{Q}_{\mathcal{T}}$ implies $bc|da \notin \mathcal{Q}_{\mathcal{T}}$. So, we must have $ab|de \in \mathcal{Q}_{\mathcal{T}}$ as $ab|cd \in \mathcal{Q}_{\mathcal{T}}$ implies that either $eb|cd \in \mathcal{Q}_{\mathcal{T}}$ or $ab|de \in \mathcal{Q}_{\mathcal{T}}$ holds while $bc|de \in \mathcal{Q}_{\mathcal{T}}$ implies that $eb|cd \in \mathcal{Q}_{\mathcal{T}}$ cannot hold.

(ii) Our assumption implies easily by induction with respect to $l - j$ that, for all $j, k, l \in \{1, 2, \dots, \ell\}$ with $i < j < k < l$, $x_0x_j|x_kx_l \in \mathcal{Q}_{\mathcal{T}}$ and $x_l \notin \{x_0, x_1, \dots, x_{l-1}\}$ must hold. Thus, given $i, j, k, l \in \{1, 2, \dots, \ell\}$ with $0 < i < j < k < l$, we have also $x_0x_i|x_kx_l \in \mathcal{Q}_{\mathcal{T}}$ and $x_0x_j|x_kx_l \in \mathcal{Q}_{\mathcal{T}}$ and, therefore, by transitivity, also $x_ix_j|x_kx_l \in \mathcal{Q}_{\mathcal{T}}$, as claimed. □

It remains to describe how *weighted* phylogenetic X-trees can be encoded in terms of quartets. Characterizations of such encodings were first established for binary phylogenetic X-trees in [54]. In the remainder of this section, we will describe the characterization for general phylogenetic X-trees that was presented in [87].

To this end, define a *weighted quartet system* to be a map $\mu : \mathcal{Q}(X) \to \mathbb{R}_{\geq 0}$. As usual, the *support* of a weighted quartet system μ is the quartet system $supp(\mu) = \{q \in \mathcal{Q}(X) : \mu(q) > 0\}$. Note that every weighted X-tree $\mathcal{T} = (V, E, \omega, \varphi)$ gives rise to a weighted quartet system $\mu_{\mathcal{T}}$ that is defined by putting $\mu_{\mathcal{T}}(q) := \sum_{e \in E_q} \omega(e)$ for every quartet $q = a_1a_2|b_1b_2 \in \mathcal{Q}(X)$ where E_q is defined to be the set of edges that separate a_1, a_2 from b_1, b_2. That is, E_q is empty in case $q \in \mathcal{Q}(X) - \mathcal{Q}_{\mathcal{T}}$, and, in case $q = a_1a_2|b_1b_2 \in \mathcal{Q}_{\mathcal{T}}$ holds, E_q is the set $E_{\mathcal{T}}[med_{\mathcal{T}}(a_1, a_2, b_1), med_{\mathcal{T}}(a_1, b_1, b_2)]$ of edges of the path in $\mathcal{T}$ connecting the median $med_{\mathcal{T}}(a_1, a_2, b_1)$ of $\varphi(a_1), \varphi(a_2)$, and $\varphi(b_1)$ with the median $med_{\mathcal{T}}(a_1, b_1, b_2)$ of $\varphi(a_1), \varphi(b_1)$, and $\varphi(b_2)$. For example, for the weighted X-tree $\mathcal{T}$ in Figure 3.1 we have $\mu_{\mathcal{T}}(a_2a_3|b_1b_3) = 2 + 3 = 5$.

Clearly, if we have two weighted phylogenetic X-trees $\mathcal{T} = (V, E, \omega, \varphi)$ and $\mathcal{T}' = (V, E, \omega', \varphi)$ that differ only with respect to their edge weightings ω and ω', we have $\mu_{\mathcal{T}} = \mu_{\mathcal{T}'}$ if and only if $\omega(e) = \omega(e')$ holds for every interior edge of the tree (V, E). Put differently, the weighted quartet system $\mu_{\mathcal{T}}$ associated with a weighted phylogenetic tree $\mathcal{T} = (V, E, \omega, \varphi)$ determines the edge weights on the interior edges of $\mathcal{T}$ and only those. More specifically, we have:

Theorem 3.9 *Suppose that μ is a weighted quartet system on $\mathcal{Q}(X)$. Then there exists a weighted phylogenetic X-tree $\mathcal{T} = (V, E, \omega, \varphi)$ with $\mu = \mu_{\mathcal{T}}$ if and only if the following holds for any four distinct elements $a_1, a_2, b_1, b_2 \in X$:*

(T1) $\mu(a_1a_2|b_1b_2) \neq 0$ *implies* $\mu(a_1b_1|a_2b_2) = \mu(a_1b_2|a_2b_1) = 0$.

(T2) $x \in X - \{a_1, a_2, b_1, b_2\}$ *and* $\mu(a_1a_2|b_1b_2) > 0$ *implies* $\mu(a_1a_2|b_1x) > 0$ *and* $\mu(a_1a_2|b_2x) > 0$*, or* $\mu(a_1x|b_1b_2) > 0$ *and* $\mu(a_2x|b_1b_2) > 0$.

(T3) $x \in X - \{a_1, a_2, b_1, b_2\}$ *and* $\mu(a_1a_2|b_1b_2) > \mu(a_1a_2|b_1x) > 0$ *implies* $\mu(a_1a_2|b_1b_2) = \mu(a_1a_2|b_1x) + \mu(a_1x|b_1b_2)$.

(T4) $x \in X - \{a_1, a_2, b_1, b_2\}$, $\mu(a_1a_2|b_1x) > 0$*, and* $\mu(a_1x|b_1b_2) > 0$ *implies* $\mu(a_1a_2|b_1b_2) = \mu(a_1a_2|b_1x) + \mu(a_1x|b_1b_2)$.

Proof It is straightforward to check that, if there exists a weighted phylogenetic X-tree $\mathcal{T} = (V, E, \omega, \varphi)$ with $\mu = \mu_{\mathcal{T}}$, then (T1)–(T4) hold.

Conversely, suppose that (T1)–(T4) hold for some weighted quartet system $\mu : \mathcal{Q}(X) \to \mathbb{R}_{\geq 0}$ and put $\underline{\mathcal{Q}} := supp(\mu)$. Note that (T1) implies immediately that $\underline{\mathcal{Q}}$ is thin. Moreover, $\underline{\mathcal{Q}}$ is also transitive: Indeed, given any five distinct elements $a_1, a_2, b_1, b_2, x \in X$, $a_1a_2|b_1x \in \underline{\mathcal{Q}}$ implies — in view of (T2) — that

$$a_1a_2|b_1b_2, a_1a_2|xb_2 \in \underline{\mathcal{Q}} \quad \text{or} \quad a_1b_2|b_1x, a_2b_2|b_1x \in \underline{\mathcal{Q}}$$

must hold while $a_1a_2|xb_2 \in \underline{\mathcal{Q}}$ implies that

$$a_1a_2|b_1b_2, a_1a_2|b_1x \in \underline{\mathcal{Q}} \quad \text{or} \quad a_1b_1|b_2x, a_2b_1|b_2x \in \underline{\mathcal{Q}}$$

must hold. Thus, if $a_1a_2|b_1x \in \underline{\mathcal{Q}}$ and $a_1a_2|xb_2 \in \underline{\mathcal{Q}}$ hold, we must have $a_1a_2|b_1b_2 \in \underline{\mathcal{Q}}$, as required, since $a_1b_2|b_1x \in \underline{\mathcal{Q}}$ and $a_1b_1|b_2x \in \underline{\mathcal{Q}}$ cannot hold simultaneously in a thin quartet system. Similarly, it follows directly from (T2) that $\underline{\mathcal{Q}}$ is saturated. So, by Theorem 3.7, there is (up to canonical isomorphism) a unique phylogenetic X-tree $\mathcal{T}_0 := (V, E, \varphi)$ such that $\mathcal{Q}_{\mathcal{T}_0} = \underline{\mathcal{Q}}$ holds.

We now show that there exists a weighting $\omega : E \to \mathbb{R}_{\geq 0}$ such that $\mu = \mu_{\mathcal{T}}$ holds for the weighted phylogenetic X-tree $\mathcal{T} = (V, E, \omega, \varphi)$. Clearly, it suffices to assign suitable weights to the interior edges of $\mathcal{T}_0$. We first claim that if $e = \{u, v\}$ is an interior edge of $\mathcal{T}_0$ such that there exist two distinct

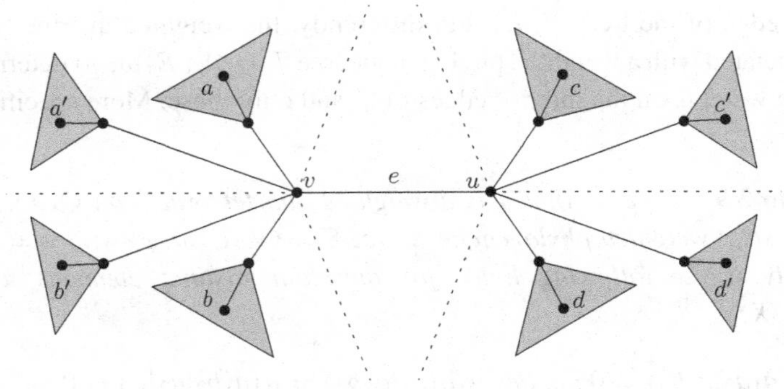

Figure 3.8 In the proof of Theorem 3.9, we may assume that the vertices labeled a, a', b, b', c, c', d, and d' lie in subtrees of $\mathcal{T}_0$ indicated by the triangles and, therefore (by subsequent replacement), even that $a = a'$, $b = b'$, $c = c'$ holds. The dotted lines mark the boundaries of regions where those vertices may be located.

quartets $q = ab|cd, q' = a'b'|c'd' \in \mathcal{Q}_{\mathcal{T}_0}$ with $\{e\} = E_q = E_{q'}$, then $\mu(q) = \mu(q')$ must hold.

Note first that, without loss of generality, we may assume that $a = a', b = b', c = c'$ and $d \neq d'$ holds (see for instance Figure 3.8). However, if say $\mu(ab|cd) > \mu(ab|cd')$ were to hold, we must have $\mu(ad'|cd) = \mu(ab|cd) - \mu(ab|cd') > 0$ in view of (T3) and, therefore, also $ad'|cd \in \mathcal{Q}_{\mathcal{T}_0}$ which is clearly impossible as there is either an edge separating $\varphi(a), \varphi(c)$ from $\varphi(d), \varphi(d')$ or there is no edge at all separating any two of $\varphi(a), \varphi(c), \varphi(d), \varphi(d')$ from the two other vertices.

Thus, noting that, for every edge e of $\mathcal{T}_0$, there is at least one quartet $q \in \mathcal{Q}_{\mathcal{T}_0}$ with $E_q = \{e\}$, we can define a well-defined edge-weighting $\omega : E \to \mathbb{R}_{>0}$ for $\mathcal{T}_0$ by choosing, for every interior edge e of $\mathcal{T}_0$, some quartet $q = q_e \in \mathcal{Q}_{\mathcal{T}}$ with $E_q = \{e\}$ and then defining $\omega(e)$ by putting $\omega(e) := \mu(q)$ while putting, say, $\omega(e) := 1$ for every pendant edge e of the tree (V, E).

To establish that $\mu(q) = \mu_{\mathcal{T}}(q)$ then holds, for every quartet $q = ab|cd \in \mathcal{Q}_{\mathcal{T}_0}$, for the map $\mu_{\mathcal{T}} : \mathcal{Q}(X) \to \mathbb{R}_{\geq 0}$ associated to the resulting weighted X-tree $\mathcal{T} := (V, E, \varphi, \omega)$, we proceed by induction on the number m of edges in E_q. If $m = 1$, then $\mu(q) = \mu_{\mathcal{T}}(q)$ holds by (our observation above and) the definition of ω. Now suppose $m > 1$ and consider the path $\mathbf{p}$ connecting the median $u := med_{\mathcal{T}}(a, b, c)$ of $\varphi(a), \varphi(b)$, and $\varphi(c)$ with the median $w := med_{\mathcal{T}}(b, c, d)$ of $\varphi(b), \varphi(c)$, and $\varphi(d)$. Let v denote the vertex on $\mathbf{p}$ that is adjacent to w, choose any element $c' \in X - \{a, b, c, d\}$ such that $v = med_{\mathcal{T}}(a, c', d)$ holds, and note that $E_{ab|c'd} = E_q - \{v, w\}$ and

$E_{ac'|cd} = \{\{v, w\}\}$ holds. By induction, μ and μ_T coincide on $ab|c'd$ and $ac'|cd$. Moreover, we have $ab|c'd, ac'|cd \in \mathcal{Q}_T = \underline{\mathcal{Q}}$ and, therefore, by (T4) and the definition of μ_T, also

$$\mu(ab|cd) = \mu(ab|c'd) + \mu(ac'|cd) = \mu_T(ab|c'd) + \mu_T(ac'|cd) = \mu_T(q),$$

as claimed. □

4
From split systems to networks

In the last chapter, we saw that X-trees can be encoded, up to canonical isomorphism, in terms of certain split systems, quartet systems, or metrics. We now start to explore how, conversely, split systems, quartet systems, or metrics can be used to (re-)construct phylogenetic relationships in terms of trees or, more generally, networks, beginning the investigation in this chapter with split systems.

Using a recursive argument, we saw already in the previous chapter that an X-tree may be encoded by — and, thus, reconstructed recursively from — the associated system of pairwise compatible X-splits. However, split systems arising in "real life" (defined, e.g., in terms of a family of naturally arising binary characters) are rarely compatible. To see how even such data may still be dealt with using "graphical" methods, consider the following example: For $X := \langle 5 \rangle$, consider the two compatible splits $S_1 = \{1, 2\}|\{3, 4, 5\}$ and $S_2 = \{1, 2, 3\}|\{4, 5\}$ that encode the X-tree depicted in Figure 4.1(a). Clearly, it is impossible to "pop-out" the split $S_3 = \{1, 2, 4\}|\{3, 5\}$ in this X-tree by introducing a single edge, reflecting of course the fact that S_2 and S_3 are incompatible. However, we could still "expand" the tree along the edge connecting the vertices that are labeled with 3 and 5 so as to obtain the labeled network in Figure 4.1(b). In this graph, the split S_1 can be "obtained" by deleting the pendant edge, the split S_3 can be "obtained" by deleting the two parallel dashed edges, whereas the split S_2 can be "obtained" by deleting the remaining two, also parallel edges.

In particular, this graphical representation generalizes the representation of compatible split systems by X-trees as follows: Each split $S = A|B$ in Σ is displayed by an appropriate *collection* of parallel edges in the X-labeled graph so that removing the edges in this collection from the network yields a graph with precisely two connected components in such a way that the vertices labeled by (the elements in) A belong to one of these two connected components, and

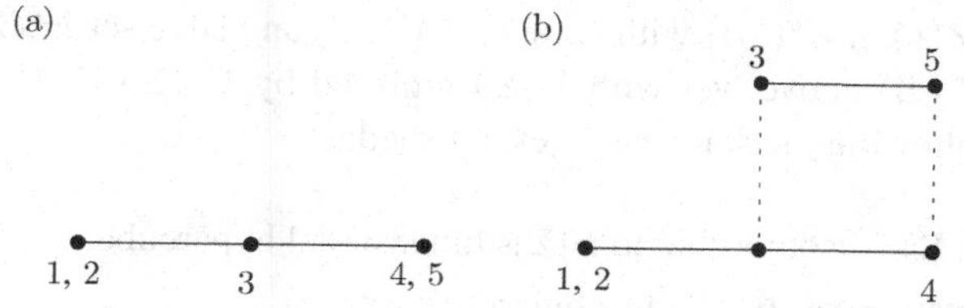

Figure 4.1 (a) An X-tree with $X = \langle 5 \rangle$ displaying two compatible splits. (b) A graph that arises from the X-tree in (a) and displays, in addition, the split $\{1, 2, 4\}|\{3, 5\}$ as described in the text.

those labeled by B belong to the other. Intriguingly, as we shall see below in the first section of this chapter, such graphs can be associated to *any* given split system Σ.

4.1 The Buneman graph

In this and all of the following sections of this chapter, we will assume that X is a fixed finite set of cardinality n and Σ an arbitrary system of X-splits. We define the *Buneman graph* $\mathcal{B}(\Sigma) = \big(V(\Sigma), E(\Sigma)\big)$ to be the graph whose vertex set $V(\Sigma)$ and edge set $E(\Sigma)$ are defined as follows: $V(\Sigma)$ is defined to consist of all maps ϕ from Σ into the power set $\mathbb{P}(X)$ of X that satisfy the following two conditions

(BG1) $\phi(S) \in S$ holds for all $S \in \Sigma$, i.e., if $S = A|B$, then $\phi(S)$ coincides either with A or with B, and

(BG2) $\phi(S) \cap \phi(S') \neq \emptyset$ holds for all $S, S' \in \Sigma$.

And the edge set $E(\Sigma)$ is defined to consist of all those 2-subsets $\{\phi, \psi\}$ of $V(\Sigma)$ for which the *difference set* $\Delta(\phi, \psi)$, i.e., the set

$$\Delta(\phi, \psi) := \{S \in \Sigma : \phi(S) \neq \psi(S)\},$$

has cardinality 1:

$$E(\Sigma) := \left\{ \{\phi, \psi\} \in \binom{V(\Sigma)}{2} : |\Delta(\phi, \psi)| = 1 \right\}.$$

This graph has appeared in the literature in various guises: As a co-pair hypergraph in [20, 21], as a special type of median graph in, e.g., [7], and in the above form in [59]. To investigate it, let us also introduce the superset $V^\star(\Sigma)$ of $V(\Sigma)$ consisting of just all maps $\phi : \Sigma \to \mathbb{P}(X)$ that satisfy (BG1) $\big($but not necessarily (BG2)$\big)$ and the corresponding *extended Buneman graph*

$\mathcal{B}^\star(\Sigma) = (V^\star(\Sigma), E^\star(\Sigma))$ with vertex set $V^\star(\Sigma)$ and edge set $E^\star(\Sigma)$ defined exactly as $E(\Sigma)$ above, yet with $V(\Sigma)$ replaced by $V^\star(\Sigma)$ in its definition. Then, the following facts are easily established:

($B^\star$-i) $\mathcal{B}^\star(\Sigma)$ is isomorphic to a $|\Sigma|$-dimensional hypercube.

($B^\star$-ii) In particular, $\mathcal{B}^\star(\Sigma)$ is a bipartite graph.

($B^\star$-iii) The graph-theoretical distance $D_{\mathcal{B}^\star(\Sigma)}(\phi, \psi)$ between any two vertices ϕ and ψ in $\mathcal{B}^\star(\Sigma)$ coincides with the cardinality $|\Delta(\phi, \psi)|$ of $\Delta(\phi, \psi)$. More precisely, there exist, for any two vertices $\phi, \psi \in V^\star(\Sigma)$ with $\ell := |\Delta(\phi, \psi)|$, altogether $\ell!$ sequences $\phi_0 := \phi, \phi_1, \dots, \phi_\ell := \psi$ of maps in $V^\star(\Sigma)$ of length $\ell + 1$ with $\{\phi_{i-1}, \phi_i\} \in E^\star(\Sigma)$ for all $i = 1, \dots, \ell$.

($B^\star$-iv) And the extended Buneman graph $\mathcal{B}^\star(\Sigma)$ coincides with $\mathcal{B}(\Sigma)$ if and only if Σ is (totally) *incompatible*, that is, any two splits in Σ are incompatible.

Furthermore, given a fixed map $\phi \in V^\star(\Sigma)$, any other map $\psi \in V^\star(\Sigma)$ is completely determined by ϕ and the difference set $\Delta(\phi, \psi)$: Indeed, defining the map $\phi^\Xi \in V^\star(\Sigma)$, for every map $\phi \in V^\star(\Sigma)$ and every subset Ξ of Σ, by

$$\phi^\Xi(S) := \begin{cases} X - \phi(S) & \text{if } S \in \Xi, \\ \phi(S) & \text{else,} \end{cases}$$

for every split $S \in \Sigma$, one has $\phi^\Xi = \psi$ for some $\phi, \psi \in V^\star(\Sigma)$ if and only if $\Xi = \Delta(\phi, \psi)$ holds. Note also that in case $\psi \in V(\Sigma)$ holds, $\Delta(\phi, \psi)$ is, in turn, completely determined by the subset

$$\Delta_{\min}(\psi|\phi) := \{S \in \Delta(\phi, \psi) : \psi(S) \in \min(\psi \backslash \phi)\} \tag{4.1}$$

of $\Delta(\phi, \psi)$ where $\min(\psi \setminus \phi) = \min_{\subseteq}(\psi \setminus \phi)$ denotes the set of inclusion-minimal subsets in the image

$$\psi \backslash \phi := \psi[\Delta(\phi, \psi)] = \{\psi(S) : S \in \Delta(\phi, \psi)\} \tag{4.2}$$

of $\Delta(\phi, \psi)$ relative to ψ where — for clarity — we write $\psi[\Xi]$ for the image $\{\psi(S) : S \in \Xi\}$ of a subset Ξ of Σ relative to any map ψ in the vertex set $V^\star(\Sigma)$ of the extended Buneman graph of Σ. Note that $\psi[\Xi] \subseteq \mathcal{C}(\Xi) \subseteq \mathcal{C}(\Sigma)$ then always holds. Note also that, in view of the fact that $\psi(S) = X - \phi(S)$ holds for all $S \in \Delta(\phi, \psi)$, the set $\min(\psi \setminus \phi)$ coincides with the collection $\{X - A : A \in \max(\phi \setminus \psi)\}$ of complements of the sets in the collection $\max(\phi \setminus \psi) = \max_{\subseteq}(\phi \setminus \psi)$ of inclusion-maximal subsets in the image

$\phi \setminus \psi = \phi[\Delta(\phi, \psi)] = \phi[\Sigma] \setminus \psi[\Sigma]$ of $\Delta(\phi, \psi)$ relative to ϕ. That is, given any split $S \in \Delta(\phi, \psi)$, we have

$$\psi(S) \in \min(\psi \setminus \phi) \iff \phi(S) \in \max(\phi \setminus \psi)$$

and, therefore, also

$$\Delta_{\min}(\psi|\phi) = \Delta_{\max}(\phi|\psi) := \{S \in \Delta(\phi, \psi) : \phi(S) \in \max(\phi \setminus \psi)\}. \quad (4.3)$$

Thus, defining the subset $\partial_\phi\, \Xi$ of Σ, for any subset Ξ of Σ and any map $\phi \in V^\star(\Sigma)$, by

$$\partial_\phi\, \Xi := \{S \in \Sigma : \exists_{S' \in \Xi}\, \phi(S) \subseteq \phi(S')\}, \quad (4.4)$$

one has $\partial_\phi \Delta_{\min}(\psi|\phi) = \Delta(\phi, \psi)$ and, therefore,

$$\psi = \phi^{\partial_\phi \Delta_{\min}(\psi|\phi)} \quad (4.5)$$

for every $\phi \in V^\star(\Sigma)$ and $\psi \in V(\Sigma)$: Indeed, $S \in \Delta(\phi, \psi)$ implies that there exists some split $S' \in \Delta_{\min}(\psi|\phi)$ with $\psi(S') \subseteq \psi(S)$ or, equivalently, $\phi(S) \subseteq \phi(S')$ and, therefore, $S \in \partial_\phi \Delta_{\min}(\psi|\phi)$ while, conversely, $S \in \partial_\phi \Delta_{\min}(\psi|\phi)$ implies that some $S' \in \Delta_{\min}(\psi|\phi)$ with $\phi(S) \subseteq \phi(S')$ exists which, in turn, implies $\psi(S') \cap \phi(S) \subseteq \psi(S') \cap \phi(S') = \emptyset$ and, therefore, $\phi(S) \neq \psi(S)$ in view of $\psi \in V(\Sigma)$, i.e., $S \in \Delta(\phi, \psi)$.

Note also that, writing ϕ^S rather than ϕ^Ξ in case Ξ consists of a single split S, only, one has $\{\phi, \psi\} \in E^\star(\Sigma)$ for some $\phi, \psi \in V^\star(\Sigma)$ if and only if $\psi = \phi^S$ holds for some (necessarily unique) split $S \in \Sigma$, i.e., the unique split $S = S_{\phi,\psi}$ in $\Delta(\phi, \psi)$.

In this section, we will present some very basic properties of Buneman graphs, some of which are illustrated in Figure 4.2. We begin with the following simple observation:

Lemma 4.1 *Given a vertex ϕ in $\mathcal{B}(\Sigma)$ and a split $S \in \Sigma$, the following three assertions are equivalent:*

(i) *$\phi(S)$ is contained in the set* $\min(\phi[\Sigma])$ *of inclusion-minimal subsets in the image $\phi[\Sigma]$ of Σ relative to ϕ.*
(ii) *The map $\phi^S \in V^\star(\Sigma)$ is, in fact, a vertex in $V(\Sigma)$ and thus forms, together with ϕ, an edge in $E(\Sigma)$.*
(iii) *There exists some vertex ψ in $V(\Sigma)$ with $\phi(S) \in \min(\phi \setminus \psi)$.*

Proof (i)⇒(ii): To see that $\phi^S \in V(\Sigma)$ holds in case $S \in \min(\phi[\Sigma])$, it suffices to note that we have (a) $\phi^S(S') \cap \phi^S(S) = \phi(S') \cap \big(X - \phi(S)\big) \neq \emptyset$ for all $S' \in \Sigma - S$ as $\phi(S') \cap \big(X - \phi(S)\big) = \emptyset$ would imply $\phi(S') \subsetneq \phi(S)$ in contradiction to $\phi(S) \in \min(\phi[\Sigma])$ and, by definition, we have (b) $\phi^S(S') \cap \phi^S(S'') = \phi(S') \cap \phi(S'') \neq \emptyset$ for all $S', S'' \in \Sigma - S$.

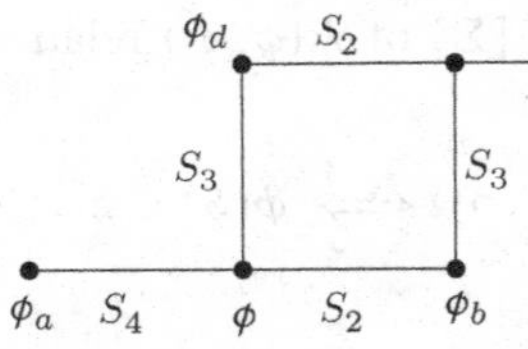

Figure 4.2 An example of a Buneman graph illustrating observations discussed in the text: The split system Σ consists of the splits $S_1 = \{c\}|\{a, b, d\}$, $S_2 = \{a, d\}|\{b, c\}$, $S_3 = \{a, b\}|\{c, d\}$, and $S_4 = \{a\}|\{b, c, d\}$, and the vertex ϕ is defined by $\phi(S_1) := \{a, b, d\}$, $\phi(S_2) := \{a, d\}$, $\phi(S_3) := \{a, b\}$, and $\phi(S_4) := \{b, c, d\}$. We see that the edges of type S_i form, for every $i = 1, 2, 3, 4$, cutsets as described in Lemma 4.5, the edges incident to ϕ correspond to the inclusion-minimal subsets in $\phi[\Sigma]$ as discussed in Corollary 4.3, and $\phi(S_i)$ corresponds to that part of $\mathcal{B}(\Sigma)$ (relative to the cutset associated with the split S_i) that contains the vertex ϕ.

(ii)$\Rightarrow$(iii): This is trivial: Just put $\psi := \phi^S$.

(iii)$\Rightarrow$(i): This follows immediately from the following, slightly more general observation using the concept of an *ideal* in a partially ordered set $M = (M, \preceq)$, that is, a subset $M' \subseteq M$ such that $m \in M'$ holds for all $m \in M$ with $m \preceq m'$ for some $m' \in M'$:

Lemma 4.2 *Given any two vertices ϕ, ψ in $\mathcal{B}(\Sigma)$, the set $\phi\backslash\psi$ is an ideal in the partially ordered set $\phi[\Sigma]$ and, hence, in $\phi\backslash\psi'$ for every map $\psi' \in V(\Sigma)$ with $\phi\backslash\psi \subseteq \phi\backslash\psi'$, that is, $A_1, A_2 \in \phi[\Sigma]$, $A_1 \in \phi\backslash\psi$, and $A_2 \subseteq A_1$ implies $A_2 \in \phi\backslash\psi$.*

Proof Let A_1, A_2 be arbitrary sets in $\phi[\Sigma]$ with $A_1 \in \phi\backslash\psi$ and $A_2 \subseteq A_1$. Then, there exist splits S_1 and S_2 in Σ with $\phi(S_1) = A_1 = X - \psi(S_1)$ and $\phi(S_2) = A_2$. Thus, our assumption $A_2 \subseteq A_1 = \phi(S_1)$ implies $\emptyset = A_2 \cap \psi(S_1)$ and, hence, $\psi(S_2) \neq A_2 = \phi(S_2)$ and, therefore, $A_2 \in \phi\backslash\psi$, as claimed.

This finishes also the proof of Lemma 4.1. □

A first simple consequence of Lemma 4.1 is the following:

Corollary 4.3 *The degree $\deg_{\mathcal{B}(\Sigma)}(\phi)$ of a vertex $\phi \in V(\Sigma)$ of the Buneman graph $\mathcal{B}(\Sigma)$ coincides with the cardinality $|\min(\phi[\Sigma])|$ of the set $\min(\phi[\Sigma])$ of inclusion-minimal subsets in $\phi[\Sigma]$.*

Proof Indeed, by definition, a vertex $\psi \in V(\Sigma)$ is incident to the vertex ϕ if and only if $|\Delta(\phi, \psi)| = 1$ holds in which case ψ is necessarily of the form $\psi = \phi^S$ for the unique split $S = S_{\phi,\psi} \in \Delta(\phi, \psi)$ and, hence, if and only if it is of this form for some split $S \in \Sigma$ with $\phi(S) \in \min(\phi[\Sigma])$.

Thus, mapping each vertex $\psi \in V(\Sigma)$ that is incident to the vertex ϕ to the set $\phi(S_{\phi,\psi}) \in \min(\phi[\Sigma])$ defines a canonical one-to-one correspondence between the set $N_{\mathcal{B}(\Sigma)}(\phi)$ of neighbors of ϕ in $\mathcal{B}(\Sigma)$ and the set $\min(\phi[\Sigma])$. □

Another consequence is the following corollary:

Corollary 4.4 (i) *$\mathcal{B}(\Sigma)$ is an isometric subgraph of $\mathcal{B}^\star(\Sigma)$.*

(ii) *More specifically, there exists, for all $\phi, \psi \in V(\Sigma)$, a canonical one-to-one correspondence between the set consisting of all sequences $\phi_0 := \phi, \phi_1, \ldots, \phi_{|\Delta(\phi,\psi)|} := \psi$ of maps in $V(\Sigma)$ with $|\Delta(\phi_{i-1}, \phi_i)| = 1$ for all $i = 1, \ldots, |\Delta(\phi, \psi)|$ and the set of all linear orders "$\preceq$" defined on $\phi \setminus \psi$ that extend the partial order of $\phi \setminus \psi$ defined by set inclusion.*

(iii) *Given any three vertices $\phi, \phi', \psi \in V(\Sigma)$, every shortest path $\phi_0 := \phi, \phi_1, \ldots, \phi_{|\Delta(\phi,\phi')|} := \phi'$ connecting ϕ and ϕ' must pass through ψ if and only if $\Delta(\phi, \phi') = \Delta(\phi, \psi) \cup \Delta(\psi, \phi')$ holds and $\phi(S) \subsetneq \phi(S')$ holds for every $S \in \Delta(\phi, \psi)$ and $S' \in \Delta(\psi, \phi')$ while, conversely, there exists a shortest path $\phi_0 := \phi, \phi_1, \ldots, \phi_{|\Delta(\phi,\phi')|} := \phi'$ in $\mathcal{B}(\Sigma)$ connecting ϕ and ϕ' not passing through ψ if and only if there exist splits $S \in \Delta(\phi, \psi)$ and $S' \in \Delta(\phi', \psi)$ with $\psi(S) \cup \psi(S') \neq X$ (or, equivalently, $\phi(S) \cap \phi'(S') \neq \emptyset$).*

Proof Clearly, (ii) implies (i) because, given any two maps $\phi, \psi \in V(\Sigma)$, there must exist, as a partial order can always be extended to a linear order, a path of length $|\Delta(\phi, \psi)|$ in $\mathcal{B}(\Sigma)$ connecting ϕ and ψ. And there can be no shorter path in $\mathcal{B}(\Sigma)$ as there is none in $\mathcal{B}^\star(\Sigma)$.

To establish (ii), assume $\phi, \psi \in V(\Sigma)$, select a split $S \in \Delta_{\min}(\phi|\psi)$ and put $\phi_1 := \phi^S$. Clearly, $\phi_1 \in V(\Sigma)$, $\{\phi, \phi_1\} \in E(\Sigma)$, $\Delta(\phi, \phi_1) = \{S\}$, and $\Delta(\phi_1, \psi) = \Delta(\phi, \psi) - S$ and, therefore, also $\phi_1 \setminus \psi = \phi \setminus \psi - \{\phi(S)\}$ holds in view of Lemma 4.1. Thus, (ii) follows easily using induction with respect to $|\Delta(\phi, \psi)|$.

The last assertion now follows from the fact that, given any three vertices $\phi, \phi', \psi \in V(\Sigma)$, every shortest path $\phi_0 := \phi, \phi_1, \ldots, \phi_{|\Delta(\phi,\phi')|} := \phi'$ connecting ϕ and ϕ' must pass through ψ if and only if (a) $\Delta(\phi, \phi') = \Delta(\phi, \psi) \cup \Delta(\psi, \phi')$ — or, equivalently, $\Delta(\phi, \psi) \cap \Delta(\psi, \phi') = \emptyset$ — holds and (b) $\phi(S) \prec \phi(S')$ holds, for all $S \in \Delta(\phi, \psi)$ and $S' \in \Delta(\psi, \phi')$, for every linear order "$\preceq$" defined on $\phi \setminus \phi'$ that extends the partial order of $\phi \setminus \phi'$ defined by set inclusion. □

It follows that the Buneman graph $\mathcal{B}(\Sigma)$ displays, as mentioned above, the underlying split system Σ in a way that directly generalizes the way X-trees display compatible split systems: Indeed, note first that, as with X-trees, there exists a canonical labeling map $\varphi = \varphi_\Sigma : X \to V(\Sigma) : x \mapsto \phi_x$ where $\phi_x = \phi_x^\Sigma$ denotes, for every $x \in X$, the map $\phi_x : \Sigma \to \mathbb{P}(X)$ from Σ into $\mathbb{P}(X)$

that associates, to any split $S \in \Sigma$, the set $S(x) \in S$ that contains the element x. Clearly, ϕ_x satisfies (BG1), and it also satisfies (BG2) since $S(x) \cap S'(x) \supseteq \{x\} \neq \emptyset$ holds for any two splits $S, S' \in \Sigma$ — more specifically, a map ϕ in $V^\star(\Sigma)$ is of the form $\phi = \phi_x$ for some $x \in X$ if and only if the intersection $\bigcap \phi[\Sigma] = \bigcap_{S \in \Sigma} \phi(S)$ of all the subsets of X in the image of ϕ is non-empty. Note that we also have

$$D_{\mathcal{B}(\Sigma)}(\phi_x, \phi_y) = |\Delta(\phi_x, \phi_y)| = |\{S \in \Sigma : S(x) \neq S(y)\}|$$

for any two elements $x, y \in X$.

Next, associating to every edge $e = \{\phi, \psi\} \in E^\star(\Sigma)$ its *type* $\kappa_\Sigma^\star(e)$, i.e., the unique split $\kappa_\Sigma^\star(e) = S_{\phi,\psi}$ in $\Delta(\phi, \psi)$, we obtain a surjective map $\kappa_\Sigma^\star : E^\star(\Sigma) \to \Sigma$ whose (also necessarily surjective) restriction to $E(\Sigma)$ we denote by κ_Σ. Clearly, one has $\Delta(\phi, \psi) = \big\{\kappa_\Sigma^\star(\{\phi_{i-1}, \phi_i\}) : i = 1, \ldots, |\Delta(\phi, \psi)|\big\}$ for every shortest path $\phi_0 := \phi, \phi_1, \ldots, \phi_{|\Delta(\phi,\psi)|} := \psi$ from a vertex ϕ to a vertex ψ in $\mathcal{B}(\Sigma)$ or $\mathcal{B}^\star(\Sigma)$. We also claim

Lemma 4.5 *The set $\kappa_\Sigma^{-1}(S)$ is, for every split $S = A|B \in \Sigma$, a cutset of $\mathcal{B}(\Sigma)$ "inducing" the split S. That is, removing the edges in $\kappa_\Sigma^{-1}(S)$ from $\mathcal{B}(\Sigma)$ yields a subgraph with precisely two connected components, one denoted by $\mathcal{B}(\Sigma|A)$ containing, among its vertices, all the vertices ϕ_a with $a \in A$, and the other, denoted by $\mathcal{B}(\Sigma|B)$, containing all the vertices ϕ_b with $b \in B$.*

Proof Let $S = A|B$ be an arbitrary split in Σ. It suffices to show that the two subsets $\mathcal{B}(\Sigma|A) := \{\phi \in V(\Sigma) : \phi(S) = A\} \supseteq \{\phi_a : a \in A\}$ and $\mathcal{B}(\Sigma|B) := \{\phi \in V(\Sigma) : \phi(S) = B\} \supseteq \{\phi_b : b \in B\}$ of $V(\Sigma)$ form (exactly the two) connected components of the subgraph of $\mathcal{B}(\Sigma)$ obtained by removing the edges in $\kappa_\Sigma^{-1}(S)$. Yet, it follows immediately from the definition of $E(\Sigma)$ that every path in $\mathcal{B}(\Sigma)$ from a vertex in $\mathcal{B}(\Sigma|A)$ to a vertex in $\mathcal{B}(\Sigma|B)$ must contain at least one edge in $\kappa_\Sigma^{-1}(S)$ while, in view of Corollary 4.4, no shortest path between any two vertices in $\mathcal{B}(\Sigma|A)$ (and, analogously, in $\mathcal{B}(\Sigma|B)$), can pass through an edge in $\kappa_\Sigma^{-1}(S)$. □

Note also that restricting the maps ϕ in $V(\Sigma)$ to a given subset Σ' of Σ induces not only a surjective *graph morphism* $\mathrm{res}_{\Sigma\to\Sigma'}^\star$ from $\mathcal{B}^\star(\Sigma)$ to $\mathcal{B}^\star(\Sigma')$, i.e., a surjective map from $V^\star(\Sigma)$ onto $V^\star(\Sigma')$ that maps every edge e in $E^\star(\Sigma)$ either onto a single vertex in $V^\star(\Sigma')$ or onto an edge in $E^\star(\Sigma')$, but also a surjective map $\mathrm{res}_{\Sigma\to\Sigma'}$ from $V(\Sigma)$ to $V(\Sigma')$, i.e., the following lemma holds:

Lemma 4.6 *Given a subset Σ' of a system Σ of X-splits, and a map $\phi' \in V(\Sigma')$, then there exists (at least) one extension $\phi \in V(\Sigma)$ of ϕ', i.e., a map $\phi \in V(\Sigma)$ with $\phi|_{\Sigma'} = \phi'$.*

Moreover, the resulting surjective graph morphism $\mathrm{res}_{\Sigma\to\Sigma'}$ *contracts every edge* $e \in E(\Sigma)$ *that is of type* $\kappa_\Sigma(e) \in \Sigma - \Sigma'$ *onto a single vertex while it maps every other edge in* $E(\Sigma)$ *onto an edge in* $E(\Sigma')$ *of the same type.*

Proof Using induction with respect to $|\Sigma|$, we may assume, without loss of generality, that $\Sigma = \Sigma' + S$ holds for some split $S = A|B$ in Σ, in which case we can extend ϕ' to a map $\phi \in V(\Sigma)$ by putting $\phi(S) := A$ if and only if $A \cap \phi'(S') \neq \emptyset$ holds for every split $S' \in \Sigma'$, and we can extend ϕ' to a map $\phi \in V(\Sigma)$ by putting $\phi(S) := B$ if and only if $B \cap \phi(S') \neq \emptyset$ holds for every split $S' \in \Sigma'$. So, we can surely always extend it in at least one of these two ways as $A \cap \phi'(S') = \emptyset$ and $B \cap \phi'(S'') = \emptyset$ for some splits $S', S'' \in \Sigma'$ would imply $\phi'(S') \cap \phi'(S'') \subseteq B \cap A = \emptyset$ in contradiction to $\phi' \in V(\Sigma')$. □

Next, we define a subset Φ of $V(\Sigma)$ (or of $V^\star(\Sigma)$, respectively) to be a *face* of $\mathcal{B}(\Sigma)$ (or of $\mathcal{B}^\star(\Sigma)$) if it is non-empty and there exists some subset $\Xi = \Xi_\Phi$ of Σ such that $\psi \in \Phi \iff \Delta(\phi, \psi) \subseteq \Xi$ holds for all maps $\phi \in \Phi$ and ψ in $V(\Sigma)$ (or in $V^\star(\Sigma)$) in which case Φ will also be called a *face of type* Ξ (of $\mathcal{B}(\Sigma)$ or of $\mathcal{B}^\star(\Sigma)$, respectively).

Remarkably, faces can be characterized in many different ways:

Lemma 4.7 *Given a non-empty subset* Φ *of* $V^\star(\Sigma)$*, the following assertions are equivalent:*

(i) Φ *is a face of* $\mathcal{B}^\star(\Sigma)$.
(ii) *There exists a subset* Σ' *of* Σ *such that* Φ *is the pre-image of a vertex of* $\mathcal{B}^\star(\Sigma')$ *relative to the graph morphism* $\mathrm{res}^\star_{\Sigma\to\Sigma'} : V^\star(\Sigma) \to V^\star(\Sigma')$.
(iii) Φ *is a convex subset of* $V^\star(\Sigma)$.
(iv) Φ *is a gated subset of* $V^\star(\Sigma)$.

Similarly, given a non-empty subset Φ *of* $V(\Sigma)$*, the following assertions are equivalent:*

(i) Φ *is a face of* $\mathcal{B}(\Sigma)$.
(ii) *There exists a subset* Σ' *of* Σ *such that* Φ *is the pre-image of a vertex of* $\mathcal{B}(\Sigma')$ *relative to the graph morphism* $\mathrm{res}_{\Sigma\to\Sigma'} : V(\Sigma) \to V(\Sigma')$.
(iii) Φ *is a convex subset of* $V(\Sigma)$.
(iv) Φ *is a gated subset of* $V(\Sigma)$.

Proof It follows immediately from the definitions that a subset Φ of $V(\Sigma)$ (or of $V^\star(\Sigma)$, respectively) is a face of type Ξ of $\mathcal{B}(\Sigma)$ (or of $\mathcal{B}^\star(\Sigma)$) for some subset Ξ of Σ if and only if Φ is the pre-image of a vertex of $\mathcal{B}(\Sigma - \Xi)$ (or of $\mathcal{B}^\star(\Sigma - \Xi)$) relative the graph morphism $\mathrm{res}_{\Sigma\to(\Sigma-\Xi)} : V(\Sigma) \to V(\Sigma - \Xi)$ or $\mathrm{res}^\star_{\Sigma\to(\Sigma-\Xi)} : V^\star(\Sigma) \to V^\star(\Sigma - \Xi)$).

Since $D_{\mathcal{B}^\star(\Sigma)}$ and $D_{\mathcal{B}(\Sigma)}$ are proper metrics, Lemma 1.4(ii) yields that every gated subset of $V^\star(\Sigma)$ or $V(\Sigma)$ is convex. Thus, it remains to show that any face is a gated subset and that any convex subset is a face.

To establish the first fact, we define, for any map $\psi \in V^\star(\Sigma)$ and any subset Φ of $V^\star(\Sigma)$, the map ψ_Φ in $V^\star(\Sigma)$ by putting

$$\psi_\Phi(S) := \begin{cases} X - \psi(S) & \text{if } \psi(S) \notin \{\phi(S) : \phi \in \Phi\}, \\ \psi(S) & \text{otherwise,} \end{cases} \tag{4.6}$$

for every split S in Σ. That is, $\psi_\Phi(S)$ coincides with $\psi(S)$ for all S in Σ for which not all maps $\phi \in \Phi$ "agree" while, if all maps $\phi \in \Phi$ map a split $S \in \Sigma$ onto the same subset A in S, $\psi_\Phi(S)$ also coincides with that subset A — implying that $\psi_\Phi(S) \in \{\psi(S), \phi(S)\}$ must hold for all $S \in \Sigma$ and $\phi \in \Phi$. Thus, if Φ is a face of type Ξ of either $\mathcal{B}(\Sigma)$ or $\mathcal{B}^\star(\Sigma)$ and ϕ is an arbitrary map in Φ, $\psi_\Phi(S) = \phi(S)$ must hold for all $S \in \Sigma - \Xi$ as all maps in Φ must coincide with ϕ on all splits not in Ξ, i.e., we have $\Delta(\psi_\Phi, \phi) \subseteq \Xi$ and, therefore, also $\psi_\Phi \in \Phi$ in case Φ is a face of $\mathcal{B}^\star(\Sigma)$, and also in case Φ is a face of $\mathcal{B}(\Sigma)$ provided $\psi_\Phi \in V(\Sigma)$ holds. Furthermore, given any split $S \in \Sigma$, we have $\psi(S) \neq \phi(S) \iff \psi_\Phi(S) \neq \phi(S)$ or $\psi(S) \neq \psi_\Phi(S)$ $\big($as $\psi(S) = \phi(S)$ implies $\psi(S) = \psi_\Phi(S) = \phi(S)\big)$ and, therefore, $\Delta(\psi, \phi) = \Delta(\psi, \psi_\Phi) \dot\cup \Delta(\psi_\Phi, \phi)$ as well as $D_{\mathcal{B}^\star(\Sigma)}(\psi, \phi) = D_{\mathcal{B}^\star(\Sigma)}(\psi, \psi_\Phi) + D_{\mathcal{B}^\star(\Sigma)}(\psi_\Phi, \phi)$ implying that any face Φ in $\mathcal{B}^\star(\Sigma)$ is indeed a gated subset of $V^\star(\Sigma)$.

Next, to show that the same holds also for faces Φ of $\mathcal{B}(\Sigma)$, it suffices to note that

$$\psi_\Phi \in V(\Sigma) \tag{4.7}$$

holds for every map $\psi \in V(\Sigma)$ and every subset Φ of $V(\Sigma)$: Indeed, given any two splits $S_1, S_2 \in \Sigma$, we have $\psi_\Phi(S_1) \cap \psi_\Phi(S_2) \neq \emptyset$ in case $\psi_\Phi(S_1) = \psi(S_1)$ and $\psi_\Phi(S_2) = \psi(S_2)$ as well as in case $\psi(S_1) \neq \psi_\Phi(S_1)$ and $\psi(S_2) \neq \psi_\Phi(S_2)$ (as the latter implies $\psi(S_1) \neq \phi(S_1)$ and $\psi(S_2) \neq \phi(S_2)$ for all $\phi \in \Phi$) while, say, $\psi(S_1) = \psi_\Phi(S_1)$ implies that $\psi(S_1) = \phi_0(S_1)$ must hold for at least one map $\phi_0 \in \Phi$ while $\psi(S_2) \neq \psi_\Phi(S_2)$ implies that $\phi(S_2) = \psi_\Phi(S_2)$ must hold for all $\phi \in \Phi$. Thus, $\psi_\Phi(S_1) \cap \psi_\Phi(S_2) = \phi_0(S_1) \cap \phi_0(S_2) \neq \emptyset$ must also hold in this case in view of $\phi_0 \in V(\Sigma)$.

Finally, to show that any non-empty convex subset Φ of $V(\Sigma)$, or of $V^\star(\Sigma)$, is a face of $\mathcal{B}(\Sigma)$ or $\mathcal{B}^\star(\Sigma)$, respectively, put $\Xi_\Phi := \bigcup_{\phi, \phi' \in \Phi} \Delta(\phi, \phi')$, and choose a map $\phi \in \Phi$ and a map $\psi \in V^\star(\Sigma)$. Clearly, if $\psi \in \Phi$ holds, then, by definition of Ξ_Φ, also $\Delta(\phi, \psi) \subseteq \Xi_\Phi$ must hold. It remains to show that, if $\Delta(\phi, \psi) \subseteq \Xi_\Phi$ holds, then $\psi \in \Phi$ holds in case Φ is a convex subset of $V^\star(\Sigma)$, and in case $\psi \in V(\Sigma)$ holds and Φ is a convex subset of $V(\Sigma)$.

To this end, we may assume by induction with respect to $|\Delta(\phi, \psi)|$ that, without loss of generality, $\Delta(\phi, \psi)$ consists of a single split S, only, and that there exists therefore, by definition of Ξ_Φ, an edge $\{\phi', \psi'\}$ in $E^\star(\Sigma)$, or in $E(\Sigma)$, with $\{\phi', \psi'\} \subseteq \Phi$ and $\Delta(\phi', \psi') = \Delta(\phi, \psi) = \{S\}$. Thus, assuming that $\phi(S) = \phi'(S) \neq \psi(S) = \psi'(S)$ and, therefore, also $\Delta(\phi, \phi') = \Delta(\psi, \psi')$ and $\Delta(\phi, \psi') = \Delta(\psi, \phi') = \Delta(\phi, \phi') + S$ holds, we have $D_{\mathcal{B}^\star(\Sigma)}(\phi, \psi') = 1 + D_{\mathcal{B}^\star(\Sigma)}(\psi, \psi') = D_{\mathcal{B}^\star(\Sigma)}(\phi, \psi) + D_{\mathcal{B}^\star(\Sigma)}(\psi, \psi')$ and, there fore, $\psi \in \Phi$ in view of the convexity of Φ and $\phi, \psi' \in \Phi$. □

Let us finally note that Corollary 4.4 suggests we consider, for any given vertex $\phi \in V(\Sigma)$, the induced subgraph $\mathcal{B}^\phi(\Sigma)$ of $\mathcal{B}(\Sigma)$ with vertex set $V(\Sigma) - \phi$ and the graph $\Gamma^\phi = \big(X^\phi, E^\phi\big)$ with vertex set $X^\phi := \{x \in X : \phi_x \neq \phi\}$ and edge set

$$E^\phi := \left\{\{x, y\} \in \binom{X^\phi}{2} : \exists_{S \in \Delta(\phi_x, \phi), S' \in \Delta(\phi_y, \phi)}\, \phi(S) \cup \phi(S') \neq X\right\}.$$

Then, Corollary 4.4 implies that the restriction of the map $\varphi_\Sigma : X \to V(\Sigma) : x \mapsto \phi_x$ to X^ϕ induces a map $\theta^\phi : \pi_0(\Gamma^\phi) \to \pi_0\big(\mathcal{B}^\phi(\Sigma)\big)$ from the set $\pi_0(\Gamma^\phi)$ of connected components of the graph Γ^ϕ into the set $\pi_0\big(\mathcal{B}^\phi(\Sigma)\big)$ of connected components of the graph $\mathcal{B}^\phi(\Sigma)$. It has been shown in [57] that this map θ^ϕ is, in fact, a bijection implying, in particular, that ϕ is a cut vertex if and only if the graph Γ^ϕ is disconnected.

More generally, we may define a face Φ of $\mathcal{B}(\Sigma)$ to be a *cut face* of $\mathcal{B}(\Sigma)$ if the induced subgraph of $\mathcal{B}(\Sigma)$ with vertex set $V(\Sigma) - \Phi$ is disconnected. It then follows easily that, given a subset Ξ of Σ, a face Φ of $\mathcal{B}(\Sigma)$ of type Ξ is a cut face if and only if the graph $\Gamma^\Phi = \big(V^\Phi, E^\Phi\big)$ is disconnected, which is defined as follows: One puts $V^\Phi := \{x \in X : \phi_x \notin \Phi\}$, picks a fixed map $\phi_0 \in \Phi$, and then puts

$$E^\Phi := \left\{\{x, y\} \in \binom{V^\Phi}{2} : \exists_{S \in \Delta(\phi_x, \phi_0) - \Xi, S' \in \Delta(\phi_y, \phi_0) - \Xi}\, \phi(S) \cup \phi(S') \neq X\right\}.$$

Indeed, to establish this fact, one applies the above observation on cut vertices to the Buneman graph $\mathcal{B}(\Sigma - \Xi)$ and its vertex $\phi_* := \mathrm{res}_{\Sigma \to (\Sigma - \Xi)}(\phi_0)$ and notes that $\Phi = \mathrm{res}^{-1}_{\Sigma \to (\Sigma - \Xi)}(\phi_*)$ must always hold.

4.2 The Buneman graph of a compatible split system

As we have seen in the previous section, we can display any split system by the corresponding Buneman graph. In this section, we will show that this representation is *consistent* in that, if the split system Σ is compatible, then the associated Buneman graph $\mathcal{B}(\Sigma)$ is (up to canonical isomorphism) the unique X-tree displaying this split system. More specifically, we shall prove the following theorem.

Theorem 4.8 *Suppose that $S = A|B$ is a split in Σ, and that ϕ, ψ are two maps in $V(\Sigma)$. Then the following hold:*

(i) *S is compatible with any other split S' in Σ if and only if there exists exactly one edge $e \in E(\Sigma)$ of type S if and only if every edge $e \in E(\Sigma)$ of type S is a bridge in $\mathcal{B}(\Sigma)$.*

(ii) *There exists exactly one shortest path in $\mathcal{B}(\Sigma)$ connecting ϕ and ψ if and only if any two splits $S_1, S_2 \in \Delta(\phi, \psi)$ are compatible.*

(iii) *Defining* $\mathrm{Incpt}(\Sigma)$ *by*

$$\mathrm{Incpt}(\Sigma) := \{\Xi \subseteq \Sigma : \textit{any two distinct splits in } \Xi \textit{ are incompatible}\},$$

and, for every $S \in \Sigma$, its subset $\mathrm{Incpt}(\Sigma|S)$ *by*

$$\mathrm{Incpt}(\Sigma|S) := \{\Xi \in \mathrm{Incpt}(\Sigma) : S \in \Xi\},$$

the number of vertices in the Buneman graph $\mathcal{B}(\Sigma)$ is given by

$$|V(\Sigma)| = |\mathrm{Incpt}(\Sigma)|, \tag{4.8}$$

the number of edges is equal to

$$|E(\Sigma)| = \sum_{\Xi \in \mathrm{Incpt}(\Sigma)} |\Xi|, \tag{4.9}$$

and, given any split $S \in \Sigma$, the number $\left|\kappa_\Sigma^{-1}(S)\right|$ of edges in $E(\Sigma)$ of type S is given by

$$\left|\kappa_\Sigma^{-1}(S)\right| = |\mathrm{Incpt}(\Sigma|S)|. \tag{4.10}$$

(iv) *Σ is compatible if and only if $\mathcal{B}(\Sigma)$ is a tree, in which case the triple $\mathcal{T} = (V(\Sigma), E(\Sigma), \varphi_\Sigma)$ is an X-tree for which, moreover, $\Sigma = \Sigma_\mathcal{T}$ holds.*

Proof (i) It follows easily from Lemma 4.5 that an edge $e = \{\phi, \psi\} \in E(\Sigma)$ is a bridge in $\mathcal{B}(\Sigma)$ if and only if it is the only edge of its type in which case it is a bridge between $\mathcal{B}(\Sigma|A)$ and $\mathcal{B}(\Sigma|B)$ in case the edge e is of type $S = A|B$. Furthermore, S is incompatible with a split $S' = \{A', B'\} \in \Sigma$ if and only if $\mathcal{B}(\{S, S'\})$ coincides with $\mathcal{B}^\star(\{S, S'\})$ and, thus, contains more than one edge of type S. So, the same must then hold as well for $\mathcal{B}(\Sigma)$ in view of Lemma 4.6.

(ii) In view of Corollary 4.4, it suffices to note that $\phi \setminus \psi$ is linearly ordered by inclusion if and only if any two splits in $\Delta(\phi, \psi)$ are compatible. Indeed, if two splits $S, S' \in \Delta(\phi, \psi)$ are compatible, one must either have $\phi(S) \subseteq \phi(S')$ or $\phi(S') \subseteq \phi(S)$ as $\phi(S) \cap \phi(S') \neq \emptyset$ and $\psi(S) \cap \psi(S') = \big(X - \phi(S)\big) \cap \big(X - \phi(S')\big) \neq \emptyset$ implies that either $\phi(S) \cap \big(X - \phi(S')\big) = \emptyset$ or $\big(X - \phi(S)\big) \cap \phi(S') = \emptyset$ must hold in case S and S' are compatible. Conversely, if the two splits $S, S' \in \Delta(\phi, \psi)$ are not compatible, neither $\phi(S) \subseteq \phi(S')$ nor $\phi(S') \subseteq \phi(S)$ can hold.

(iii) To prove that (4.8), (4.9), and (4.10) hold, we use a construction that was originally presented in [55] and yields, for every $\phi \in V(\Sigma)$, a bijective map H_ϕ from $V(\Sigma)$ onto $\text{Incpt}(\Sigma)$ which readily implies (4.8). Based on this map, we will then characterize those 2-subsets of $\text{Incpt}(\Sigma)$ that correspond to edges in $E(\Sigma)$ of given type to obtain (4.9) and (4.10).

We begin with noting that, given any two vertices ϕ, ψ in $\mathcal{B}(\Sigma)$, any two distinct splits $S_1, S_2 \in \Delta_{\min}(\psi|\phi)$ must be incompatible as neither $\psi(S_1) \cap \psi(S_2) = \emptyset$ nor $\phi(S_1) \cap \phi(S_2) = \emptyset$ nor $\psi(S_1) \subseteq \psi(S_2)$ nor $\psi(S_2) \subseteq \psi(S_1)$ can hold. Consequently, associating, given a fixed vertex $\phi \in V(\Sigma)$, to any vertex $\psi \in V(\Sigma)$ the set $\Delta_{\min}(\psi|\phi) \in \text{Incpt}(\Sigma)$ yields a map

$$H_\phi : V(\Sigma) \to \text{Incpt}(\Sigma) : \psi \mapsto \Delta_{\min}(\psi|\phi)$$

that is injective in view of Equation (4.5). And it is surjective because, given any subset $\Xi \in \text{Incpt}(\Sigma)$, there exists a map $\phi_\Xi \in V(\Sigma)$ with $H_\phi(\phi_\Xi) = \Xi$, *viz.* the map $\phi_\Xi := \phi^{\partial_\phi \Xi} \in V^\star(\Sigma)$ defined by

$$\phi_\Xi(S) := \begin{cases} X - \phi(S) & \text{if } S \in \partial_\phi \Xi, \\ \phi(S) & \text{otherwise.} \end{cases}$$

Indeed, one has $\phi_\Xi \in V(\Sigma)$ because $\phi_\Xi(S_1) \cap \phi_\Xi(S_2) \neq \emptyset$ holds for all S_1, S_2 in $\partial_\phi \Xi$ as $S_1', S_2' \in \Xi$, $\phi(S_1) \subseteq \phi(S_1')$, and $\phi(S_2) \subseteq \phi(S_2')$ implies

$$\begin{aligned} \phi_\Xi(S_1) \cap \phi_\Xi(S_2) &= \big(X - \phi(S_1)\big) \cap \big(X - \phi(S_2)\big) \\ &\supseteq \big(X - \phi(S_1')\big) \cap \big(X - \phi(S_2')\big) = \phi_\Xi(S_1') \cap \phi_\Xi(S_2') \neq \emptyset, \end{aligned}$$

the latter because S_1' and S_2' are contained in Ξ and, thus, incompatible. It also holds for all S_1, S_2 in $\Sigma - \partial_\phi \Xi$ as this implies $\phi_\Xi(S_1) \cap \phi_\Xi(S_2) = \phi(S_1) \cap \phi(S_2) \neq \emptyset$. And it holds in case $S_1 \in \partial_\phi \Xi$ and $S_2 \in \Sigma - \partial_\phi \Xi$ as $\emptyset = \phi_\Xi(S_1) \cap \phi_\Xi(S_2) = \big(X - \phi(S_1)\big) \cap \phi(S_2)$ would imply $\phi(S_2) \subseteq \phi(S_1)$ and, hence, $\phi(S_2) \subseteq \phi(S')$ for some $S' \in \Xi$ and, therefore, $S_2 \in \partial_\phi \Xi$, in contradiction to our assumption $S_2 \notin \partial_\phi \Xi$. And one has

$$\begin{aligned}H_\phi(\phi_\Xi) &= \Delta_{\min}(\phi_\Xi|\phi) = \Delta_{\max}(\phi|\phi_\Xi)\\ &= \{S \in \Delta(\phi_\Xi, \phi) : \phi(S) \in \max(\phi \backslash \phi_\Xi)\}\\ &= \{S \in \partial_\phi \Xi : \phi(S) \in \max(\phi[\partial_\phi \Xi])\}\\ &= \{S \in \Xi : \phi(S) \in \max \phi[\Xi]\} = \Xi\end{aligned}$$

where the first line follows from the definition of the map H_ϕ and (4.3), and the second one from the definition of the map ϕ_Ξ and the fact that, by definition of $\partial_\phi \Xi$, one has $\max(\phi[\partial_\phi \Xi]) = \max(\phi[\Xi])$ for every map $\phi \in V^\star(\Sigma)$ and every subset Ξ of Σ, and that $\max(\phi[\Xi]) = \phi[\Xi]$ holds for every map $\phi \in V^\star(\Sigma)$ and every subset Ξ of Σ that consists of pairwise incompatible splits. Together, this clearly implies Equation (4.8).

Next, note that Equation (4.9) follows from Equation (4.10): Indeed, if $\left|\kappa_\Sigma^{-1}(S)\right| = |\mathrm{Incpt}(\Sigma|S)|$ holds for every split $S \in \Sigma$, one has also

$$\begin{aligned}|E(\Sigma)| &= \sum_{S\in\Sigma} \left|\kappa_\Sigma^{-1}(S)\right| = \sum_{S\in\Sigma} |\mathrm{Incpt}(\Sigma|S)|\\ &= \sum_{S\in\Sigma}\ \sum_{\Xi\in\mathrm{Incpt}(\Sigma)} \delta_{S,\Xi} = \sum_{\Xi\in\mathrm{Incpt}(\Sigma)}\ \sum_{S\in\Sigma} \delta_{S,\Xi} = \sum_{\Xi\in\mathrm{Incpt}(\Sigma)} |\Xi|\end{aligned}$$

where $\delta_{S,\Xi}$ is defined, as usual, by

$$\delta_{S,\Xi} := \begin{cases} 1 & \text{in case } S \in \Xi,\\ 0 & \text{otherwise.}\end{cases}$$

Finally, to establish Equation (4.10), it suffices to note that, given a fixed map $\phi \in V(\Sigma)$ and a split $S \in \Sigma$, the bijection H_ϕ induces a bijection from the subset $\{\psi \in V(\Sigma) : \{\psi, \psi^S\} \in E(\Sigma), \psi(S) \neq \phi(S)\}$ of $V(\Sigma)$ onto the subset $\mathrm{Incpt}(\Sigma|S)$ of $\mathrm{Incpt}(\Sigma)$: Indeed, if $\psi \in V(\Sigma)$, $\{\psi, \psi^S\} \in E(\Sigma)$, and $\psi(S) \neq \phi(S)$ holds, by Lemma 4.1, we must have $\psi(S) \in \min(\psi[\Sigma])$ and, therefore $S \in \Delta_{\min}(\psi|\phi) = H_\phi(\psi)$, i.e., $H_\phi(\psi) \in \mathrm{Incpt}(\Sigma|S)$. Conversely, $\Xi \in \mathrm{Incpt}(\Sigma|S)$ implies $\psi := \phi_\Xi \in V(\Sigma)$, $S \in \Delta_{\min}(\psi|\phi)$ and, therefore, $\{\psi, \psi^S\} \in E(\Sigma)$, and $\psi(S) \neq \phi(S)$ in view of $S \in \Xi \subseteq \Delta(\phi, \psi)$.

We note in passing that there are various further results in the literature concerning the number of vertices in Buneman graphs (see e.g., [17, 42, 129]).

(iv) The last assertion follows quite easily from **each** of the Assertions (i), (ii), and (iii): It follows from Assertion (i) as a connected graph $G = (V, E)$ with vertex set V and edge set $E \subseteq \binom{V}{2}$ is a tree if and only every edge in E is a bridge. It follows from Assertion (ii) as a connected bipartite graph $G = (V, E)$ with V and E as above is a tree if and only if there exists exactly one shortest path in G between any two vertices $v, w \in V$, and there exists, conversely, for any two splits $S_1, S_2 \in \Sigma$ a pair of vertices $\phi, \psi \in V(\Sigma)$ with $S_1, S_2 \in \Delta(\phi, \psi)$. And it follows from Assertion (iii) as a finite connected graph $G = (V, E)$ with vertex set V and edge set $E \subseteq \binom{V}{2}$ is a tree if and only if $|V| = 1 + |E|$ holds. So, $\mathcal{B}(\Sigma)$ is a tree if and only if $|\text{Incpt}(\Sigma)| = \sum_{\Xi \in \text{Incpt}(\Sigma)} 1$ coincides with $1 + \sum_{\Xi \in \text{Incpt}(\Sigma)} |\Xi|$ and, hence, if and only if $|\Xi| = 1$ holds for every non-empty set $\Xi \in \text{Incpt}(\Sigma)$, that is, if and only if any two splits in Σ are compatible.

Moreover, if Σ is compatible, it follows readily from Lemma 4.5 that $\mathcal{T} = (V(\Sigma), E(\Sigma), \varphi_\Sigma)$ is an X-tree and that $\Sigma = \Sigma_{\mathcal{T}}$ holds. □

4.3 Median networks

As mentioned before, Buneman graphs have been studied in various guises and, in particular, as special instances of *median graphs*, that is, connected graphs $G = (V, E)$ for which the induced metric space (V, D_G) is a median space. Note that median graphs are natural generalizations of trees and hypercubes, and have been much investigated in the literature, see [108] for a review.

Here, we shall show that the Buneman graph of a split system is not only a median graph, but that it results by taking medians of medians of medians, and so on, of the vertices in

$$V(X) = V_\Sigma(X) := \{\phi_x : x \in X\},$$

i.e., the vertices in $V(\Sigma)$ that are labeled by elements of X. More specifically, given a subset $V' \subseteq V$ of the point set V of any metric space (V, D), define the *median hull* of V', denoted by $\mathbf{Med}(V') = \mathbf{Med}_D(V')$, either as the smallest subset of V containing V' and the set $Med_D(a, b, c)$ for any three points a, b, c it contains or, recursively, by putting

$\mathbf{Med}_0(V') := V'$,
$\mathbf{Med}_{i+1}(V') := \bigcup_{u,v,w \in \mathbf{Med}_i(V')} Med(u, v, w)$ for all $i \in \mathbb{N}$, and
$\mathbf{Med}(V') := \bigcup_{i \in \mathbb{N}} \mathbf{Med}_i(V')$.

Note that, since V is a finite set, there always exists some $i_0 \in \mathbb{N}$ with $\mathbf{Med}_{i_0}(V') = \mathbf{Med}_{i_0+1}(V')$ and, hence, with $\mathbf{Med}_i(V') = \mathbf{Med}(V')$ for all $i \geq i_0$. Furthermore, the following holds:

Theorem 4.9 *With X and Σ as above, both, the Buneman graph $\mathcal{B}(\Sigma)$ and the extended Buneman graph $\mathcal{B}^\star(\Sigma)$ are median graphs, and the vertex set $V(\Sigma)$ of $\mathcal{B}(\Sigma)$ coincides with the median hull* $\mathbf{Med}(V(X))$ *of $V(X)$ relative to $D_{\mathcal{B}^\star(\Sigma)}$ as well as $D_{\mathcal{B}(\Sigma)}$.*

Proof It is obvious that $\mathcal{B}^\star(\Sigma)$, being a hypercube, is a median graph and that, for all $\psi, \phi_1, \phi_2 \in V(\Sigma)$, we must have

$$med(\psi, \phi_1, \phi_2)(S) = \phi_\Phi(S) = \begin{cases} X - \psi(S) & \text{if } \psi(S) \notin \{\phi_1(S), \phi_2(S)\}, \\ \psi(S) & \text{otherwise,} \end{cases}$$

for $\Phi := \{\phi_1, \phi_2\}$ (cf.(4.6)) and any split $S \in \Sigma$, implying that also $\mathcal{B}(\Sigma)$ is a median graph in view of Equation (4.7).

This implies in particular that $\mathbf{Med}(V(X)) \subseteq V(\Sigma)$ holds. To show that, actually, $\mathbf{Med}(V(X)) = V(\Sigma)$ holds, it clearly suffices to note that $\phi \in \mathbf{Med}(V(X))$ holds in case there exists some $\psi \in \mathbf{Med}(V(X))$ with $\{\phi, \psi\} \in E(\Sigma)$. To establish this fact, note first that, given any $\ell + 1$ maps $\psi, \phi_1, \ldots, \phi_\ell \in V^\star(\Sigma)$, one has $\psi_{\{\phi_1,\ldots,\phi_\ell\}} = med(\psi, \psi_{\{\phi_1,\ldots,\phi_{\ell-1}\}}, \phi_\ell)$ for any $\ell \geq 2$ and all $\psi, \phi_1, \ldots, \phi_\ell \in V^\star(\Sigma)$ as $med(\psi, \psi_{\{\phi_1,\ldots,\phi_{\ell-1}\}}, \phi_\ell)(S) \neq \psi(S) \iff \psi_{\{\phi_1,\ldots,\phi_{\ell-1}\}}(S) = \phi_\ell(S) \neq \psi(S) \iff \phi_1(S) = \cdots = \phi_{\ell-1}(S) = \phi_\ell(S) \neq \psi(S) \iff \psi_{\{\phi_1,\ldots,\phi_\ell\}}(S) \neq \psi(S)$ holds for every split $S \in \Sigma$ implying, by induction relative to ℓ, that

$$\psi_{\{\phi_1,\ldots,\phi_\ell\}} \in \mathbf{Med}(V(X))$$

holds for all $\psi, \phi_1, \ldots, \phi_\ell \in \mathbf{Med}(V(X))$. Thus, given any two maps $\phi, \psi \in V(\Sigma)$ with $\{\psi, \phi\} \in E(\Sigma)$ and $\psi \in \mathbf{Med}(V(X))$, also $\phi \in \mathbf{Med}(V(X))$ must hold as, denoting the unique split in $\Delta(\phi, \psi)$ by S_0 and the, say, ℓ elements in $\phi(S_0)$ by $x_1, x_2, \ldots, x_\ell$, one has $\phi = \psi_{\{\phi_{x_1},\ldots,\phi_{x_\ell}\}}$ in view of $\Delta(\psi, \psi_{\{\phi_{x_1},\ldots,\phi_{x_\ell}\}}) = \{S_0\} = \Delta(\psi, \phi)$: Indeed, one has $\psi_{\{\phi_{x_1},\ldots,\phi_{x_\ell}\}}(S) \neq \psi(S) \iff x_1, x_2, \ldots, x_\ell \notin \psi(S) \iff \phi(S_0) \cap \psi(S) = \emptyset \iff \psi(S) \subseteq \psi(S_0) \iff \psi(S) = \psi(S_0) \iff S = S_0$ as (cf. Lemma 4.1) our assumptions imply $\psi(S_0) \in \min(\psi[\Sigma])$. So, we have indeed $\Delta(\psi, \psi_{\{\phi_{x_1},\ldots,\phi_{x_\ell}\}}) = \{S_0\} = \Delta(\psi, \phi)$ and, therefore, $\psi_{\{\phi_{x_1},\ldots,\phi_{x_\ell}\}} = \phi$, as claimed. □

It was essentially these observations that first motivated the use of median networks to analyze DNA sequence data. Although we will not go into further details here, the basic idea that was proposed for computing median networks works as follows: First align and recode the sequences of the organisms in question to obtain a collection of binary sequences, all of the same length k. Then, considering this as a collection of vertices within a k-dimensional hypercube, take their median hull to obtain their median network (see [14] for more details). This approach is now commonly used to analyze mitochondrial data sets for population data [69, 77] as, for such data sets, not much information tends to get lost in the process of recoding four-state DNA sequences (namely A,T,C,G) into binary ones [8]. An example of a median network computed for DNA data collected from some New Zealand buttercups is presented in

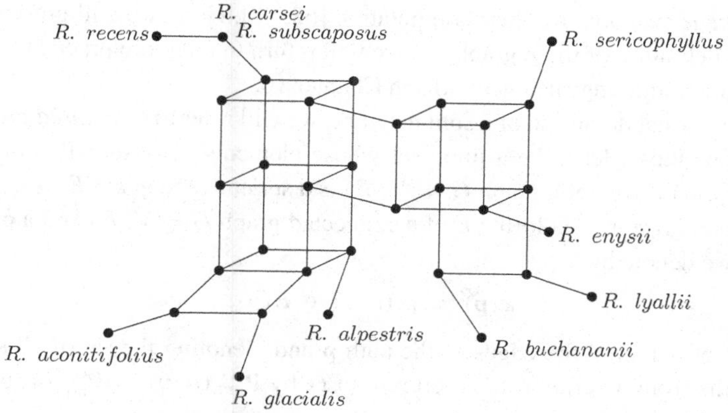

Figure 4.3 The median network computed in [97] from DNA sequences of 10 species of New Zealand buttercups (i.e., New Zealand flowers from the genus *Ranunculus*) illustrating the radiation of New Zealand buttercups during their evolution after New Zealand re-emerged well above sea level about 30 million years ago.

Figure 4.3. However, in cases where this recoding could be problematic, progress has been made by using *quasi-median graphs* [16, 115, 133] instead: These are natural generalizations of median networks that can be generated from non-binary sequences as subgraphs of Hamming graphs (see e.g., [13] and [15]).

4.4 Split networks

Even though Buneman graphs can be useful tools for analyzing DNA sequence data, they may quite often have a large number of vertices and edges which makes them somewhat impractical. For example, as we have noted above, the Buneman graph of a totally incompatible split system is a hypercube and, thus, has an exponential number of vertices in terms of the number of splits. Some methods have therefore been proposed to "reduce" [14] or "prune" [97] large Buneman graphs to produce subgraphs that are easier to visualize. The resulting subgraphs have an important property in common with Buneman graphs in that they still display splits via cutsets of edges, as Buneman graphs do according to Lemma 4.5.

Graphs having this property are known as *split networks*, and their usage for visualizing split systems arising in evolutionary studies was proposed, for example, in [7, 11]. They are now commonly used in phylogenetics, and software is available for their computation [65, 98]. Here, we will present the formal definition of these graphs — we will return to the more practical application of computing such networks in Chapter 10.

For a formal definition of a split network, we will refer to C*-colored graphs*: To define those, let C be a finite set whose elements we also call *colors*. A C*(-edge)-coloring* of a graph $G = (V, E)$ is a surjective map $\kappa : E \to C$. Furthermore, given a C-coloring κ of a connected graph $G = (V, E)$ and a path $\mathbf{p}$ in G, we denote by

$$\kappa(\mathbf{p}) := \{\kappa(e) : e \in E(\mathbf{p})\}$$

the set of colors of the edges of the path $\mathbf{p}$ and, denoting the set of all shortest paths from a vertex u to a vertex v in G by $\mathbf{P}_{\min}(u, v) = \mathbf{P}^G_{\min}(u, v)$, we denote by

$$\kappa(u, v) := \bigcap_{\mathbf{p} \in \mathbf{P}_{\min}(u,v)} \kappa(\mathbf{p})$$

the set of all colors that appear on *every* shortest path in G from u to v, implying that $|\kappa(u, v)| \leq |\kappa(\mathbf{p})| \leq D_G(u, v)$ holds, for every C-coloring κ of a connected graph (V, E), for any two vertices u, v of G, and every path $\mathbf{p} \in \mathbf{P}_{\min}(u, v)$, and that $\kappa(u, v) = \{\kappa(\{u, v\})\}$ holds for every edge $\{u, v\} \in E$. A *split graph* can now be defined as a triple (V, E, κ) such that $G = (V, E)$ is a connected and bipartite graph and $\kappa : E \to C$ is a C-coloring of (V, E) for which $D_G(u, v) = |\kappa(u, v)|$ holds for all $u, v \in V$ — implying that also $\kappa(u, v) = \kappa(\mathbf{p})$ must hold for any two vertices u, v of a split graph (V, E, κ) and every path $\mathbf{p} \in \mathbf{P}^G_{\min}(u, v)$.

Split graphs have been studied in [101, 134], and various characterizations are known. The following theorem establishes the crucial property of split graphs used in the definition of split networks below. The proof is adapted from [64].

Theorem 4.10 *Suppose that $G = (V, E)$ is a graph, C is a set, $\kappa : E \to C$ is a C-coloring of G, and (V, E, κ) is a split graph. Then, $\kappa(u, v) \subseteq \kappa(\mathbf{p})$ holds for every (and not only every shortest) path $\mathbf{p}$ from a vertex u to a vertex v in G and, given any $c \in C$, the graph $G_c := (V, E - \{e \in E : \kappa(e) = c\})$ consists of precisely two connected components while $\kappa(e) = c$ holds for an edge $e = \{u, v\} \in E$ if and only if the connected components $G_c(u)$ and $G_c(v)$ of G_c containing its two endpoints u, v are distinct. In particular:*

(i) *$\kappa(u, w) = \kappa(u, v) \,\triangle\, \kappa(v, w)$ holds for all $u, v, w \in V$.*

(ii) *A path $\mathbf{p}$ in G is a shortest path from its starting point to its end point if and only if the length of $\mathbf{p}$ coincides with the cardinality $|\kappa(\mathbf{p})|$ of $\kappa(\mathbf{p})$, i.e., if and only if all the edges of $\mathbf{p}$ have distinct colors.*

(iii) *More specifically, if* $\mathbf{p} = u_0 := u, u_1, \ldots, u_l := v$ *is any path from a vertex* u *to a vertex* v *in* G*, then* $\kappa(u, v)$ *consists of all* $c \in C$ *for which the number of all* $i \in \{1, \ldots, l\}$ *with* $\kappa(\{u_{i-1}, u_i\}) = c$ *is odd, i.e.,* $\kappa(u, v)$ *is the "symmetric difference" (or "mod-2 sum") of the family* $(\kappa(u_{i-1}, u_i))_{i=1,\ldots,l}$ *of one-color sets, and* $\mathbf{p}$ *is a shortest path if and only if it "crosses every color" at most once.*

Proof First, we show that $\kappa(u, v) \subseteq \kappa(\mathbf{p})$ holds for every (and not only every shortest) path $\mathbf{p}$ from a vertex u to a vertex v in G. We prove this by induction on the length ℓ of the path $\mathbf{p}$. If $\ell = 1$, then $\kappa(u, v) = \kappa(\mathbf{p})$ clearly holds. Now, assume $\ell \geq 2$ and let $\mathbf{p} = u_0 =: u, u_1, \ldots, u_\ell := v$ be any path in G from u to v. Note that, since G is bipartite, $D_G(u, u_{\ell-1}) = D_G(u, v) \pm 1$ holds. Put $\mathbf{p}' := u_0, u_1, \ldots, u_{\ell-1}$ and consider the two cases $D_G(u, u_{\ell-1}) = D_G(u, v) + 1$ and $D_G(u, u_{\ell-1}) = D_G(u, v) - 1$. In the first case, we have

$$\kappa(u, v) = \kappa(u_0, u_\ell) \subseteq \kappa(u_0, u_{\ell-1}) \subseteq \kappa(\mathbf{p}') \subseteq \kappa(\mathbf{p}),$$

where the first inclusion follows from the fact that extending any shortest path $\mathbf{p}_1$ from u_0 to u_ℓ by the vertex $u_{\ell-1}$ yields a shortest path $\mathbf{p}_1'$ from u_0 to $u_{\ell-1}$ and, hence, a path for which $\kappa\left(\mathbf{p}_1'\right) = \kappa(u_0, u_{\ell-1})$ as well as $\kappa\left(\mathbf{p}_1'\right) = \kappa(\mathbf{p}_1) \cup \kappa(u_{\ell-1}, u_l) = \kappa(u_0, u_\ell) \cup \kappa(u_{\ell-1}, u_\ell)$ holds. The second inclusion follows from the induction hypothesis, and the last inclusion is obvious. Similar arguments yield

$$\begin{aligned}\kappa(u, v) = \kappa(u_0, u_\ell) &= \kappa(u_0, u_{\ell-1}) + \kappa(\{u_{\ell-1}, u_\ell\}) \\ &\subseteq \kappa(\mathbf{p}') + \kappa(\{u_{\ell-1}, u_\ell\}) = \kappa(\mathbf{p})\end{aligned}$$

in the second case. This finishes the proof of the claim.

Now, with c as above, consider an arbitrary edge $e = \{u, v\}$ of G with $\kappa(e) = c$. Since κ is surjective, such an edge must exist. Furthermore, $\pi_0(G_c) = \{G_c(u), G_c(v)\}$ or, equivalently, $G_c(u) \neq G_c(v)$ and $G_c(v) \cup G_c(u) = V$ must hold: Clearly, $G_c(u) \neq G_c(v)$ follows immediately from the fact that, as we have just seen, $\{c\} = \kappa(\{u, v\}) \subseteq \kappa(\mathbf{p})$ must hold for every path $\mathbf{p}$ from u to v in G. And we have $G_c(v) \cup G_c(u) = V$ as, given any vertex w in V, we must have $|D_G(u, w) - D_G(v, w)| = 1$ in view of the fact that G is bipartite. So, we may assume — without loss of generality — that, say, $D_G(u, w) = D_G(v, w) + 1$ holds. Then, any shortest path from w to v can be extended by the edge e to a shortest path from w to u implying that $\kappa(u, w) = \kappa(v, w) + c$ and $c \notin \kappa(v, w)$ must hold in view of $|\kappa(u, w)| = D_G(u, w) = D_G(v, w) + 1 = |\kappa(v, w)| + 1$. So, $w \in G_c(v)$ must hold in this case.

Assertions (i)–(iii) now are simple consequences as, given any $c \in C$, every path $\mathbf{p}$ from a vertex u to a vertex v in G must cross c-colored edges an odd number of times in case $c \in \kappa(u, c)$ holds, and an even number of times otherwise. □

Corollary 4.11 *Continuing with the notations and assumptions introduced in Theorem* 4.10, *let U denote a subset of the vertex set V of G, let $C_U := \kappa(E_U)$ denote the κ-image of the edge set $E_U = E \cap \binom{U}{2}$ of the induced graph $G[U] = (U, E_U)$ with vertex set U, and denote by κ_U the restriction of the map κ to E_U, considered as a map from E_U onto C_U. Then, the triple (U, E_U, κ_U) is a split graph if and only if $G[U]$ is an isometric subgraph of G.*

Proof Clearly, since G is a bipartite graph, the induced subgraph $G[U]$ is necessarily bipartite, too, and the map $\kappa_U : E_U \to C_U$ is surjective by definition of C_U. To establish that (U, E_U, κ_U) is a split graph if and only if the graph $G[U]$ is an isometric subgraph of G, note first that both assertions imply that $G[U]$ is connected and, hence, that $\mathbf{P}_{\min}^{G[U]}(u, v) \neq \emptyset$ holds for any $u, v \in U$. It remains to show that $D_{G[U]}(u, v) = |\kappa_U(u, v)|$ holds for all $u, v \in U$ if and only if $D_{G[U]}(u, v) = D_G(u, v)$ holds for all $u, v \in U$.

However, we have $\kappa(u, v) \subseteq \kappa_U(u, v)$ in view of the above theorem and therefore $D_G(u, v) = |\kappa(u, v)| \leq |\kappa_U(u, v)| \leq D_{G[U]}(u, v)$. Thus, $D_G(u, v) = D_{G[U]}(u, v)$ implies $|\kappa_U(u, v)| = D_{G[U]}(u, v)$, as required.

Conversely, $|\kappa_U(u, v)| = D_{G[U]}(u, v)$ implies that no two edges in a path $\mathbf{p} \in \mathbf{P}_{\min}^{G[U]}(u, v)$ can have the same color, which, in view of Theorem 4.10(iii), implies that every path $\mathbf{p}$ in $\mathbf{P}_{\min}^{G[U]}(u, v)$ must also be a shortest path from u to v in G and, therefore, that $D_G(u, v) = D_{G[U]}(u, v)$ must hold for all $u, v \in U$. This establishes the corollary. □

With these results in hand, we now define a *split network displaying a split system* $\Sigma \subseteq \Sigma(X)$ to be a quadruple $\mathcal{N} = (V, E, \kappa, \varphi)$ such that (V, E, κ) is a split graph for a map $\kappa : E \to \Sigma$ from the edge set E onto the set Σ considered as a color set, and a labeling map $\varphi : X \to V$ such that, for each split $S \in \Sigma$, $S = \{\varphi^{-1}(V_1), \varphi^{-1}(V_2)\}$ holds where V_1 and V_2 denote the two connected components of the graph $\left(V, E - \kappa^{-1}(S)\right)$. Furthermore, we define $\mathcal{N}$ to be a *proper* split network displaying Σ if and only if it satisfies the following condition

(P) Given any two splits S and S' in Σ and any vertex $v \in V$, there exists some $x \in X$ with $S, S' \notin \kappa(\varphi(x), v)$.

We now verify that the extended Buneman graph as well as the Buneman graph can be viewed as split networks, the latter one even as a proper split network. Indeed, this follows immediately from the following, considerably more general result:

Theorem 4.12 *Given a subset U of $V^\star(\Sigma)$, consider the induced edge-colored graph (U, E_U, κ_U) with $(U, E_U) := \big(\mathcal{B}^\star(\Sigma)\big)[U]$ denoting the graph induced by U in $\mathcal{B}^\star(\Sigma)$ while the edge-coloring κ_U is given by restricting the map $\kappa_\Sigma^\star : E^\star(\Sigma) \to \Sigma$ to the edge set E_U of $\big(\mathcal{B}^\star(\Sigma)\big)[U]$ and replacing Σ by the image $\Sigma_U := \kappa_\Sigma^\star(E_U)$ of E_U relative to $\kappa_\Sigma^\star$. Then the following hold:*

(i) *(U, E_U, κ_U) is a split graph if and only if any two vertices $\phi, \psi \in U$ can be connected by a shortest path $\phi_0 := \phi, \phi_1, \ldots, \phi_{|\Delta(\phi,\psi)|} := \psi$ in $\mathcal{B}^\star(\Sigma)$ all of whose vertices are contained in U — that is, if and only if the induced graph (U, E_U) is an isometric subgraph of $\mathcal{B}^\star(\Sigma)$.*

(ii) *If, in addition, U contains $V(X)$, then adding the map φ_Σ, considered as a map from X into U, to the split graph (U, E_U, κ_U) gives rise to a split network $\mathcal{N}_U := (U, E_U, \kappa_U, \varphi_\Sigma)$ displaying Σ.*

(iii) *Conversely, given any split network $\mathcal{N} = (V, E, \kappa, \varphi)$ with vertex set V and edge set E displaying Σ, the associated map $\iota_\mathcal{N} : V \to V^\star(\Sigma)$ that maps any v in the vertex set V of $\mathcal{N}$ onto the map*

$$\phi_v = \phi_v^\mathcal{N} : \Sigma \to \mathbb{P}(X) : S \mapsto S(v) := \{x \in X : S \notin \kappa(\varphi(x), v)\}$$

maps V injectively onto a subset $U = U_\mathcal{N} := \iota_\mathcal{N}(V)$ of $V^\star(\Sigma)$ that contains $V(X)$ and induces an isometric subgraph (U, E_U) of $\mathcal{B}^\star(\Sigma)$ and, thus, gives rise to a split graph $\mathcal{N}^\star := (U, E_U, \kappa_U)$ onto which the split graph $\mathcal{N}$ is mapped isomorphically by $\iota_\mathcal{N}$. In particular, one has $\kappa(u, v) = \Delta(\phi_v, \phi_u)$ for all $u, v \in V$.

(iv) *Finally, $U_\mathcal{N}$ is actually a subset of $V(\Sigma)$ if and only if $\mathcal{N}$ is a* proper *split network displaying Σ.*

Proof Clearly, (i) follows directly from Corollary 4.11.

(ii) Note first that, if U contains $V(X)$, then $\Sigma_U = \Sigma$ must hold since, for every split $S \in \Sigma$, there exist $x, y \in X$ such that $S \in \Delta(\phi_x, \phi_y) = \kappa_U(\phi_x, \phi_y)$ holds. And it follows from Lemma 4.5 that, for every split $S = A|B \in \Sigma$, the graph $\big(U, E_U - \kappa_U^{-1}(S)\big)$ has precisely two connected components $\overline{A}$ and $\overline{B}$ so that $S = \left\{\varphi_U^{-1}\big(\overline{A}\big), \varphi_U^{-1}\big(\overline{B}\big)\right\}$ holds, as required.

(iii) Clearly, $S(v) \in S$ holds for every vertex v of $\mathcal{N}$ and every split $S \in \Sigma$: By definition, the graph $\big(V, E - \kappa^{-1}(S)\big)$ has two connected components V_1 and V_2 and $S = \varphi^{-1}(V_1)|\varphi^{-1}(V_2)$ holds. Thus, assuming that, say, $v \in V_1$ holds, we have $\varphi(x) \in V_1$ for some $x \in X$ if and only if there is no shortest path from v to $\varphi(x)$ in $\mathcal{N}$ that contains an edge in $\kappa^{-1}(S)$ or, equivalently, if and only if

$S \notin \kappa(\varphi(x), v)$ holds. So, we have $S(v) = \{x \in X : S \notin \kappa(\varphi(x), v)\} = \{x \in X : \varphi(x) \in V_1\} = \varphi^{-1}(V_1)$ and, hence, $S(v) \in \mathcal{S}$, as required.

Moreover, we have $V(X) \subseteq U$ in view of the fact that $\phi^{\mathcal{N}}_{\varphi(x)} = \iota_{\mathcal{N}}(\varphi(x)) = \phi_x$ holds for all $x \in X$ as, obviously, $x \in \varphi^{-1}(V_x) = \phi^{\mathcal{N}}_{\varphi(x)}(S)$ holds, for every split $S \in \Sigma$, for the connected component V_x of $(V, E - \kappa^{-1}(S))$ that contains $\varphi(x)$.

Further, we have $\kappa(u, v) = \Delta(\phi_v, \phi_u)$ for any two vertices $u, v \in V$ as, given any split $S \in \Sigma$, we have

$$S \in \Delta(\phi_u, \phi_v) \iff \phi_u(S) \neq \phi_v(S) \iff S(u) \neq S(v)$$
$$\iff \exists_{x \in X} S \in \kappa(\phi_x, u) \,\Delta\, \kappa(\phi_x, v) \iff S \in \kappa(u, v).$$

In particular, $\{u, v\}$ is an edge of $\mathcal{N}$ with $\kappa(\{u, v\}) = S$ if and only if $\{\phi_u, \phi_v\}$ is an edge of $\mathcal{B}^\star(\Sigma)$ and, hence, of $\mathcal{N}^\star$ with $\Delta(\phi_u, \phi_v) = \{S\}$.

Note that this immediately yields that

- the map $\iota_{\mathcal{N}}$ is injective,
- $\mathcal{N}^\star$ is a split graph,
- and $\iota_{\mathcal{N}}$ induces a canonical isomorphism from $\mathcal{N}$ onto $\mathcal{N}^\star$.

The last assertion now follows immediately from the definition of $\mathcal{B}(\Sigma)$ — more specifically, from property (BG2). □

Corollary 4.13 *There exists, for any system Σ of X-splits, a canonical one-to-one correspondence between subsets U of $V^\star(\Sigma)$ that* (a) *contain $V(X)$ and* (b) *induce an isometric subgraph $G_U = (U, E_U)$ of $\mathcal{B}^\star(\Sigma)$ and isomorphism classes of split networks $\mathcal{N} = (V, E, \kappa, \varphi)$ displaying Σ. Similarly, there exists a canonical one-to-one correspondence between subsets U of $V(\Sigma)$ that satisfy the two conditions* (a) *and* (b) *above and isomorphism classes of proper split networks $\mathcal{N} = (V, E, \kappa, \varphi)$ displaying Σ.*

It follows also that there may be many proper split networks besides the Buneman graph that display a split system Σ — see Figure 4.4 for an example. In particular, as we can see in this figure, (proper) split networks may use a much smaller number of vertices and edges for displaying a split system than the Buneman graph. An important class of split systems for which one can systematically construct split networks having considerably fewer vertices and edges than the Buneman graph are the *circular split systems* [10]: A split system $\Sigma \subseteq \Sigma(X)$ is called circular if there exists a linear order "$\leq$" of X with

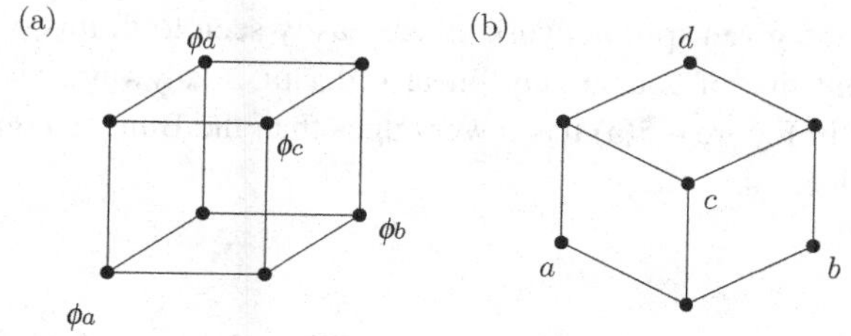

Figure 4.4 (a) The Buneman graph for the three splits $\{a, b\}|\{c, d\}$, $\{a, c\}|\{b, d\}$, and $\{a, d\}|\{b, c\}$. (b) A split network with only nine edges displaying the same three splits.

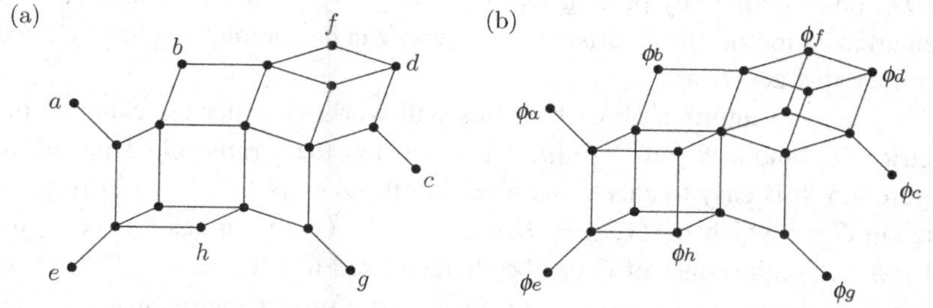

Figure 4.5 (a) A split network displaying a circular split system Σ on $X = \{a, b, \dots, g\}$. (b) The Buneman graph $\mathcal{B}(\Sigma)$.

the property that, for every split $S = A|B \in \Sigma$, there are elements $x, y \in X$ with $x \leq y$ and either $A = \{z \in X : x \leq z \leq y\}$ or $B = \{z \in X : x \leq z \leq y\}$. In this case, one may also say that the split S *fits* that linear order of X. It was shown in [64] that, for every circular split system Σ on a set X with n elements, there exists a split network displaying Σ that has $O(n^4)$ vertices and edges. Moreover, and possibly more importantly for visualization purposes, these split networks are "planar", that is, they can be drawn in the plane in such a way that no two edges cross each other.

In Figure 4.5(a), we depict a split network $\mathcal{N}$ that displays a circular split system Σ. As can be seen, no two edges in the drawing of $\mathcal{N}$ cross each other. Note also that all vertices that are labeled by an element in X are located on the "outside" of the split network and that we can obtain a linear order of X to which every split in Σ fits by "walking around" the network in a clockwise direction: Indeed, ordering the elements of X in the order in which they are met in this fashion, starting at, say, the vertex "a" yields a linear order of X for which $a < b < f < d < c < g < h < e$ holds, and the six non-trivial splits displayed by the given split network all are easily seen to fit this order (while any trivial split fits, of course, any linear order of X anyway). Note that the split network in Figure 4.5(a) has fewer edges than the Buneman graph $\mathcal{B}(\Sigma)$ depicted in Figure 4.5(b).

4.5 Split graphs and metrics: The theory of X-nets

In this section, we will discuss how split graphs relate to metrics. Note first that, given any split graph (V, E, κ) with vertex set V, edge set E, and color set C, associating to each color $c \in C$ a positive weight $\omega(c) \in \mathbb{R}_{>0}$ induces a metric D_ω on V defined by putting $D_\omega(u, v) := \sum_{c \in \kappa(u,v)} w(c)$ for all $u, v \in V$. Remarkably, metric spaces arising in this way can be completely characterized in rather abstract terms.

To get some intuition about how this will work, consider for example the metric D_G induced on $X = \{a, b, c, d, e\}$ by the graph G depicted in Figure 4.6. It is easy to check that precisely those pairs $\{x, z\} \subseteq \binom{X}{2}$ form an edge in G for which $D_G(x, z) = D_G(x, y) + D_G(y, z)$ implies $y \in \{x, z\}$ for all $y \in X$. So, the edges of G can be characterized in terms of D_G. This motivates the following definitions, all referring to a proper metric space (V, D) with a finite point set V:

(i) Two elements $u, v \in V$ are said to form a *primitive pair* $\{u, v\}$ in V if they are distinct and $[u, v]_D = \{u, v\}$ holds (or, equivalently, if the interval $[u, v]_D$ has cardinality 2). The set of primitive pairs in V is denoted by $\mathrm{Prim}(V, D)$.

(ii) The metric space (V, D) is defined to be *bipartite* if every primitive pair $\{u, v\} \in \mathrm{Prim}(V, D)$ is a gated subset.

(iii) We denote by "$\|$" the binary relation on V^2 defined, for all (u, u'), $(v, v') \in V^2$, by

$$(u, u') \| (v, v') \iff \{u', v\} \subseteq [u, v']_D \;\&\; \{u, v'\} \subseteq [u', v]_D.$$

(iv) And we say that (V, D) is an (*abstract*) L_1-*space* if "$\|$" is an equivalence relation on V^2 (as this clearly holds for the L_1-metric defined on $\mathbb{R}^n$).

Note that, given any two primitive pairs e, e' in a bipartite space (V, D), one can always label the elements in e and e' as u, v and u', v' such that $D(u, u') = \max\{D(w, w') : w \in e, w' \in e'\}$ holds, in which case $v, v' \in [u, u']_D$ must hold, and one has either $(u, v) \| (v', u')$ or $D(u, u') = D(u, v) + D(v, v') + D(v', u')$ (or, equivalently, $v \in [u, v']_D$ and $v' \in [v, u']_D$). Using

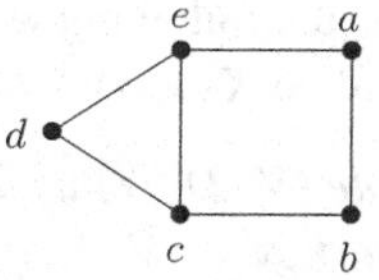

Figure 4.6 The graph G is completely determined by the induced metric D_G.

this terminology, the following holds (for a proof of this and many related results, see [52]):

Theorem 4.14 *Given a finite metric space (V, D), there exists a split graph (V, E, κ) with vertex set V, edge set E, and color set C and a positive weighting $\omega : C \to \mathbb{R}_{>0}$ such that D coincides with the induced metric D_ω if and only if (V, D) is a bipartite L_1-space in which case the split graph (V, E, κ) is uniquely determined, up to relabeling, by V and D.*

Next, consider a split network $\mathcal{N} = (V, E, \kappa, \varphi)$ displaying a system Σ of X-splits as defined in the previous section, and let $\omega : \Sigma \to \mathbb{R}_{>0}$ be an arbitrary positive weighting of Σ. It can be shown that $\mathcal{N}$ is a proper split network if and only if the labeling map φ has the following property (P′) relative to the induced metric D_ω:

(P′) For any two primitive pairs $\{u_1, v_1\}, \{u_2, v_2\}$ in V for which some $u \in V$ with $u_1 \in [u, v_1]_{D_\omega}$ and $u_2 \in [u, v_2]_{D_\omega}$ exists, there exists also some $x \in X$ with $u_1 \in [\varphi(x), v_1]_{D_\omega}$ and $u_2 \in [\varphi(x), v_2]_{D_\omega}$.

In general, φ will always satisfy the following, slightly weaker property:

(P″) For any two primitive pairs $\{u_1, v_1\}, \{u_2, v_2\}$ in V for which neither $(u_1, v_1)\|(u_2, v_2)$ nor $(u_1, v_1)\|(v_2, u_2)$ holds, there exist three elements $x, y, z \in X$ such that the gates of $\varphi(x)$ and $\varphi(y)$ relative to $\{u_1, v_1\}$ coincide and are distinct from the gate of $\varphi(z)$ relative to $\{u_1, v_1\}$ while the gates of $\varphi(z)$ and $\varphi(y)$ relative to $\{u_2, v_2\}$ coincide and are distinct from the gate of $\varphi(x)$ relative to $\{u_2, v_2\}$.

We therefore define a triple (V, D, ψ) consisting of a finite set V, a metric D defined on D, and a map $\psi : X \to V$, to be a *proper X-net* if the metric space (V, D) is a bipartite L_1-space and the map ψ satisfies the condition (P′) relative to D (with φ replaced by ψ, of course) and we define it to be just an *X-net* if (V, D) is a bipartite L_1-space and the map ψ satisfies the condition (P″) relative to D.

Referring to these definitions, the following was shown in [52] (where, however, proper X-nets were just called X-nets):

Theorem 4.15 *For every triple (V, D, ψ) as above, there exists a system Σ of X-splits, a proper split network $\mathcal{N} = (V, E, \kappa, \varphi)$, and a map $\omega : \Sigma \to \mathbb{R}_{>0}$ such that the associated triple (V, D_ω, φ) coincides with the given triple (V, D, ψ) if and only if that triple is a proper X-net — in which case φ must, of course, coincide with ψ, Σ is uniquely determined by the triple (V, D, ψ), and so is — up to canonical isomorphism — the whole network $\mathcal{N} = (V, E, \kappa, \varphi)$.*

Note that Theorem 4.15 remains true if the term "proper" is dropped everywhere. We leave the details to the reader.

5

From metrics to networks: The tight span

In the previous chapter, we saw how to obtain a network from a split system. We now explore how, using the tight-span construction, we can obtain a network from a metric.

5.1 The tight span

Given an arbitrary set X and an arbitrary symmetric bivariate map $D : X \times X \to \mathbb{R}$,

- let $\mathbb{R}^X$ denote the set of all maps from X into $\mathbb{R}$,
- for any map $f \in \mathbb{R}^X$, let $\|f\| := \sup_{x \in X} |f(x)|$ denote its L_∞-norm,
- for each $a \in X$, let $k_a = k_a^D \in \mathbb{R}^X$ denote the associated *Kuratowski map* $X \to \mathbb{R} : x \mapsto D(a, x)$,
- let $P(X, D)$ or , simply, $P(D)$ denote the subset of $\mathbb{R}^X$ that consists of those maps $f \in \mathbb{R}^X$ that satisfy the following condition:

  (P) $D(x, y) \leq f(x) + f(y)$ holds for all $x, y \in X$,
- and let $T(X, D)$ or, simply, $T(D)$, denote the *tight span* of D that is defined to consist of all those maps in $P(D)$ that are minimal with respect to the canonical "pointwise" partial ordering of $\mathbb{R}^X$ defined, for all $f, g \in \mathbb{R}^X$, by

$$f \leq g \Leftrightarrow g \geq f \Leftrightarrow f(x) \leq g(x) \text{ for all } x \in X,$$

  i.e., we put $T(X, D) := \min_{\leq} \big(P(X, D)\big)$.

The tight span is of interest in particular when D is a metric in which case it may also be called the tight span of the associated metric space (X, D). In this case, it was first studied in a ground-breaking paper by John Isbell from 1964 concerning *injective metric spaces* [103]. Subsequently, the tight span was

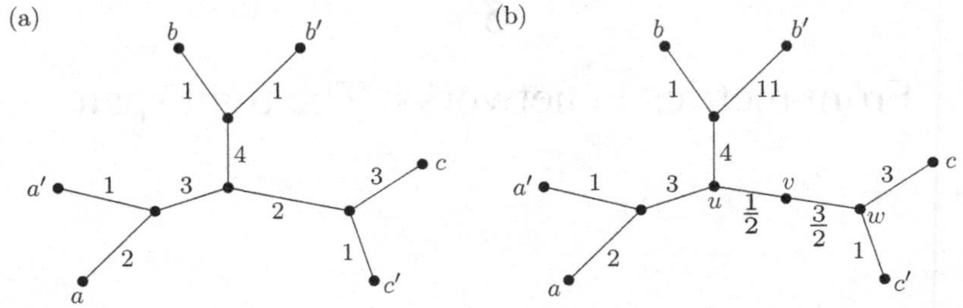

Figure 5.1 (a) An edge-weighted phylogenetic tree on $X = \{a, a', b, b', c, c'\}$ used to illustrate Kuratowski's construction and how to obtain maps in the tight span of a treelike metric D on X. For example, k_a maps a onto 0, a' onto 3, c' onto 8, and all other elements in X onto 10. (b) The map f_u associated to the central vertex u maps a' onto 4, c' onto 3, and all other elements in X onto 5, and it is easily seen to be contained in $T(D)$. Additional vertices like the vertex v of degree 2 also yield maps in the tight span of the induced metric.

rediscovered in 1984 [50] where it was linked to treelike spaces, networks, and the cohomological dimension of groups. Intriguingly, it was rediscovered yet again by Chrobak and Lamore in 1994 in the context of the so-called 3-server problem [43].

Although the definition of the tight span may appear a bit strange at first glance, tight spans arise quite naturally when considering phylogenetic trees. For example, consider the metric $D := D_{\mathcal{T}}$ associated to the simple edge-weighted phylogenetic X-tree $\mathcal{T} = (V, E, \omega)$ depicted in Figure 5.1(a) with leaf set $X := \{a, a', b, b', c, c'\}$. To every vertex w of $\mathcal{T}$, we can associate the map $f_w : X \to \mathbb{R}$ in $\mathbb{R}^X$ defined by putting $f_w(x) := D_{(V,E,\omega)}(w, x)$ for every $x \in X$. It is easy to check that any such map is contained in the tight span $T(D)$ of D. Indeed, this even holds for any arbitrarily inserted vertex v of degree 2: That is, we can arbitrarily insert a vertex v into any edge $\{u, w\}$ of $\mathcal{T}$ by adding a new vertex v to $\mathcal{T}$, replacing the edge $\{u, w\}$ by the two edges $\{u, v\}$ and $\{v, w\}$, and assigning positive edge weights $\omega(\{u, v\})$ and $\omega(\{v, w\})$ so that $\omega(\{u, v\}) + \omega(\{v, w\}) = \omega(\{u, w\})$ holds (thus "splitting" this edge into two). In this way, we still get maps of the form f_v that are contained in $T(D)$ $\big($cf. Figure 5.1(b)$\big)$. And with some more effort, one can also check that, conversely, there exists, for every map $f \in T(D)$, either a vertex v of $\mathcal{T}$ or a way to split some edge in $\mathcal{T}$ into two by inserting an additional vertex v as just described so that $f = f_v$ holds, that is, $T(D)$ parameterizes exactly "all" the points in $\mathcal{T}$.

The following result provides a useful characterization of the maps in the tight span that holds for arbitrary symmetric bivariate maps D:

Lemma 5.1 *For any symmetric bivariate map D defined on a set X, a map $f \in \mathbb{R}^X$ is contained in the associated tight span $T(D)$ if and only if*

$$f(x) = \sup\{D(x, y) - f(y) : y \in X\} \tag{5.1}$$

holds for every $x \in X$. In particular,

$$\sup_{x \in X} \big(f(x) - g(x)\big) = \sup_{x,y \in X} \big(D(x, y) - f(y) - g(x)\big) = \sup_{y \in X} \big(g(y) - f(y)\big) \tag{5.2}$$

and, therefore, also

$$\sup_{x \in X} \big(f(x) - g(x)\big) = \sup_{x \in X} |f(x) - g(x)| = \|f - g\| = \|f, g\| \tag{5.3}$$

holds for the L_∞-distance $\|f, g\|$ for any two maps $f, g \in T(D)$ (i.e., the L_∞-norm $\|f - g\|$ of their difference $f - g$), $k_a \in T(D)$ holds for every $a \in X$ if and only if D is a metric, and a map $f \in T(D)$ coincides with the Kuratowski map k_a for some $a \in X$ in case D is a metric if and only if $f(a) = 0$ holds.

Proof Let f be an arbitrary map in $T(D)$ and assume, for a contradiction, that there exists some $x \in X$ with $f(x) \neq \sup_{y \in X} \big(D(x, y) - f(y)\big)$. Since $f \in P(D)$ and, therefore, $f(x) + f(y) \geq D(x, y)$ for all $y \in X$, we must have $f(x) > \sup_{y \in X} \big(D(x, y) - f(y)\big)$ and therefore, in particular, $f(x) > D(x, x) - f(x)$ or, equivalently, $f(x) > \frac{D(x,x)}{2}$.

Now define $f' : X \to \mathbb{R}$ by $f'(z) := f(z)$ for all $z \in X - x$ and $f'(x) := \max\left\{\sup_{y \in X} \big(D(x, y) - f(y)\big), \frac{D(x,x)}{2}\right\}$. Clearly, we have $f' \lneqq f$. However, we also have $f' \in P(D)$ which yields the required contradiction: Since $f \in P(D)$, it suffices to show that $f'(x) + f'(z) \geq D(x, z)$ holds for all $z \in X$. Yet, if $z = x$, this follows immediately from the definition of f', and if $z \in X - x$, then $f'(x) + f'(z) \geq \big(D(x, z) - f(z)\big) + f(z) = D(x, z)$ will also hold.

Conversely, suppose that $f(x) = \sup_{y \in X} \big(D(x, y) - f(y)\big)$ holds for every $x \in X$. Then, $f \in P(D)$ must hold as $f(x) = \sup_{y \in X} \big(D(x, y) - f(y)\big) \geq D(x, z) - f(z)$ holds for all $x, z \in X$. To show that $f \in T(D)$ also holds, consider an arbitrary map $g \in P(D)$ with $g \leq f$. Then, $f(x) = \sup_{y \in X} \big(D(x, y) - f(y)\big) \leq \sup_{y \in X} \big(D(x, y) - g(y)\big) \leq g(x)$ holds for all $x \in X$ and, hence, $f \leq g$ which implies $f = g$, as required. This establishes Equations (5.1), (5.2), and (5.3).

It is also obvious that $k_a = k_a^D \in P(D)$ holds for some $a \in X$ if and only if the inequality $D(x, y) \leq D(a, x) + D(a, y)$ holds for all $x, y \in X$. And it is easy to see that D is a metric if and only if $k_a \in T(D)$ holds for every $a \in X$:

Indeed, if D is a metric, we have

$$\begin{aligned} k_a(x) &= D(a,x) \geq \sup{}_{y\in X}\big(D(x,y) - D(a,y)\big) \\ &= \sup{}_{y\in X}\big(D(x,y) - k_a(y)\big) \geq D(a,x) - k_a(a) = k_a(x) \end{aligned}$$

for all $a, x, y \in X$. Conversely, if $k_a \in T(D)$ holds for some $a \in X$, not only must $D(x, y) \leq D(a, x)+D(a, y)$ hold for all $x, y \in X$ and, hence, in particular $0 \leq D(a, a)$ in view of $D(a, a) \leq D(a, a)+D(a, a)$, but also $D(a, a) = 0$ as required: $D(a, a) > 0$ would imply that $k'_a \lneqq k_a$ would hold for the map

$$k'_a : X \to \mathbb{R} : x \mapsto \begin{cases} \frac{1}{2}D(a,a) & \text{if } x = a, \\ D(a,x) & \text{otherwise,} \end{cases}$$

that would also be contained in $P(D)$ in contradiction to the definition of $T(D)$.

Finally, if D is a metric and $f(a) = 0$ holds for some $f \in T(D)$, we have $f(x) = f(x) + f(a) \geq D(x, a) = k_a(x)$ for every $x \in X$ and, therefore, $f = k_a$ in view of $f, k_a \in T(D)$. □

We will now go on to further investigate the tight span in case D is a metric while returning to considering more general maps D in Chapter 9. To begin with, we note

Proposition 5.2 *Assume that (X, D) is a metric space. Then, the following hold:*

(i) *$P(D)$ is contained in $\mathbb{R}^X_{\geq 0}$.*

(ii) $\sup\{f(y) - D(x, y) : y \in X\} = \sup\{D(x, y) - f(y) : y \in X\} = \|f, k_x\| = f(x)$ *holds for all $x \in X$ and $f \in T(D)$.*

(iii) *$T(D)$ is a proper metric space relative to the restriction L^D_∞ of the map $L_\infty : \mathbb{R}^X \times \mathbb{R}^X \to \mathbb{R} \cup \{+\infty\} : (f, g) \mapsto \|f, g\|$ to $T(D)$, i.e., $0 < \|f, g\| < +\infty$ holds for any two distinct maps $f, g \in T(D)$.*

(iv) *The* Kuratowski embedding

$$\iota = \iota_D : X \to T(D) : x \mapsto k_x$$

of X into its tight span (considered as a metric space relative to L^D_∞) is a canonical isometric embedding of (X, D) into $T(D)$, i.e., $\|k_x, k_y\| = D(x, y)$ holds for all x, y in X, while its image $\iota_D(X)$ coincides with the set of all $f \in T(D)$ for which $supp(f) := \{x \in X : f(x) \neq 0\}$ differs from X.

Proof (i) Clearly, $f(x) + f(x) \geq D(x, x) = 0$ must hold for all $f \in P(D)$ and $x \in X$ in case D is a metric.

(ii) $f(x) = \sup_{y \in X}\big(D(x, y) - f(y)\big) = \sup_{y \in X}\big(k_x(y) - f(y)\big) = \|k_x, f\|$ holds for all $f \in T(D)$ and $x \in X$ by definition and in view of Lemma 5.1.

(iii) Choosing any $x \in X$, (ii) implies that $\|f, g\| \leq \|f, k_x\| + \|k_x, g\| = f(x) + g(x) < \infty$ holds for all $f, g \in T(D)$.

(iv) Assertion (ii) implies that $\|k_x, k_y\| = k_y(x) = D(x, y)$ holds for all $x, y \in X$, i.e., that the map $\iota_D : X \to T(D)$ is indeed an isometric embedding of X into $T(D)$ (and therefore injective if and only if D is proper). That its image $\iota_D(X)$ coincides with the set of all $f \in T(D)$ for which $supp(f)$ differs from X follows immediately from the last assertion in Lemma 5.1. □

Remarkably, L^D_∞ satisfies the 4-point condition characterizing treelike metrics whenever (and, hence, if and only if) D does. More generally, defining D to be Δ-*hyperbolic* for some $\Delta \in \mathbb{R}$ (cf. [82]) if and only if

$$D(x, y) + D(u, v) \leq \Delta + \max\{D(x, u) + D(y, v), D(x, v) + D(y, u)\}$$

holds for all $x, y, u, v \in X$, we have

Theorem 5.3 *A metric $D : X \times X \to \mathbb{R}$ is Δ-hyperbolic for some $\Delta \in \mathbb{R}$ if and only if the induced metric L^D_∞ is Δ-hyperbolic.*

Proof Indeed, given any $f_1, f_2, g_1, g_2 \in T(D)$, we have (cf. Lemma 5.1)

$$\begin{aligned}
&\|f_1, f_2\| + \|g_1, g_2\| \\
&= \sup\big(D(x, y) - f_1(x) - f_2(y)\big) + \sup\big(D(u, v) - g_1(v) - g_2(u)\big) \\
&= \sup\big(D(x, y) + D(u, v) - f_1(x) - f_2(y) - g_1(u) - g_2(v)\big) \\
&\leq \Delta + \sup\left(\max\begin{Bmatrix} D(x, v) + D(y, u) \\ D(x, u) + D(y, v) \end{Bmatrix} - f_1(x) - f_2(y) - g_1(u) - g_2(v)\right) \\
&= \Delta + \max\begin{Bmatrix} \sup\big(D(x, u) + D(y, v) - f_1(x) - g_1(u) - f_2(y) - g_2(v)\big) \\ \sup\big(D(x, u) + D(y, v) - f_1(x) - g_2(v) - f_2(y) - g_1(u)\big) \end{Bmatrix} \\
&= \Delta + \max\{\|f_1, g_1\| + \|f_2, g_2\|, \|f_1, g_2\| + \|f_2, g_1\|\}
\end{aligned}$$

with the supremum always taken over all variables x, y, u, v mentioned (while "$\max\{\alpha, \beta\}$" always refers to the maximum of the two numbers α and β, of course). □

In particular, replacing f_1, f_2, g_1, and g_2 by f, g, h, k_x, k_y, and k_z appropriately, we get

Corollary 5.4 *If D is Δ-hyperbolic, one has*

$$\|f, g\| + h(x) \leq \Delta + \max\{\|f, h\| + g(x), \|g, h\| + f(x)\},$$

$$f(x) + g(y) \leq \Delta + \max\{f(y) + g(x), \|f, g\| + D(x, y)\},$$

$$\|f, g\| + D(x, y) \leq \Delta + \max\{f(x) + g(y), f(y) + g(x)\}$$

or, equivalently,

$$\begin{aligned}\|f, g\| &\leq \Delta + \max\{f(x) + g(y), f(y) + g(x)\} - D(x, y)\\ &= \Delta + (f(x) + f(y) - D(x, y)) + \max\{g(y) - f(y), g(x) - f(x)\},\end{aligned}$$

and

$$f(x) + D(y, z) \leq \Delta + \max\{D(x, y) + f(z), D(x, z) + f(y)\}$$

or, equivalently,

$$\begin{aligned}&\min\{f(x) + f(y) - D(x, y), f(x) + f(z) - D(x, z)\}\\ &\leq \Delta + \big(f(y) + f(z) - D(y, z)\big)\end{aligned}$$

for all $f, g, h \in T(D)$ and all $x, y, z \in X$. In particular, the binary relation "$\vee_f$", defined for any $f \in \mathbb{R}^X$ on X by

$$x \vee_f y \Leftrightarrow f(x) + f(y) > D(x, y)$$

for all $x, y \in X$, is an equivalence relation on $\{x \in X : x \vee_f x\} = supp(f)$ in case D is treelike, and we will denote the corresponding partition of $supp(f)$ by Π_f in this case.

One may also wonder what happens if we take the "tight span of the tight span of a metric", i.e., what can we say about $T^2(D) := T\left(T(D), L_\infty^D\right)$? Remarkably, we just get back $T(D)$:

Theorem 5.5 *Suppose that (X, D) is a metric space. Then, the isometric embedding*

$$\iota_{L_\infty^D} : T(D) \to T^2(D) : f \mapsto \left(f^* := k_f^{L_\infty^D} : T(D) \to \mathbb{R} : g \mapsto \|f, g\|\right)$$

is a bijective isometry.

Proof Since $\iota_{L_\infty^D}$ is an isometric embedding and, therefore, injective, it suffices to show that $\iota_{L_\infty^D}$ is surjective. To this end, consider an arbitrary map

$F \in T^2(D)$ and the associated map $f : X \to \mathbb{R} : x \mapsto F(k_x)$. It clearly suffices to show that $f \in T(D)$ and $f^* = F$ holds. To see that $f \in T(D)$ holds, note first that

$$\begin{aligned} f(x) = F(k_x) &= \sup_{g \in T(D)} \big(\|k_x, g\| - F(g)\big) = \sup_{g \in T(D)} \big(g(x) - F(g)\big) \\ &= \sup_{g \in T(D)} \big(\sup_{y \in X} \big(D(x,y) - g(y)\big) - F(g)\big) \\ &= \sup_{y \in X} \big(D(x,y) - \inf_{g \in T(D)} (F(g) + \|g, k_y\|)\big) \\ &= \sup_{y \in X} \big(D(x,y) - F(k_y)\big) = \sup_{y \in X} \big(D(x,y) - f(y)\big) \end{aligned}$$

holds for all $x \in X$ as $F(k_y) \geq \inf_{g \in T(D)}(F(g) + \|g, k_y\|) \geq F(k_y)$ and, therefore,

$$\inf_{g \in T(D)} (F(g) + \|g, k_y\|) = F(k_y)$$

holds for all $F \in T^2(D)$ and $y \in X$: Indeed, putting $g := k_y$, we see that $F(k_y) \geq \inf_{g \in T(D)}(F(g) + \|g, k_y\|)$ must always hold while the inequality $F(g) + \|g, k_y\| \geq F(k_y)$ follows from the triangle inequality applied to F, g^*, and k_y^* as the respective distances between these maps relative to the canonical metric defined on $T^2(D)$ coincide, in view of the above results, with $F(g)$, $\|g, k_y\|$, and $F(k_y)$, respectively. Thus, $f \in T(D)$ holds in view of Lemma 5.1, as claimed.

To show that also $\iota_{L_\infty^D}(f) = F$ holds, it suffices — in view of $\iota_{L_\infty^D}(f), F \in T^2(D)$ — to note that $\big(\iota_{L_\infty^D}(f)\big)(g) \leq F(g)$ holds for every map $g \in T(D)$ which is easily done:

$$\begin{aligned} \big(\iota_{L_\infty^D}(f)\big)(g) = \|f, g\| &= \sup_{x \in X} \big(g(x) - f(x)\big) = \sup_{x \in X} \big(g(x) - F(k_x)\big) \\ &= \sup_{x \in X} \big(g(x) - k_x(x) - F(k_x)\big) \\ &\leq \sup_{x \in X} \big(\sup_{h \in T(D)} \big(g(x) - h(x) - F(h)\big)\big) \\ &= \sup_{h \in T(D)} \big(\sup_{x \in X} \big(g(x) - h(x)\big) - F(h)\big) \\ &= \sup_{h \in T(D)} \big(\|g, h\| - F(h)\big) = F(g). \end{aligned}$$ □

As pointed out in [50], we can think of the tight span of a metric space as being an abstract convex hull of that space, and so, referring to this analogy, the last result may be interpreted as asserting that this convex hull coincides with its own convex hull. This analogy can be made more formal in terms of category theory where tight spans play the role of *injective objects*. This was already done by John Isbell in his 1964 paper, and we refer the interested reader to his paper [103], and [50], for more details.

5.2 A canonical contraction from *P(D)* onto *T(D)*

In this section, we will show that the tight span $T(D)$ of a metric D is a retract of $P(D)$, that is, there exists a (canonical) contraction $\rho = \rho_D$ from $P(D)$ onto $T(D)$. Amongst other things, this implies in particular that $T(D)$ is *contractible*.

To define ρ_D, note first that it follows immediately from our definitions that $f \in P(D)$ holds for some $f \in \mathbb{R}^X$ if and only if $f_\# \leq f$ holds for the map

$$f_\# : X \to \mathbb{R} \cup \{\infty\} : x \mapsto \sup\{D(x, y) - f(y) : y \in X\},$$

and that $f \in T(D) \Leftrightarrow f = f_\#$ holds for all $f \in \mathbb{R}^X$. Note also that, in view of the fact that

$$\begin{aligned} f_\#(x) &= \sup\{D(x, y) - f(y) : y \in X\} \\ &= \sup\{D(x, y) - D(y, z) + D(y, z) - f(y) : y \in X\} \\ &\leq \sup\{D(x, y) - D(y, z) : y \in X\} + \sup\{D(y, z) - f(y) : y \in X\} \\ &= D(x, z) + f_\#(z) \end{aligned}$$

holds for all $x, z \in X$, one has either $f_\# \in \mathbb{R}^X$ or $f_\#(x) = +\infty$ for all $x \in X$. We denote the set $\{f \in \mathbb{R}^X : f_\# \in \mathbb{R}^X\}$ by $P_\#(D)$. Clearly, $f \in P_\#(D)$ holds for some $f \in \mathbb{R}^X$ if and only if $\sup\{g(y) - f(y) : y \in X\} < +\infty$ holds for some or, equivalently, for all maps $g \in T(D)$ as, given any $f \in \mathbb{R}^X$, we have

$$\begin{aligned} &\sup\{g(y) - f(y) : y \in X\} \\ &\quad = \sup\{g(y) - D(x, y) + D(x, y) - f(y) : y \in X\} \\ &\quad \leq \sup\{g(y) - D(x, y) : y \in X\} + \sup\{D(x, y) - f(y) : y \in X\} \\ &\quad = \|g, k_x\| + f_\#(x) = g(x) + f_\#(x) \end{aligned}$$

and

$$\begin{aligned} f_\#(x) &= \sup\{D(x, y) - f(y) : y \in X\} \\ &= \sup\{D(x, y) - g(y) + g(y) - f(y) : y \in X\} \\ &\leq \sup\{D(x, y) - g(y) : y \in X\} + \sup\{g(y) - f(y) : y \in X\} \\ &= g(x) + \sup\{g(y) - f(y) : y \in X\} \end{aligned}$$

for all $x \in X$ and $g \in T(D)$.

Furthermore, we have

$$f_\#(x) + f(y) = \sup\{D(x,z) - f(z) : z \in X\} + f(y)$$
$$\geq \big(D(x,y) - f(y)\big) + f(y) = D(x,y)$$

for all $f \in \mathbb{R}^X$ and $x, y \in X$ and therefore, denoting for every map $f \in \mathbb{R}^X$ the map $\frac{f+f_\#}{2}$ by ∂f, also $\partial f \in P(D)$ for all $f \in P_\#(D)$: Indeed, we have

$$\begin{aligned}(\partial f)(x) + (\partial f)(y) &= \frac{f(x) + f_\#(x)}{2} + \frac{f(y) + f_\#(y)}{2} \\ &= \frac{f_\#(x) + f(y)}{2} + \frac{f_\#(y) + f(x)}{2} \\ &\geq \frac{D(x,y)}{2} + \frac{D(x,y)}{2} = D(x,y)\end{aligned}$$

for all $f \in \mathbb{R}^X$ and $x, y \in X$. Moreover, "$f_\# \leq f \Leftrightarrow f \in P(D)$" implies that also

$$f \in P(D) \Leftrightarrow f_\# \leq f \Leftrightarrow \partial f \leq f$$

and, therefore, also $\partial^2 f := \partial(\partial f) \leq \partial f$ holds for every $f \in P_\#(D)$ as well as

$$f \in T(D) \Leftrightarrow f_\# = f \Leftrightarrow \partial f = f$$

holds for every $f \in \mathbb{R}^X$. In particular, ∂ defines a map from $P_\#(D)$ into itself that decreases every map $f \in P(D)$ and whose fix-point set is $T(D)$.

Next, note that $g \leq f$ implies $f_\# \leq g_\#$ for all $f, g \in \mathbb{R}^X$. In consequence, $f \in P(D)$ or, equivalently, $f_\# \leq \partial f = \frac{f+f_\#}{2} \leq f$ implies that $f_\# \leq (\partial f)_\# \leq \partial^2 f \leq \partial f \leq f$ and, thus, in view of

$$\partial f - f_\# = \frac{f + f_\#}{2} - f_\# = \frac{f - f_\#}{2} = f - \frac{f + f_\#}{2} = f - \partial f,$$

also

$$\begin{aligned}\|\partial f, \partial^2 f\| &= \left\|\partial f - \frac{\partial f + (\partial f)_\#}{2}\right\| = \left\|\frac{\partial f - (\partial f)_\#}{2}\right\| \\ &\leq \left\|\frac{\partial f - f_\#}{2}\right\| = \left\|\frac{f - \partial f}{2}\right\|\end{aligned}$$

holds. So, defining the operator $\partial^k : P_\#(D) \to P_\#(D)$ for all $k \in \mathbb{N}$ recursively by putting $\partial^0 := Id_{P_\#(D)}$ and $\partial^{k+1} := \partial \circ \partial^k$ for all $k \in \mathbb{N}$, the above facts imply that

$$(f^{(k)})_\# \leq (f^{(k+1)})_\# \leq f^{(k+1)} \leq f^{(k)} \in P(D)$$

as well as

$$\|f^{(k+1)}, (f^{(k+1)})_\#\| \leq \frac{1}{2}\|f^{(k)}, (f^{(k)})_\#\|$$

holds, for all $k \in \mathbb{N}_{>0}$ and any map $f \in P_\#(D)$, for the sequence $(f^{(k)})_{k\in\mathbb{N}}$ defined by $f^{(k)} := \partial^k f$ — or, in recursive form, by $f^{(0)} := f$ and $f^{(k+1)} := \partial f^{(k)}$ — for all $k \in \mathbb{N}$. More precisely, the above arguments show that

$$|f^{(k+1)}(x) - (f^{(k+1)})_\#(x)| \leq \frac{1}{2}|f^{(k)}(x) - (f^{(k)})_\#(x)|$$

holds for every $f \in P_\#(D)$ and $x \in X$, implying that (even in case $\|f_\#, f\| = \infty$) the sequence $(f^{(k)})_{k\in\mathbb{N}}$ converges pointwise quite quickly and monotonically decreases for all $f \in P_\#(D)$ from $k = 1$ onwards $\big($and even from $k = 0$ onwards in case $f \in P(D)\big)$ to a map $f^{(\infty)} := \lim_{k\to\infty} f^{(k)}$ for which, in view of

$$(f^{(k)})_\# \leq (f^{(k+1)})_\# \leq f^{(k+1)} \leq f^{(k)},$$

also

$$(f^{(k)})_\# \leq (f^{(\infty)})_\# \leq f^{(\infty)} \leq f^{(k)}$$

and, therefore, $(f^{(\infty)})_\# = f^{(\infty)}$ or, equivalently, $f^{(\infty)} \in T(D)$ must hold.

Furthermore, we have evidently $f = f^{(\infty)} \Leftrightarrow f \in T(D)$ and $f^{(\infty)} \leq f \Leftrightarrow f \in P(D)$ for all $f \in P_\#(D)$. Also, we have $\|f^{(\infty)}, g^{(\infty)}\| \leq \|f, g\|$ for all $f, g \in P_\#(D)$ — and, hence, in particular, for all $f, g \in P(D)$. Indeed, we have

$$\begin{aligned} f_\#(x) &= \sup\{D(x, y) - f(y) : y \in X\} \\ &= \sup\{D(x, y) - g(y) + g(y) - f(y) : y \in X\} \\ &\leq \sup\{D(x, y) - g(y) : y \in X\} + \sup\{g(y) - f(y) : y \in X\} \\ &= g_\#(x) + \sup\{g(y) - f(y) : y \in X\} \end{aligned}$$

for all $f, g \in \mathbb{R}^X$. Thus, $f, g \in P_\#(D)$ implies that

$$\sup\{f_\#(x) - g_\#(x) : x \in X\} \leq \sup\{g(x) - f(x) : x \in X\}$$

and, therefore, also

$$\begin{aligned}\|\partial f, \partial g\| &= \left\| \frac{f+f_\#}{2} - \frac{g+g_\#}{2} \right\| = \left\| \frac{f-g}{2} + \frac{f_\# - g_\#}{2} \right\| \\ &\le \left\| \frac{f-g}{2} \right\| + \left\| \frac{f_\# - g_\#}{2} \right\| = \frac{\|f,g\|}{2} + \frac{\|f_\#, g_\#\|}{2} \\ &\le \frac{\|f,g\|}{2} + \frac{\|f,g\|}{2} = \|f,g\|\end{aligned}$$

must hold.

In other words, we have

Theorem 5.6 *Given any metric space (X, D), the map $\rho_\# : P_\#(D) \to T(D) : f \mapsto f^{(\infty)}$ is a contraction from $P_\#(D)$ onto $T(D)$ for which*

$$\rho_\#(f) = f \Leftrightarrow f \in T(D)$$

and

$$\|\rho_\#(f), \rho_\#(g)\| \le \|f, g\|$$

holds for all $f, g \in P_\#(D)$. In particular, the restriction of $\rho_\#$ to $P(D)$ yields a contraction $\rho = \rho_D$ from $P(D)$ onto $T(D)$.

Recall that a subset $Y \subseteq \mathbb{R}^X$ is (affinely) *convex* if, for all $y, y' \in Y$ and all $\alpha \in [0,1]$, $\alpha y + (1-\alpha)y' \in Y$ holds. The following facts follow easily from combining Theorem 5.6 with the fact that $P(D)$ is, for every metric space (X, D), a convex subset of $\mathbb{R}^X$:

Corollary 5.7 *Given any metric space (X, D), the following hold:*

(i) *$T(D)$ is a retract of $P(D)$.*
(ii) *$T(D)$ is contractible.*
(iii) *$T(D)$ is a geodesic metric space.*

Proof (i) It follows immediately from Theorem 5.6 that putting $\tau_D(f,t) := (1-t)f + t\,\rho_D(f)$ for all $f \in P(D)$ and $t \in [0,1]$ yields the required homotopy.

(ii) Similarly, choosing one fixed map $g \in T(D)$ and putting $\tau_D^{(g)}(f,t) := \rho_D\big((1-t)f + t\,g\big)$ for all $f \in T(D)$ and $t \in [0,1]$, this follows again immediately from Theorem 5.6.

(iii) Assume $f, g \in T(D)$, put $L := \|f, g\|$, and consider the straight line connecting f and g in $P(D) \subseteq \mathbb{R}^X$, i.e., the map $\tau^*_{f,g} : [0, L] \to P(D) : t \mapsto \frac{L-t}{L} f + \frac{t}{L} g$. Concatenating this map with the contraction ρ_D, Theorem 5.6 now implies that the resulting map $\tau_{f,g} := \rho_D \circ \tau^*_{f,g}$ is a geodesic from f

to g in $T(D)$, as claimed: Indeed, $\tau_{f,g}$ is an isometry as $\left\|\tau^*_{f,g}(t'), \tau^*_{f,g}(t)\right\| = \frac{t'-t}{L}\|g, f\| = t' - t$ holds for all $t, t' \in [0, L]$ with $t \leq t'$. And this, in turn, implies that $\|\tau_{f,g}(t), \tau_{f,g}(0)\| \leq t$, $\|\tau_{f,g}(t'), \tau_{f,g}(t)\| \leq t' - t$ and $\|\tau_{f,g}(L), \tau_{f,g}(t')\| \leq L - t'$ and, therefore, also

$$\begin{aligned} L &= \|f, g\| \\ &\leq \|\tau_{f,g}(L), \tau_{f,g}(t')\| + \|\tau_{f,g}(t'), \tau_{f,g}(t)\| + \|\tau_{f,g}(t), \tau_{f,g}(0)\| \\ &\leq (L - t') + (t' - t) + t = L \end{aligned}$$

holds for all t, t' as above which, in turn, implies that equality must hold throughout. So, we have $\|\tau_{f,g}(t'), \tau_{f,g}(t)\| = t' - t$, as claimed. □

Note also that a map $f \in T(D)$ is contained in the image of a geodesic from a point of the form k_x to a point of the form k_y for some $x, y \in X$ if and only if $f(x) + f(y) = D(x, y)$ holds. Thus, given $f \in T(D)$ and two elements $x, y \in X$, the map f is contained in the image of a geodesic from k_x to k_y if and only if the supremum $\sup_{z \in X} \big(D(x, z) - f(z)\big)$ is actually attained at y, a fact that was one of the reasons for naming $T(D)$ the "tight span" of D.

Remark 5.8 *The above considerations suggest we define $P^{(2)}(D)$ to denote the set of all pairs $(f, g) \in \mathbb{R}^X \times \mathbb{R}^X$ for which*

$$f(x) + g(y) \geq D(x, y)$$

holds for all $x, y \in X$. Clearly, given some $f \in \mathbb{R}^X$, one has $f \in P(D)$ if and only if $(f, f) \in P^{(2)}(D)$ holds, and there exists some $g \in \mathbb{R}^X$ with $(f, g) \in P^{(2)}(D)$ if and only if $f_\# \in \mathbb{R}^X$ holds in which case $(f, g) \in P^{(2)}(D)$ holds for some $g \in \mathbb{R}^X$ if and only if $f_\# \leq g$ holds. In particular, using the standard convention $\inf A := \infty$ in case A is the empty subset of $\mathbb{R}$, one has $f_\#(x) = \inf\{g(x) : (f, g) \in P^{(2)}(D)\}$ for all $f \in \mathbb{R}^X$ and $x \in X$.

Note also that restricting the map $f \mapsto f_\#$ to the set $P_\#(D)$ defines a map from $P_\#(D)$ into itself, and that iterating this map, we have $(f_\#)_\# \leq f$ and "$(f_\#)_\# = f \iff f = g_\#$ for some $g \in \mathbb{R}^X$" for all $f \in P_\#(D)$. That is, associating the map $f_\#$ to a map f in $P_\#(D)$ defines some sort of "Galois connection" (cf. also Chapter 8) on this set, considered as a partially ordered set.

Of course, one has $P_\#(D) = \mathbb{R}^X$ for one or, as well, for all symmetric bivariate maps $D : X \times X \to \mathbb{R}$ if and only if X is finite. So, in this case, only $P^{(2)}(D)$ might have some interesting structure. Yet in case X is infinite, it might even be of some interest to study not only the set $P^{(2)}(D)$, but also the set $P_\#(D)$ in some more detail, e.g., for (X, D) the real line with its standard L_2-metric or its subspace, the interval $[0, 1]$.

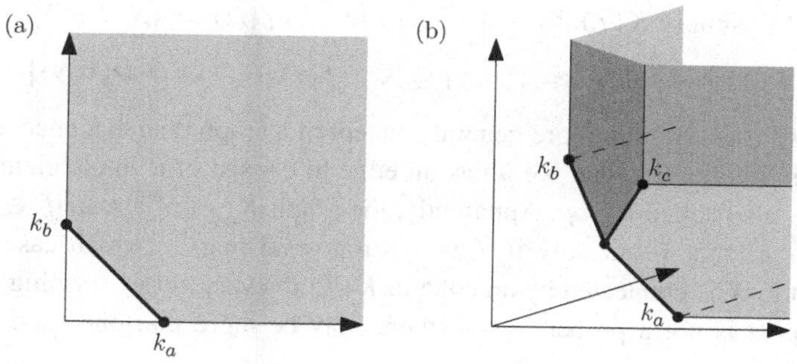

Figure 5.2 (a) The points in the shaded region belong to $P(D)$ for the metric D on $\{a, b\}$ with $D(a, b) = 1$. The bold straight line segment connecting the points k_a and k_b forms the tight span of D. (b) The convex polytope $P(D)$ for the metric D on $\{a, b, c\}$ with $D(a, b) = D(a, c) = D(b, c) = 1$. This polytope has six one-dimensional unbounded faces. Two of them are drawn with a dashed line because they are "hidden" by some of the six two-dimensional unbounded faces, four of which are indicated by shading. The three bold straight line segments form the tight span of D.

5.3 The tight span of a finite metric space

In case X is finite and D is a metric on X, we shall now give a more explicit geometrical description of $T(D)$.

First, we note that the set $P(D)$ introduced in the last section is a non-compact *convex polytope* in $\mathbb{R}^X$ in this case since it is nothing other than the intersection of a finite collection of half-spaces, and it is non-compact as it contains every sufficiently large map $f \in \mathbb{R}^X$. In contrast, $T(D)$ is a bounded subset of $\mathbb{R}^X$ if X is finite as $0 \leq f(x) = \sup_{y \in X} \big(D(x, y) - f(y)\big) \leq \sup_{y \in X} D(x, y) \leq \max(D) := \max_{z,y \in X} D(z, y)$ holds for all $x \in X$.

To be more specific, we define, for all $x, y \in X$, the half-space

$$H_{x,y}^{\geq D} := \{f \in \mathbb{R}^X \ : \ f(x) + f(y) \geq D(x, y)\}.$$

Then,

$$P(D) = \bigcap_{x,y \in X} H_{x,y}^{\geq D}$$

clearly holds. Two simple, yet instructive examples of $P(D)$ are given in Figure 5.2(a) and (b) for metric spaces consisting of two and three points, respectively.

The tight span of D has a rather natural description in terms of the polytope $P(D)$. For any $f \in \mathbb{R}^X$, we define the *tight-equality graph of f*, denoted by

$K(f|D)$ or simply $K(f)$, by putting $K(f) = K(f|D) := \big(X, E(f)\big)$ with

$$E(f) = E(f|D) := \big\{\{x, y\} \subseteq X : f(x) + f(y) = D(x, y)\big\}$$

where we use a slightly more general concept of a graph than that encountered so far in this book in that we allow an edge to consist of a single element of X, only, also called a *loop*. Apparently, the graph $K(f)$ of a map $f \in T(D)$ contains a loop if and only if f is a Kuratowski map in which case every element in X is connected by an edge in $K(f)$ to every vertex forming a loop (in case D is not a proper metric, there may be more that one such vertex in $K(f)$).

Note also that the edges of $K(f)$ are clearly formed by exactly those pairs of elements $\{x, y\}$ in X for which $x \vee_f y$ does not hold. In particular, as, by Corollary 5.4, "$\vee_f$" is an equivalence relation on $supp(f)$ in case D is treelike, we obtain

Lemma 5.9 *If D is treelike, then the graph $K(f)$ coincides with the complete multi-partite graph $\Gamma(\Pi_f)$ associated with the partition Π_f of $supp(f)$, i.e., it connects any two elements in X by an edge if and only if they do not belong to the same subset in Π_f — and thus any element in $X - supp(f)$ to all elements including itself. In particular, $K(f)$ is connected for every map $f \in T(D)$ if D is treelike.*

It is also obvious that, given any $f \in \mathbb{R}^X$, two elements $x, y \in X$ are connected by an edge in $K(f)$ if and only if f is contained in the boundary hyperplane

$$H^D_{x,y} := \{g \in \mathbb{R}^X : g(x) + g(y) = D(x, y)\}$$

of $H^{\geq D}_{x,y}$. Hence, the smallest face $[f] = [f]_D$ of $P(D)$ containing some map $f \in P(D)$, i.e., the intersection of all hyperplanes $H^D_{x,y}$ $(x, y \in X)$ containing f, is given by

$$[f] = \{g \in P(D) : E(f) \subseteq E(g)\}. \tag{5.4}$$

We now characterize those faces of $P(D)$ that belong to $T(D)$:

Lemma 5.10 *For any $f \in P(D)$, the following assertions are equivalent:*

(i) $f \in T(D)$,
(ii) *$[f]_D$ is a subset of $T(D)$,*
(iii) *$[f]_D$ is bounded,*
(iv) *the support*

$$supp\big(K(f)\big) := \{x \in X : \textit{there exists some } y \in X \textit{ with } \{x, y\} \in E(f)\}$$

of $K(f)$ coincides with X.

In particular, $T(D)$ is a compact contractible cell complex formed by all compact faces of the convex polytope $P(D)$.

Proof (i) $\Leftrightarrow$ (iv): For $f \in P(D)$, we have

$$f \in T(D) \Leftrightarrow \forall_{x \in X} f(x) = \sup_{y \in X}\big(D(x, y) - f(y)\big)$$

and therefore, as X is finite, $f \in T(D) \Leftrightarrow \forall_{x \in X} \exists_{y \in X} f(x) + f(y) = D(x, y)$, that is, if and only if $supp\big(K(f)\big) = X$, as required.

(i) $\Rightarrow$ (ii): Clearly, if $f \in T(D)$ and $g \in [f]_D$, we have $supp\big(K(f)\big) = X$ by (iv) and, therefore, also $supp\big(K(g)\big) = X$ in view of $E(f) \subseteq E(g)$. So, $g \in T(D)$ must hold, as required.

(ii) $\Rightarrow$ (iii): This is trivial.

(iii) $\Rightarrow$ (i): If $f \in P(D) - T(D)$ holds, "(i) $\Leftrightarrow$ (iv)" implies the existence of some $x \in X$ such that $f(x) + f(y) > D(x, y)$ holds for all $y \in X$ and, hence, $E(f) = E(g)$ as well as $g \in [f]_D$ for every $g \in \mathbb{R}^X$ with $g(y) := f(y)$ for all $y \in X - x$ and $g(x) \geq f(x)$. So, $[f]_D$ is not bounded.

It follows in particular that $T(D)$ coincides precisely with the union of all compact faces of $P(D)$. □

So, $T(D)$ — being a compact contractible cell complex — has a well-defined dimension which is bounded from above at least by the dimension of $\mathbb{R}^X$, that is, by $|X|$. In fact, we can say more about this dimension:

Lemma 5.11 *With (X, D) as above, the dimension $\dim[f]_D$ of the convex cell $[f]_D$ coincides, for any f in $P(D)$, with the dimension of the linear space*

$$W(f) := \{v \in \mathbb{R}^X : v(x) + v(y) = 0 \textit{ for all } \{x, y\} \in E(f)\}$$

and, hence, with the number b_f of bipartite connected components of $K(f)$ (counting every isolated vertex not connected to itself by a loop as a single such component, of course). In particular,

(i) $\dim[f]_D \leq \lfloor |supp\big(K(f)\big)|/2 \rfloor + |X - supp\big(K(f)\big)|$ *holds for all* $f \in P(D)$,

(ii) $\dim[f]_D \leq |X|/2$ *holds for all* $f \in T(D)$,

(iii) *the 0-dimensional faces or "vertices" of $T(D)$ are precisely the 0-dimensional faces of $P(D)$,*

(iv) $\dim[k_x]_D = 0$ *holds for every $x \in X$, i.e., every element of X gives rise to a vertex of $T(D)$.*

(v) *If D is treelike and $f \in T(D)$ holds, then f is either a vertex in $T(D)$ or Π_f is a bipartition of X in which case $[f]_D$ is an edge of $T(D)$ (as well as of $P(D)$). In particular, $T(D)$ is a one-dimensional cell complex in case D is treelike.*

Proof The assertion that $\dim[f]_D = \dim W(f)$ holds for any $f \in P(D)$ follows immediately, essentially by the definition of $[f]_D$, from the observation that $\dim[f]_D = \dim \bigcap_{\{x,y\}\in E(f)} H^D_{x,y}$ holds for any $f \in P(D)$ as any sufficiently small variation of f within $\bigcap_{\{x,y\}\in E(f)} H^D_{x,y}$ remains in $P(D)$ in view of the fact that $f(x) + f(y) > D(x, y)$ holds for all $x, y \in X$ with $\{x, y\} \notin E(f)$. Moreover, it is obvious that

$$\bigcap_{\{x,y\}\in E(f)} H^D_{x,y} = \{g \in \mathbb{R}^X : \{x, y\} \in E(f) \Rightarrow g(x) + g(y) = f(x) + f(y)\}$$

$$= \{g \in \mathbb{R}^X : g - f \in W(f)\} = \{f + v : v \in W(f)\}$$

and, therefore, also $\dim \bigcap_{\{x,y\}\in E(f)} H^D_{x,y} = \dim W(f)$ always holds. So, also $\dim[f]_D = \dim W(f)$ holds for all $f \in P(D)$.

Next, recall that a connected component C of $K(f)$ is non-bipartite if and only if the following holds: There exists, for any $x \in C$, some $\ell \in \mathbb{N}$ and a sequence $x_0 := x, x_1, \ldots, x_{2\ell}, x_{2\ell+1} := x$ of vertices in C with $\{x_i, x_{i-1}\} \in E(f)$ for all $i = 1, \ldots, 2\ell+1$. This implies that $2\,v(x) = \sum_{i=0}^{2\ell}(-1)^i\big(v(x_i)+ v(x_{i+1})\big) = 0$ must hold for every $v \in W(f)$ and every $x \in X$ that is contained in a non-bipartite connected component of $K(f)$. In contrast, one can freely choose the value $v(x)$ of a map $v \in W(f)$ on one element $x \in C$ in case C is a bipartite component C of $K(f)$. However, one is then forced to put $v(y) = \pm v(x)$ for any other element $y \in C$, the sign depending on whether the paths connecting x and y in $K(f)$ have even or odd length (recall that, in a bipartite graph, the lengths of all such paths must be of the same parity!). So, if Y denotes any subset of the union of all bipartite connected components of $K(f)$ that intersects any such component in exactly one vertex, the restriction homomorphism $\mathbb{R}^X \to \mathbb{R}^Y : v \mapsto v|_Y$ induces a canonical $\mathbb{R}$-linear isomorphism between $W(f)$ and $\mathbb{R}^Y$ which immediately yields $\dim[f]_D = \dim W(f) = |Y| = b_f$, as required.

(i) To see that $\dim[f] \le \lfloor |supp(K(f))|/2 \rfloor + |X - supp(K(f))|$ holds, note that every bipartite connected component that is a subset of $supp(K(f))$ (and, thus, in particular does not consist of a single isolated vertex in X, not even connected to itself by a loop) must have at least two vertices.

(ii) Further, we have $supp(K(f)) = X$ for every $f \in T(D)$ and, therefore, $\dim[f]_D \le |X|/2$ for every such f in view of Lemma 5.10, (i) $\Rightarrow$ (iv).

(iii) This follows from Lemma 5.10, (iii) $\Rightarrow$ (i).

(iv) Indeed, given any element $x \in X$, every element y in X, including the element x, is connected to x by an edge in $K(k_x)$.

(v) The last assertion follows immediately from the former assertions in view of the fact that $K(f)$ coincides, in case D is treelike and f is a map in

$T(D)$, with the complete multi-partite graph $\Gamma(\Pi_f)$ associated with the partition Π_f of $supp(f)$ defined for any map f in the tight span of a treelike metric (cf. Lemma 5.9) and that, therefore, $b_f \leq 1$ always holds while equality holds if and only if Π_f is a bipartition of X. □

It is worth noting that the facts collected above imply the following result which actually was the principal motivation for considering the tight-span construction and inspired the work presented in [50]:

Corollary 5.12 *If D is a treelike metric, then $T(D)$ consists, as a sub-complex of the cell complex formed by the faces of $P(D)$, of its vertices and (compact) edges, only, and must therefore be a tree when considered as just a graph whose vertices are the vertices of $P(D)$ (and, hence, of $T(D)$) while its edges are (represented by) the pairs $\{f, g\}$ of vertices of $P(D)$ that form the endpoints of an edge in $P(D)$ (and, hence, of $T(D)$).*

Proof We have seen in Lemma 5.11(v) that $\dim[f]_D \leq 1$ holds for every map $f \in T(D)$ in case D is a treelike metric and that, in consequence, $T(D)$ consists indeed, considered as a sub-complex of the cell complex formed by the faces of $P(D)$, of vertices and edges of $P(D)$, only. Actually, it is the "compact 1-skeleton" of $P(D)$, i.e., the union of all vertices and compact 1-cells contained in $P(D)$. So, considered as just a graph, basic combinatorial topology implies that it must be a tree as it is contractible in view of Corollary 5.7(ii). □

However, we can establish the fact that this graph must be a tree also in a direct way without any reference to combinatorial topology: If it were not a tree, there would exist a sequence of $\ell \geq 3$ distinct vertices $f_1, \ldots, f_\ell$ in $T(D)$ such that, with $f_0 := f_\ell$, the ℓ pairs $\{f_{i-1}, f_i\}$ ($i \in \{1, \ldots, \ell\}$) are the endpoints of edges in $T(D)$. However, for any $x \in X$, there must exist some $i \in \{1, \ldots, \ell\}$ with $\max\{f_1(x), \ldots, f_\ell(x)\} = f_i(x)$ which is impossible in view of the following lemma which is also of some interest in itself as it indicates how to explicitly construct a path from an arbitrary vertex f in $T(D)$ to a Kuratowski map:

Lemma 5.13 *Assume that D is a treelike metric defined on a finite set X, x is an element in X, and f is a vertex of $T(D)$ that is distinct from the Kuratowski map k_x. Then, there exists exactly one vertex g in $T(D)$ such that $f(x) \geq g(x)$ holds and f and g are the endpoints of an edge in $T(D)$ in which case also $f(x) > g(x)$ and $\|f, g\| = f(x) - g(x)$ must hold.*

Proof Assume that $\{f, g\}$ is an edge in $T(D)$ and that $f(x) \geq g(x)$ holds. We first show that g is determined by x and f. To this end, note that $h := \frac{f+g}{2}$

must be a map in $T(D)$ that is not a vertex. So, the partition Π_h must be a proper bipartition of X, i.e., a bipartition of X into exactly two distinct, non-empty subsets A and B. Thus, we have $h(x_1) + h(x_2) = D(x_1, x_2)$ for some $x_1, x_2 \in X$ if and only if we have $\{x_1, x_2\} = \{a, b\}$ for some $a \in A$ and some $b \in B$. However, in view of $f(x_1) + f(x_2), g(x_1) + g(x_2) \geq D(x_1, x_2)$, we have $h(x_1) + h(x_2) = D(x_1, x_2) \Leftrightarrow f(x_1) + f(x_2) = g(x_1) + g(x_2) = D(x_1, x_2) \Leftrightarrow \{x_1, x_2\} \in E(f) \cap E(g)$ for all $x_1, x_2 \in X$. So, $f(a) + f(b) = g(a) + g(b) = D(a, b)$ and, therefore, $f(a) - g(a) = g(b) - f(b)$ must hold for all $a \in A$ and $b \in B$.

In consequence, if we put $\rho := f(x) - g(x)$ and if, say $x \in A$ holds, we must have $g(b) = f(b) + \rho$ for all $b \in B$ and $g(a) = f(a) - \rho$ for all $a \in A$, that is, g is completely determined by f, A, B, and ρ, and $\rho > 0$ must hold in view of $f \neq g$ and $f(x) \geq g(x)$. However, x and f determine ρ, A, and B as our assumptions imply that the set $\{x' \in X : x' \vee_f x\} = \{x' \in X : f(x) + f(x') > D(x, x')\}$ must coincide with A because $b \in B$ implies $f(x) + f(b) = D(x, b)$ and $a \in A$ implies $f(x) + f(a) = g(x) + g(a) + 2\rho > D(x, a)$, and $\dim[g] = 0$ implies that ρ must coincide with $\frac{1}{2}\min\{f(a) + f(a') - D(a, a') : a, a' \in A\}$. Indeed, if $2\rho < f(a) + f(a') - D(a, a')$ were to hold for all $a, a' \in A$, we would have $g(a) + g(a') = f(a) + f(a') - 2\rho > D(a, a')$ for all $a, a' \in A$ and, hence, $E(g) = E(h)$, i.e., g would not be a *vertex* of $T(D)$. And if $2\rho > f(a) + f(a') - D(a, a')$ were to hold for some $a, a' \in A$, we would have $g(a) + g(a') = f(a) + f(a') - 2\rho < D(a, a')$ for these two elements $a, a' \in A$; so, we would not even have $g \in P(D)$.

It remains to show that, for every vertex f in $T(D)$, there exists a suitable vertex g as above in $T(D)$. But if f is a vertex in $T(D)$ and x is an element in X with $f(x) > 0$, we may put $A := \{x' \in X : x' \vee_f x\}$, $B := X - A$, and $\rho := \frac{1}{2}\min\{f(a) + f(a') - D(a, a') : a, a' \in A\}$, and define $g \in \mathbb{R}^X$ by putting

$$g(y) := \begin{cases} f(y) - \rho & \text{if } y \in A, \\ f(y) + \rho & \text{if } y \in B, \end{cases}$$

for all $y \in X$. It is then easily verified that g is a vertex in $T(D)$, that f and g are the endpoints of an edge in $T(D)$, *viz.* the edge formed by all $h \in T(D)$ for which $\{a, b\} \in E(h)$ holds for all $a \in A$ and $b \in B$, and that $f(x) \geq g(x)$ holds, as required. □

Thus, considering $T(D)$ as a graph in case D is treelike, this graph cannot contain any cycle and must, hence, be a tree as claimed in Corollary 5.12.

While these are all rather general and abstract considerations, what does the tight span actually look like at least for small X? In Figure 5.3, we depict the

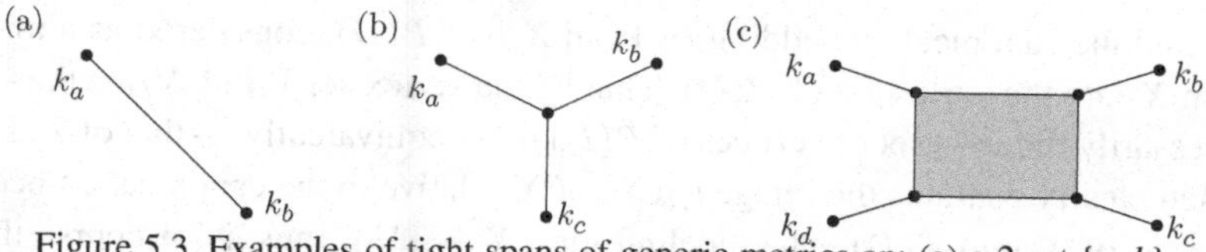

Figure 5.3 Examples of tight spans of generic metrics on: (a) a 2-set $\{a, b\}$, (b) a 3-set $\{a, b, c\}$, and (c) a 4-set $\{a, b, c, d\}$. The dots represent 0-dimensional faces, straight line segments one-dimensional faces, and the shaded rectangle in (c) represents a/the face of dimension 2.

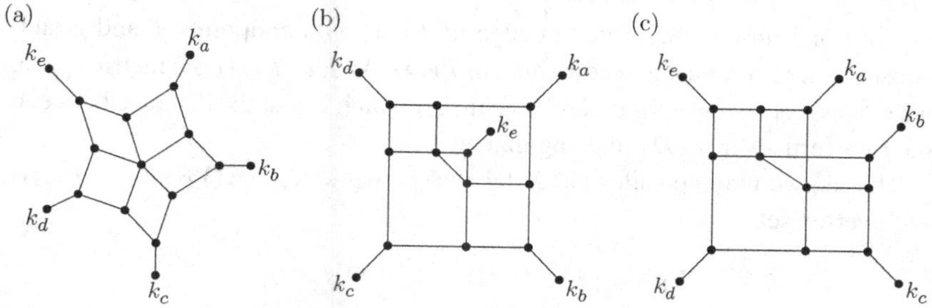

Figure 5.4 The three types of networks $\mathcal{N}_D$ (without the edge weights) that arise from generic metrics D on a 5-point set $X = \{a, \ldots, e\}$ (up to relabeling of the vertices).

tight span for *generic* metrics with $|X| = 2, 3, 4$ (cf. also Figure 5.2 for the cases $|X| = 2, 3$). In [50], the tight spans of generic 5-point metrics were classified — there are three "combinatorial types" of such metrics (see Figure 5.4). More recently, all generic 6-point metrics have been computed in [125] of which there are 339 types. Note finally that some software packages have been developed that allow the computation of tight spans for arbitrary metrics either directly (see e.g., [80]) or via general-purpose routines for manipulating intersections of half-spaces (see e.g., [39]).

5.4 Networks from tight spans

We now describe how we can use the tight span to obtain an *X-labeled weighted graph* or network $\mathcal{N} = (V, E, \omega, \varphi)$ representing a finite metric space (X, D). Clearly, there is a natural X-labeled network $\mathcal{N}_D$ to consider, *viz.* the — necessarily connected (see e.g., [105]) — "compact" 1-skeleton of the convex polytope $P(D) \subseteq \mathbb{R}^X$ together with the weighting induced by the L_∞-metric on

$\mathbb{R}^X$ and the canonical embedding ι_D from X into $P(D)$, considered as a map from X into the vertex set of $P(D)$. That is, the vertex set V_D of $\mathcal{N}_D$ is the — necessarily finite — set of vertices of $P(D)$ or — equivalently — that of $T(D)$ which clearly contains the image $\iota_D(X)$ of X relative to the canonical embedding ι_D of X into $T(D)$. The labeling map $X \to V_D$, thus, is of course the map ι_D considered as a map from X into V_D. The edge set E_D of $\mathcal{N}_D$ consists of precisely those 2-subsets $\{f, g\}$ of V_D that form the two vertices of some bounded one-dimensional face or "edge" of $P(D)$ — that is, those 2-subsets $\{f, g\}$ of V_D for which the face $\left[\frac{f+g}{2}\right]$ of $P(D)$ containing the map $\frac{f+g}{2}$ has dimension 1 and is, therefore, an edge of $P(D)$ with endpoints f and g and, consequently, compact and contained in $T(D)$. And, as $T(D)$ is a metric space, we can associate a weight to any edge in E_D which is just the distance between the two vertices in $T(D)$ forming that edge.

That is, we may consider the X-labeled network $\mathcal{N}_D = (V_D, E_D, \omega_D, \iota_D)$ with vertex set

$$V_D := \{f \in P(D) \,:\, \dim[f]_D = 0\},$$

edge set

$$E_D := \left\{\{f, g\} \in \binom{V_D}{2} \,:\, \dim[(f+g)/2] = 1\right\},$$

labeling map ι_D (considered as a map from X into V_D), and weighting

$$\omega_D : E_D \to \mathbb{R}_{>0} : \{f, g\} \mapsto \|f, g\| = L_\infty^D(f, g).$$

Obviously, if D is treelike or, equivalently, $T(D)$ is a one-dimensional cell complex, the graph (V_D, E_D) coincides exactly with the graph considered already in Corollary 5.12 in the previous section.

In Figure 5.3, the networks that come from the tight span of 2-, 3-, and 4-point metric spaces have as vertices the black dots, and the edges correspond to the straight line segments connecting them, while, in Figure 5.4, we present the three possible networks arising for generic 5-point metric spaces as determined in [50].

We have already seen in Corollary 5.12 in the previous section that $\mathcal{N}_D$ is an edge-weighted X-tree whenever D is treelike. Here, we want to supplement this by providing the non-inductive proof of the fact that a weighted X-tree $\mathcal{T}$ is determined — up to canonical isomorphism — by the induced metric $D_\mathcal{T}$ referred to already in Section 3.1. To this end, we define the metric $D_\mathcal{N}$ induced by an X-labeled weighted graph $\mathcal{N} = (V, E, \omega, \varphi)$ by putting $D_\mathcal{N}(x, y) := D_{(V,E,\omega)}(\varphi(x), \varphi(y))$ for all $x, y \in X$ — a metric that, once again, just as in the case of X-trees, needs to be clearly distinguished from the metric $D_{(V,E,\omega)}$ defined on V by the underlying weighted graph (V, E, ω). Regarding this metric, the following can be established:

Theorem 5.14 *Suppose that D is a metric on a finite set X. Then, D is tree-like — or, equivalently (cf. Section 3.1), it satisfies the 4-point condition — if and only if $\mathcal{N}_D$ is an edge-weighted X-tree relative to the map $\iota_D : X \to V_D \subseteq T(D) : x \mapsto k_x$ for which $D = D_{\mathcal{N}_D}$ holds.*

More specifically, given any edge-weighted X-tree

$$\mathcal{T} = (V, E, \omega : E \to \mathbb{R}_{>0}, \varphi : X \to V)$$

with $D = D_{\mathcal{T}}$, let $D' := D_{(V,E,\omega)}$ denote the metric induced by ω on the vertex set V of $\mathcal{T}$. Then, associating to each $v \in V$ the map

$$f_v : X \to \mathbb{R} : x \mapsto D'\big(v, \varphi(x)\big),$$

induces a canonical isomorphism between $\mathcal{T}$ and $\mathcal{N}_D$, i.e.,

- *it induces a bijection $\gamma_D : V \to V_D : v \mapsto f_v$,*
- *two vertices $u, v \in V$ form an edge in E if and only if the two maps f_u and f_v form an edge in E_D,*
- *one has $D'(u, v) = \|f_u, f_v\|$ for all $u, v \in V$, implying in particular that $\omega(\{u, v\}) = \omega_D(\{f_u, f_v\})$ holds for every edge $\{u, v\} \in E$, and*
- *one has $\gamma_D\big(\varphi(x)\big) = f_{\varphi(x)} = k_x = \iota_D(x)$ for all $x \in X$.*

Proof It suffices to show that, with the notations and assumptions introduced above, the map $V \to \mathbb{R}^X : v \mapsto f_v$ induces a canonical isomorphism from $\mathcal{T}$ onto $\mathcal{N}_D$ — implying in particular that $D = D_{\mathcal{N}_D}$ must hold in this case.

Clearly, we have $D(x, y) = D_{\mathcal{T}}(x, y) = D'\big(\varphi(x), \varphi(y)\big)$ and $D'\big(v, \varphi(x)\big) = f_v(x) = \|f_v, k_x\| = \|f_v, f_{\varphi(x)}\|$ for all $x, y \in X$ and $v \in V$. Further, $D'(v, u) = \|f_v, f_u\|$ must hold for all $v, u \in V$: Indeed, continuing the unique path in $\mathcal{T}$ from v to u until one meets a leaf and, hence, a vertex of the form $\varphi(x)$ for some $x \in X$, $D'(\varphi(x), v) = D'(\varphi(x), u) + D'(u, v)$ or, equivalently, $\|f_v, f_{\varphi(x)}\| = D'(\varphi(x), v) = D'(\varphi(x), u) + D'(u, v) = \|f_u, f_{\varphi(x)}\| + D'(u, v)$ must hold for this element $x \in X$ implying that $D'(v, u) = \|f_v, f_{\varphi(x)}\| - \|f_u, f_{\varphi(x)}\| \leq \|f_v, f_u\|$ holds. Conversely, we have $\|f_v, f_u\| = \sup_{x \in X}\big(f_v(x) - f_u(x)\big) = \sup_{x \in X}\big(D'(\varphi(x), v) - D'(\varphi(x), u)\big) \leq D'(u, v)$, so equality must hold, as claimed.

Next, we show that the map $\gamma_D : V \to \mathbb{R}^X : v \mapsto f_v$ indeed induces a bijection from V onto V_D: It is injective as $\|f_u, f_v\| = D'(u, v)$ holds for all $u, v \in V$. Further, we have $f_v \in P(D)$ for every $v \in V$ in view of the triangle inequality (applied to D') and our assumption that $D = D_{\mathcal{T}}$ and, therefore, $D(x, y) = D'(\varphi(x), \varphi(y))$ holds for all $x, y \in X$. And $f_v \in V_D$ holds for every $v \in V$: It holds surely for every vertex $v \in \varphi(X)$ as $f_{\varphi(x)} = k_x \in V_D$ holds for every $x \in X$. And it holds also for every vertex of degree $d \geq 3$ and, hence, for every $v \in V - \varphi(X)$: Indeed, denoting by $X_1, \ldots, X_d$ the pre-images (relative to φ) of the d subsets of V that form the connected components of the forest $T^{(v)} := (V - v, E - E_T(v))$, one must have $f_v(x_i) + f_v(x_j) = D(x_i, x_j)$ for all $x_i \in X_i$ and $x_j \in X_j$ with $1 \leq i < j \leq d$ implying that $K(f_v)$ is connected and not bipartite (as it contains a cycle of length 3 in view of $d \geq 3$) and that, therefore, $f_v \in V_D$ must hold.

Conversely, there exists, for every $f \in V_D$, some $v \in V$ with $f_v = f$: If f is of the form k_x for some $x \in X$, then $f = f_{\varphi(x)}$ holds. Otherwise, since $K(f)$ has no bipartite connected component in view of $f \in V_D$ and Lemma 5.9, $K(f)$ is a complete r-partite graph for some $r \geq 3$. Hence, we can choose three elements $x_1, x_2, x_3 \in X$ that are pairwise not $\vee_f$-equivalent. We claim that $f = f_v$ must hold for the median $v \in V$ of $\varphi(x_1), \varphi(x_2)$, and $\varphi(x_3)$ in (V, D'), which must exist by Lemma 1.5. Indeed, we have $f(x_i) + f(x_j) = D(x_i, x_j) = D'(\varphi(x_i), \varphi(x_j)) = D'(\varphi(x_i), v) + D'(v, \varphi(x_j))$ for all $1 \leq i < j \leq 3$ and, therefore, $f(x_i) = \frac{D(x_i,x_j)+D(x_i,x_k)-D(x_j,x_k)}{2} = D'(\varphi(x_i), v) = f_v(x_i)$ for all i, j, k with $\{i, j, k\} = \{1, 2, 3\}$.

Hence, it remains to show that, for any element $x \in X - \{x_1, x_2, x_3\}$, $f(x) = f_v(x)$ holds. But, as x_1, x_2, and x_3 are pairwise not $\vee_f$-equivalent, any such x can be $\vee_f$-equivalent to at most one element in $\{x_1, x_2, x_3\}$, and it can also, in view of $f_v(x_i) = f(x_i), i \in \{1, 2, 3\}$, be $\vee_{f_v}$-equivalent to at most one element in $\{x_1, x_2, x_3\}$. Thus, there must exist at least one index $i \in \{1, 2, 3\}$ such that $f(x_i)+f(x) = D(x_i, x)$ and $f_v(x_i)+f_v(x) = D(x_i, x)$ holds as there must be two distinct indices in $\{1, 2, 3\}$ for which the first, and two for which the second identity holds. But this implies $f(x) = D(x_i, x) - f(x_i) = f_v(x)$, as required.

Finally, we want to show that $\{f_u, f_v\} \in E_D$ holds for all $u, v \in V$ with $e := \{u, v\} \in E$. That is, we want to show that, putting $h := \frac{1}{2}(f_u + f_v)$, we have $\dim[h]_D = 1$ in this case: Indeed, let $S_e = \varphi^{-1}(T^{(e)}(u)) | \varphi^{-1}(T^{(e)}(v))$ denote the split of X associated with e and note that, essentially by definition, we have $f_v(a)+f_v(b) = f_u(a)+f_u(b) = D(a, b)$ as well as $f_u(a)+f_u(a') > D(a, a')$ for all $a, a' \in A := \varphi^{-1}(T^{(e)}(v))$ and $f_v(b) + f_v(b') > D(b, b')$ for all $b, b' \in B := \varphi^{-1}(T^{(e)}(u))$. Thus, we have $h(x)+h(y) = D(x, y)$ for some $x, y \in X$ if and only if we have $f_v(x) + f_v(y) = f_u(x) + f_u(y) = D(x, y)$ and, hence, if and only if we have $x \in A$ and $y \in B$ or $y \in A$ and $x \in B$ implying that $K(h)$ coincides with the complete bipartite graph $\Gamma(S_e)$ associated with the bipartition $S_e = A|B$ associated with e and that, consequently, $\dim[h]_D = 1$ holds, as claimed.

It follows that $\{f_u, f_v\} \in E_D \Leftrightarrow \{u, v\} \in E$ must hold for all $u, v \in V$ because $|V| = |V_D|$ together with the fact that both graphs, (V, E) and (V_D, E_D), are trees (the first one by assumption and the second one in view of Corollary 5.12) implies that also $|E| = |E_D|$ must hold.

Thus, the map $V \to V_D : v \mapsto f_v$ indeed induces a canonical isomorphism between $\mathcal{T}$ and $\mathcal{N}_D$, as claimed. □

We conclude this section by noting that, to every finite edge-weighted tree $T = (V, E, \omega)$, one can associate an (in general infinite) metric space (X_T, D_T) that captures not only the distances between the vertices of T, but between all "points" that lie on T. This can be made more precise by referring to "$\mathbb{R}$-trees": Indeed, it has been shown in [50] that a metric space (X, D) satisfies the 4-point condition if and only if its tight span is an $\mathbb{R}$-tree, and that (X, D) is itself an $\mathbb{R}$-tree if and only if it is a complete and connected metric space that satisfies the 4-point condition if and only if D satisfies the 4-point condition and the natural isometric embedding $\iota_D : X \to T(D)$ is a bijective isometry.

5.5 Network realizations of metrics

In the previous section, we have seen that, for D a treelike metric, the network $\mathcal{N}_D$ is the unique X-tree associated to D. But which properties does this network have in case D is an arbitrary metric? Since we are trying to represent metrics by networks, it is natural to ask whether $\mathcal{N}_D$ is always a *realization* of D, that is, an X-labeled network $\mathcal{N} = (V, E, \omega, \varphi)$ with $D = D_{\mathcal{N}}$. In this section, we will show that this and even better results do indeed hold.

First note that, in general, the number of vertices of $\mathcal{N}_D$ can be exponential in the number of elements of X. More precisely, it was shown in [94] that $2^{n-2k-1}\frac{n}{n-k}\binom{n-k}{k}$ is a tight upper bound on the number of k-dimensional faces of $T(D)$ for a metric D on a set X with n elements. To give a simple example of a family of metrics D where it is easily seen that the number of vertices of $T(D)$ is not bounded by a polynomial in $|X|$, consider a set X with m^2 elements, $m \geq 3$, and partition it arbitrarily into subsets $X_1, \ldots, X_m$ such that $|X_1| = \cdots = |X_m| = m$ holds. Define the metric D_m on X by

$$D_m(x, x') = \begin{cases} 0 & \text{if } x = x', \\ 4 & \text{if } x \neq x' \text{ and } \{x, x'\} \subseteq X_i \text{ for some } i \in \{1, \ldots, m\}, \\ 2 & \text{otherwise.} \end{cases}$$

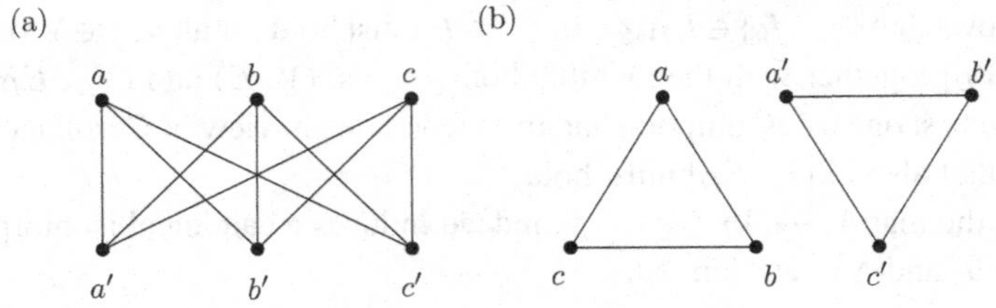

Figure 5.5 (a) A graph G for which the network $\mathcal{N}'_{D_G}$ associated to the metric D_G induced by G is a proper subgraph of $\mathcal{N}_{D_G}$ (see text for details). (b) The graph $K(f)$ for some $f \in V_D - V'_D$.

Then, for every choice of elements $a_i \in X_i$, $1 \le i \le m$, the map $f \in \mathbb{R}^X$ defined by $f(x) = 1$ if $x = a_i$ for some $i \in \{1, \dots, m\}$ and $f(x) = 3$ otherwise is a vertex of $\mathcal{N}_{D_m}$: Indeed, we clearly have $f \in P(D_m)$ and

$$E(f) = \big\{\{a_i, a_j\} : 1 \le i < j \le m\big\} \cup \bigcup_{i=1}^{m} \big\{\{a_i, x\} : x \in X_i - a_i\big\}.$$

Hence, $K(f)$ is connected, yet — in view of $m \ge 3$ — it is not bipartite which implies $\dim[f]_{D_m} = 0$, as required. It follows that $\mathcal{N}_{D_m}$ has at least $m^m = |X|^{m/2}$ vertices.

We therefore consider, for any finite metric space (X, D) as above, the weighted subgraph $\mathcal{N}'_D = \big(V'_D, E'_D, \omega'_D, \iota'_D\big)$ of $\mathcal{N}_D$ that has vertex set

$$V'_D := \{f \in V_D \;:\; K(f) \text{ is connected}\},$$

edge set

$$E'_D := \{\{f, g\} \in E_D \;:\; K\big((f + g)/2\big) \text{ is connected}\},$$

weighting $\omega'_D : E'_D \to \mathbb{R}_{>0}$ given by restricting ω_D to E'_D, and labeling map ι'_D given by considering ι_D as a map from X into V'_D (which is possible as $k_x \in V'_D$ holds for every $x \in X$).

Note that $\mathcal{N}'_D$ is, in general, a proper subnetwork of $\mathcal{N}_D$. For example, $\mathcal{N}'_D$ differs from $\mathcal{N}_D$ for the metric $D := D_G$ induced on the set $X := \{a, a', b, b', c, c'\}$ by the graph G depicted in Figure 5.5(a): Indeed, the graph $K(f)$ associated to the map $f \in \mathbb{R}^X$ with $f(x) = 1$ for all $x \in X$, which is depicted in Figure 5.5(b), has no bipartite connected components, but is not connected. Hence, we have $f \in V_D - V'_D$ (see [61] for more details). Yet, the example of a family of metrics D_m presented above illustrates that, even so, $\mathcal{N}'_D$ can be pretty large.

We now establish that $\mathcal{N}'_D$ and, therefore, also $\mathcal{N}_D$ is a realization of D.

Theorem 5.15 *If (X, D) is a finite metric space, then both, $\mathcal{N}'_D$ and $\mathcal{N}_D$, are connected realizations of D.*

Proof We have already noted above that $\mathcal{N}_D$, being the "compact 1-skeleton" of a convex polytope in $\mathbb{R}^X$, is necessarily connected. And it follows immediately from the definitions that $k_x, k_y \in V'_D \subseteq V_D$ and $\|k_x, k_y\| \le D_{\mathcal{N}_D}(k_x, k_y) \le D_{\mathcal{N}'_D}(k_x, k_y)$ holds for all $x, y \in X$. To show that also $\mathcal{N}'_D$ is connected and that equality must hold above, it suffices to show that, for every $x \in X$ and every $f \in V'_D - k_x$, there exists some $g \in V'_D$ with $\{f, g\} \in E'_D$ and $\|f, g\| + \|g, k_x\| = \|f, k_x\|$: Indeed, this implies that, given any two elements $x, y \in X$, there exists a sequence $f_0 := k_y, f_1, \dots, f_\ell := k_x$ of maps in V'_D with $\{f_{i-1}, f_i\} \in E'_D$ and $\|f_{i-1}, f_i\| + \|f_i, k_x\| = \|f_{i-1}, k_x\|$ for all $i = 1, \dots, \ell$ and, therefore, $D_{\mathcal{N}'_D}(k_x, k_y) = \|k_y, k_x\|$ and $D_{\mathcal{N}_D}(k_x, k_y) = D_{\mathcal{N}'_D}(k_x, k_y) = \|k_y, k_x\| = D(x, y)$.

So, assume $x \in X$ and $f \in V'_D - k_x$. To construct a map g as required, consider subsets A of X containing x such that

(i) no edge $e \in E(f)$ of $K(f)$ (whether it is of cardinality 1 or 2) is contained in A, and
(ii) the (by construction bipartite) graph $K_A(f)$ with edge set $E_A(f) := \{e \in E(f) : |e \cap A| = 1\}$ $(= \{e \in E(f) : e \cap A \ne \emptyset\}$ if (i) holds) and vertex set $V_A(f) := A \cup \bigcup_{e \in E_A(f)} e$ is connected.

Clearly, $A := \{x\}$ is one such subset; so, such subsets exist. Note also that $z \in V_{\{x\}}(f)$ must hold for any $z \in X$ with $\{z\} \in E(f)$ (or, equivalently, $f(z) = 0$). Moreover, if A is any such subset and $X \ne V_A(f)$ holds, it follows from our assumption that $K(f)$ is connected that some $a' \in X - V_A(f)$ and some $b' \in V_A(f)$ must exist with $\{a', b'\} \in E(f)$ and, of course, $\{a'\} \notin E(f)$ in which case $b' \in V_A(f) - A$ must hold and $A' := A + a'$ would be an even larger such subset. Thus, $X = V_A(f)$ must hold in case A is an inclusion-maximal subset of X of this form.

Now, to construct the required map g, choose just one such inclusion-maximal subset $A = A_{\max}$ of X arbitrarily, put $B := X - A$, and define the map $g_\eta \in \mathbb{R}^X$, for any positive real number η, by putting $g_\eta(a) := f(a) - \eta$ for all $a \in A$ and $g_\eta(b) := f(b) + \eta$ for all $b \in B$.

Clearly, we have $g_\eta(b_1) + g_\eta(b_2) = f(b_1) + f(b_2) + 2\eta > D(b_1, b_2)$ for all $b_1, b_2 \in B$, we have $g_\eta(a) + g_\eta(b) = f(a) + f(b) \ge D(a, b)$ for all $a \in A$ and $b \in B$ — with equality if and only if $\{a, b\} \in E_A(f)$ holds, and we have $g_\eta(a_1) + g_\eta(a_2) = f(a_1) + f(a_2) - 2\eta$ for all $a_1, a_2 \in A$.

So, we have $g_\eta(a_1) + g_\eta(a_2) \ge D(a_1, a_2)$ for all $a_1, a_2 \in A$ if and only if $\eta \le \alpha := \frac{1}{2}\min\{f(a_1) + f(a_2) - D(a_1, a_2) : a_1, a_2 \in A\}$ holds while, in case $\eta < \alpha$, we have $g_\eta(a_1) + g_\eta(a_2) > D(a_1, a_2)$ for all $a_1, a_2 \in A$. In consequence, we have

- $g_\eta \in P(D)$ for all $\eta \in [0, \alpha]$,
- $K(g_\eta)$ coincides with $K_A(f)$ for some $\eta > 0$ if and only if $\eta < \alpha$ holds,
- so, $K(g_\eta)$ is also connected and bipartite in this case,
- while, in case $\eta = \alpha$, $E(g_\eta)$ contains, in addition, one edge of the form $\{a_1, a_2\}$ with $a_1, a_2 \in A$ implying that $K(g_\eta)$ cannot be bipartite in this case.

Thus, we have $g := g_\alpha \in V'_D$ as well as $\{f, g\} \in E_D$ in view of $K\left(\frac{f+g}{2}\right) = K\left(g_{\frac{\alpha}{2}}\right) = K_A(f)$.

To finish the proof, it remains to show that $\|f, g\| + \|g, k_x\| = \|f, k_x\|$ holds, which is equivalent to $\|f, g\| = f(x) - g(x)$. But the latter equality clearly holds, since $\|f, g\| = \sup_{x \in X} |f(x) - g(x)| = \alpha = f(x) - g(x)$ holds by construction. □

Since, by Theorem 5.15, $\mathcal{N}'_D$ is a connected X-labeled subnetwork of $\mathcal{N}_D$ that realizes D, it follows from Theorem 5.14 that D is treelike if and only if $\mathcal{N}'_D$ is isomorphic to the corresponding edge-weighted X-tree, that is, we have the following corollary:

Corollary 5.16 *Suppose that D is a metric on a finite set X. Then, D is treelike if and only if $\mathcal{T} := \mathcal{N}'_D$ is an edge-weighted X-tree in which case $D = D_\mathcal{T}$ necessarily holds. In particular, if D is treelike, then $\mathcal{N}'_D$ coincides with $\mathcal{N}_D$.*

5.6 Optimal and hereditarily optimal realizations

The graph $\mathcal{N}'_D$ introduced in the last section has some further remarkable properties in addition to those presented in Theorem 5.15 and Corollary 5.16. To explain them, we first need to recall some facts concerning *optimal* realizations. An *optimal realization* of a metric D is a realization $\mathcal{N} = (V, E, \omega, \varphi)$ of D for which the *total weight*

$$\|\mathcal{N}\| := \sum_{e \in E} \omega(e)$$

is as small as possible.

For any given metric D on a finite set X, optimal realizations always exist [50, 102]. Furthermore, if D is treelike, every edge-weighted X-tree $\mathcal{T}$ representing D is actually an optimal realization of D [90], and $\mathcal{T}$ is uniquely determined by this property up to isomorphism. Even so, optimal realizations are hard to compute [3, 135] and, perhaps more importantly, they are in general not even uniquely determined by D (see e.g., [50, (A 3.3)] and [122]).

Taking this into account, one possible way out of this dilemma, discussed in [50, 58, 61], is to define a realization $\mathcal{N} = (V, E, \omega, \varphi)$ of D to be *hereditarily optimal* — or, for short, *h-optimal* — if

(i) $|X| \leq 2$ and $\mathcal{N}$ is an optimal realization, or
(ii) $|X| \geq 3$ and $\mathcal{N}$ has minimal total weight among all realizations that have the property that, for every proper subset $Y \subset X$, there exists a subset $E_Y \subseteq E$ such that $\varphi(Y) \subseteq V_Y := \bigcup_{e \in E_Y} e$ holds, and $\mathcal{N}_Y = (V_Y, E_Y, \omega|_{E_Y}, \varphi|_Y)$ is an h-optimal realization of the restriction $D|_Y$ of D to Y.

Remarkably, even though, at first sight, one may expect that the task of finding h-optimal realizations might be even more forbidding than the task of finding optimal ones, and that such realizations are even less likely to be unique, the converse is actually true. In fact, it can be shown (cf. [50, Theorem 7]) that the graph $\mathcal{N}'_D = (V'_D, E'_D, \omega'_D, \iota'_D)$ defined in the last section is an h-optimal realization of D and that every other h-optimal realization $\mathcal{N} = (V, E, \omega, \varphi)$ of D is essentially isomorphic to $\mathcal{N}'_D$ in the sense that it becomes isomorphic to $\mathcal{N}'_D$ by *fusing* (one at a time) certain pairs of edges that share a vertex of degree 2. More precisely, when fusing two distinct edges $e := \{u, v\}$ and $e' := \{v, w\}$, with $v \in V - \varphi(X)$ an unlabeled vertex of degree 2, we remove the vertex v and both edges e and e' from the graph, then add the single edge $\{u, w\}$ to the graph, and define its weight to be the sum of the weights of e and e'. In particular, $\mathcal{N}'_D$ can be characterized as being the — up to canonical isomorphism — unique hereditarily optimal realization of D without unlabeled vertices of degree 2.

As we have pointed out, there exist metrics D such that $\mathcal{N}_D \neq \mathcal{N}'_D$ holds. Thus, it is of some interest to characterize those metrics D for which $\mathcal{N}_D$ is an h-optimal realization of D or, equivalently, for which $\mathcal{N}'_D = \mathcal{N}_D$ holds. We conclude this section with a result from [58] that presents such a characterization in terms of tight-equality graphs:

Theorem 5.17 *For any metric D on X, the following two assertions are equivalent:*

(i) *$\mathcal{N}'_D$ and $\mathcal{N}_D$ coincide, that is, $V_D = V'_D$ and $E_D = E'_D$ hold.*
(ii) *The tight-equality graph $K(h)$ of every $h \in T(D)$ with $\dim[h]_D = 1$ is connected.*

Remark 5.18 *Intriguingly, each of the two assertions in this theorem is equivalent also to asserting that the metric D satisfies a certain 5-point condition which, as we shall see in Chapter 7, arises naturally when relating metrics to split systems (cf. Theorem 7.11).*

Proof We first show that (ii) implies (i). Note that, since $\mathcal{N}'_D$ is a subgraph of $\mathcal{N}_D$, it suffices to show that $V_D \subseteq V'_D$ and $E_D \subseteq E'_D$ both hold in case $K(h)$ is connected for every $h \in T(D)$ with $\dim[h]_D = 1$.

To this end, let $\{f, g\}$ be an arbitrary edge of $\mathcal{N}_D$ and consider the map $h := \frac{1}{2}(f + g)$. By assumption, the face $[h]$ of $P(D)$ containing h is an edge of $P(D)$ with endpoints f and g that is contained in $T(D)$, we have $E(h) \subseteq E(f) \cap E(g)$ in view of (5.4), and $K(h)$ is connected $\big($in view of (ii)$\big)$ and bipartite (in view of $\dim[h]_D = 1$) . Hence, $K(f)$ and $K(g)$ are both connected implying that $f, g \in V'_D$ and $\{f, g\} \in E'_D$ hold, as claimed.

To show that (i) implies (ii), suppose that $V'_D = V_D$ and $E'_D = E_D$ hold, and consider an arbitrary map $f \in T(D)$ with $\dim[f]_D = 1$. Let $Y \subseteq X$ be the bipartite connected component of $K(f)$ and let A and B denote the bipartition of Y induced by $K(f)$. Note that, in view of $f \in T(D)$, Lemma 5.10(iv) implies that both, A and B, are non-empty. For every $\varepsilon \in \mathbb{R}$, we define a map $f_\varepsilon \in \mathbb{R}^X$ by

$$f_\varepsilon = \begin{cases} f(x) & \text{if } x \in X - Y, \\ f(x) + \varepsilon & \text{if } x \in A, \\ f(x) - \varepsilon & \text{if } x \in B, \end{cases}$$

for all $x \in X$. Let $\varepsilon_{\min}$ and $\varepsilon_{\max}$ denote the minimum and maximum number ε, respectively, such that $f_\varepsilon \in T(D)$ holds. Note that $\varepsilon_{\min}$ and $\varepsilon_{\max}$ are well defined, since $g(x) \geq 0$ holds for all $g \in T(D)$ and all $x \in X$. Moreover, it follows immediately from the construction that $\varepsilon_{\min} < 0 < \varepsilon_{\max}$ holds, and that $K(f) = K(f_\varepsilon)$ holds for all ε with $\varepsilon_{\min} < \varepsilon < \varepsilon_{\max}$. Also by construction, $K(f)$ is a subgraph of $K(f_{\varepsilon_{\min}})$ and there must exist some edge of $K(f_{\varepsilon_{\min}})$ that is not an edge of $K(f)$ and the endpoints of this edge are either both contained in A, or one endpoint is in A and the other endpoint is in $X - Y$. Analogous properties hold for the graph $K(f_{\varepsilon_{\max}})$. Hence, both $K(f_{\varepsilon_{\min}})$ and $K(f_{\varepsilon_{\max}})$ have no bipartite connected component, which yields that $f_{\varepsilon_{\min}}$ and $f_{\varepsilon_{\max}}$ are contained in V_D.

To conclude the proof of the theorem, consider the map $h := \frac{1}{2}(f_{\varepsilon_{\min}} + f_{\varepsilon_{\max}})$. Note that $h = f_{\varepsilon_0}$ holds, where $\varepsilon_0 := \frac{1}{2}(\varepsilon_{\min} + \varepsilon_{\max})$. Since we have $\varepsilon_{\min} < \varepsilon_0 < \varepsilon_{\max}$ this implies that $K(f) = K(h)$ and, thus, $\dim[f]_D = \dim[h] = 1$. Hence, $\{f_{\varepsilon_{\min}}, f_{\varepsilon_{\max}}\}$ is an edge of $\mathcal{N}_D$ and, in view of $E_D = E'_D$, also an edge of $\mathcal{N}'_D$. This yields that $K(h) = K(f)$ is connected, as claimed. □

6

From quartet and tree systems to trees

We continue referring to a finite set X that we now assume to have cardinality $n \geq 4$. In the previous two chapters, we have seen how to obtain trees and/or networks from splits and metrics, respectively. Here, we will consider how to obtain trees from quartet or, more generally, *tree systems*. As we shall see, the situation is quite different in this case, mainly due to the fact that splits and metrics provide *global* information whereas quartet and tree systems only provide (overlapping) *local* information.

To illustrate this fact, recall that we showed in Chapter 2 that a phylogenetic X-tree $\mathcal{T}$ is determined by the set $\mathcal{Q}_\mathcal{T}$ of *all* quartets it displays, and in Chapter 3 that a quartet system $\mathcal{Q}$ coincides with the set $\mathcal{Q}_\mathcal{T}$ of all the quartets displayed by a phylogenetic X-tree $\mathcal{T}$ if and only if it is thin, transitive, and saturated. However, there are quartet systems $\mathcal{Q}$ that do not satisfy these conditions, but that still "encode" a unique X-tree in the sense that there is, up to canonical isomorphism, a *unique* phylogenetic X-tree $\mathcal{T}$ for which $\mathcal{Q} \subseteq \mathcal{Q}_\mathcal{T}$ holds. We shall call such quartet systems *definitive*, and we will say that $\mathcal{Q}$ *defines* $\mathcal{T}$ if this holds.

For example, it can be shown that the quartet system

$$\mathcal{Q} = \{12|37, 16|45, 15|34, 23|67\}$$

defines the tree in Figure 6.1. A phylogenetic X-tree $\mathcal{T}$ that is defined by a quartet system must be binary as a non-binary X-tree $\mathcal{T}$ can always be modified in more than one way to yield a binary X-tree $\mathcal{T}'$ with $\mathcal{Q}_\mathcal{T} \subset \mathcal{Q}_{\mathcal{T}'}$. Since a phylogenetic X-tree $\mathcal{T}$ encoded by a thin, transitive, and saturated quartet system need not be binary, it follows that such quartet systems need not be definitive.

It is an open problem though, whether or not there exists a polynomial time algorithm to decide whether a given quartet system is definitive. However, it is known that the related problem to decide whether a given quartet system

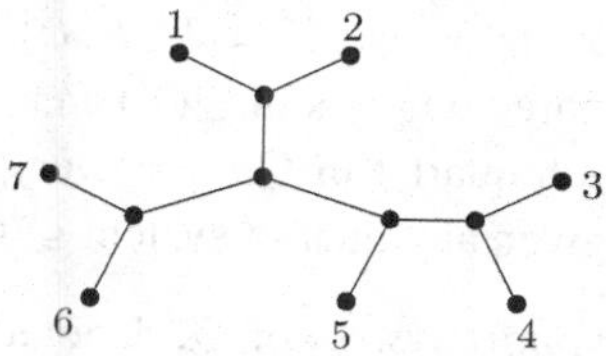

Figure 6.1 A phylogenetic $\langle 7 \rangle$-tree defined by the quartet system $\mathcal{Q} = \{12|37, 16|45, 15|34, 23|67\}$.

is *compatible*, that is, whether there exists a phylogenetic X-tree $\mathcal{T}$ such that $\mathcal{Q} \subseteq \mathcal{Q}_{\mathcal{T}}$ holds, is NP-complete [123]. This hints at the inherent computational complexity involved in this problem. Therefore, although it is possible to develop heuristics that would, at least in specific instances, always generate a tree (or network) from a quartet system (using, for example, ideas presented in [54]), we will concentrate in this chapter rather on *sparse* quartet systems, that is, quartet systems of size $n - 3$, and investigate which of them are definitive.

Our main result will be stated in Section 6.3 (Theorem 6.4). It provides, given a binary phylogenetic X-tree $\mathcal{T}$, a necessary and sufficient condition for sparse quartet systems $\mathcal{Q} \subseteq \mathcal{Q}_{\mathcal{T}}$ to define $\mathcal{T}$. Sufficiency will be established also in this section. Proving necessity is quite a bit harder. Some preparatory results will be given in Section 6.4. In particular, we will demonstrate a remarkable result recently established by Stefan Grünewald concerning the so-called "super-tree problem". It appears to be the — so far — only "positive" result established in this context, and provides a rather useful sufficient condition for the existence of super-trees for certain collections of phylogenetic trees with overlapping (but not too strongly overlapping!) support called "slim tree systems". Based on this, the altogether eight steps needed for proving the crucial direction of Theorem 6.4 will then be presented in the last section (Section 6.5).

The next two sections collect definitions and simple facts that will be needed in this context.

6.1 On quartet systems

We begin by introducing some more definitions and collecting some simple observations regarding quartet systems:

($\mathcal{Q}$1) Recall first that, given a quartet $q = ab|cd$, the "underlying" set $\{a, b, c, d\}$ is also called the support of q and denoted by $supp(q)$. Recall also that, given a phylogenetic X-tree $\mathcal{T}=(V, E, \varphi)$ and a 4-subset Y of X, at most one quartet of the three quartets in $\mathcal{Q}(Y)$ can be contained in $\mathcal{Q}_{\mathcal{T}}$ and there exists exactly one such quartet in $\mathcal{Q}_{\mathcal{T}}$ for every 4-subset Y of X if and only if $\mathcal{T}$ is binary. Next, given any quartet system $\mathcal{Q} \subseteq \mathcal{Q}(X)$,

(Q1–i) we define the *support system* of $\mathcal{Q}$, denoted by $supp(\mathcal{Q})$, to be the $\mathcal{Q}$-indexed family $supp(\mathcal{Q}) := \big(supp(q)\big)_{q\in\mathcal{Q}}$ of 4-subsets of X,

(Q1–ii) we put $\bigcup supp(\mathcal{Q}) := \bigcup_{q\in\mathcal{Q}} supp(q)$,

(Q1–iii) we define the *excess* of $\mathcal{Q}$, denoted by $exc(\mathcal{Q})$, by

$$exc(\mathcal{Q}) := \Big|\bigcup supp(\mathcal{Q})\Big| - 3 - |\mathcal{Q}|,$$

in case $\mathcal{Q}$ is not empty and by $exc(\mathcal{Q}) := 0$ else, and

(Q1–iv) we define $\mathcal{Q}$ to be *excess free* if $exc(\mathcal{Q}) = 0$ holds.

Note that $exc(\mathcal{Q})$ only depends the number $|\mathcal{Q}|$ of quartets in $\mathcal{Q}$ and the cardinality of the union $\bigcup supp(\mathcal{Q})$ and not on the actual quartets q in $\mathcal{Q}$. In particular, if $\mathcal{Q}$ contains, for every 4-subset Y of X, at most one quartet q with support Y, it only depends on the collection $\mathcal{X}_{\mathcal{Q}} := \{supp(q) : q \in \mathcal{Q}\}$ of subsets Y of X that are of the form $Y = supp(q)$ for some $q \in \mathcal{Q}$.

(Q2) Further, given a phylogenetic X-tree $\mathcal{T}=(V, E, \varphi)$, we'll say that a quartet $q = ab|cd \in \mathcal{Q}(X)$ is *gap free* relative to $\mathcal{T}$ if $q \in \mathcal{Q}_{\mathcal{T}}$ holds and the two interior vertices in the tree $\mathcal{T}_q := \mathcal{T}|_{supp(q)}$ obtained by restricting $\mathcal{T}$ to the support $supp(q)$ of q are adjacent in $\mathcal{T}$ in which case we will also say that q *distinguishes* the (necessarily) interior edge e of $\mathcal{T}$ that connects these two vertices, i.e., the edge $e \in E$ that coincides — as an element of $\binom{V}{2}$ — with the unique interior edge in the tree $\mathcal{T}_q$.

Clearly, given a quartet $q = ab|cd \in \mathcal{Q}_{\mathcal{T}}$ and an interior edge $e = \{u, v\} \in E_{int}(\mathcal{T})$ with $a, b \in \mathcal{T}^{(e)}(u)$ and $c, d \in \mathcal{T}^{(e)}(v)$, the following assertions are equivalent:

(dstg–i) The quartet q distinguishes e.

(dstg–ii) The unique edge $e^{u\to a} \in E(u)$ that separates u from a is distinct from the unique edge $e^{u\to b} \in E(u)$ that separates u from b, and the unique edge $e^{v\to c} \in E(v)$ that separates v from c is distinct from the unique edge $e^{v\to d} \in E(v)$ that separates v from d.

(dstg–iii) The path $\mathbf{p}_{\mathcal{T}}(a, b)$ passes through u, but not through v while the path $\mathbf{p}_{\mathcal{T}}(c, d)$ passes through v, but not through u.

(dstg–iv) One has $u = med_{\mathcal{T}}(a, b, c) = med_{\mathcal{T}}(a, b, d)$ and $v = med_{\mathcal{T}}(b, c, d) = med_{\mathcal{T}}(a, c, d)$.

(dstg–v) The path $\mathbf{p}_{\mathcal{T}}(a, c)$ shares exactly the edge e with the path $\mathbf{p}_{\mathcal{T}}(b, d)$.

(dstg–vi) The path $\mathbf{p}_{\mathcal{T}}(a, d)$ shares exactly the edge e with the path $\mathbf{p}_{\mathcal{T}}(b, c)$.

The following simple observation will be of some use in establishing the crucial direction of the main theorem in the last section:

Lemma 6.1 *Assume that $n \geq 5$ holds and that we are given a binary X-tree $\mathcal{T} = (V, E, \varphi)$, two distinct cherries x, y and x', y' of X, and two quartets $q = xy|ab$ and $q' = x'y'|a'b'$ that, respectively, distinguish the two (necessarily interior and distinct) edges $e := e_{\mathcal{T}}(x, y)$ and $e' := e_{\mathcal{T}}(x', y')$ that — in turn — are, respectively, incident with the two vertices $v := v_{\mathcal{T}}(x) = v_{\mathcal{T}}(y)$ and $v' := v_{\mathcal{T}}(x') = v_{\mathcal{T}}(y')$. Then the following hold:*

(Q2–i) *If the two edges e and e' share a vertex, one must have $\{x, y\} \cap \{a', b'\} \neq \emptyset$ and $\{x', y'\} \cap \{a, b\} \neq \emptyset$.*

(Q2–ii) *If $e \cap e' = \emptyset$ as well as $a = x'$ and $b = b'$ holds, one has $D_{\mathcal{T}}(b, x) + D_{\mathcal{T}}(a, a') < D_{\mathcal{T}}(x, a) + D_{\mathcal{T}}(b, a')$ and, therefore, $xb|aa' \in \mathcal{Q}_{\mathcal{T}}$ and, by symmetry, also $yb|aa' \in \mathcal{Q}_{\mathcal{T}}$.*

Proof Note first that we may assume, without loss of generality, that $\mathcal{T}$ is a simple binary X-tree. Let $u := med(a, b, x)$ denote the median of a, b, and x so that $e = \{u, v\}$, $D_{\mathcal{T}}(u, x) = 2$ as well as

$$D_{\mathcal{T}}(b, x) = D_{\mathcal{T}}(b, u) + D_{\mathcal{T}}(u, x), \quad D_{\mathcal{T}}(x, a) = D_{\mathcal{T}}(x, u) + D_{\mathcal{T}}(u, a),$$

and

$$D_{\mathcal{T}}(a, b) = D_{\mathcal{T}}(a, u) + D_{\mathcal{T}}(u, b)$$

hold. In addition, let $u' := med(a', b', x')$ denote the median of a', b', and x' so that $e' = \{u', v'\}$ as well as $D_{\mathcal{T}}(b', x') = D_{\mathcal{T}}(b', u') + D_{\mathcal{T}}(u', x')$, $D_{\mathcal{T}}(x', a') = D_{\mathcal{T}}(x', u') + D_{\mathcal{T}}(u', a')$, and $D_{\mathcal{T}}(a', b') = D_{\mathcal{T}}(a', u') + D_{\mathcal{T}}(u', b')$ holds.

Now consider the path $\mathbf{p}_{\mathcal{T}}(x, x') = w_0, w_1, \ldots, w_\ell$ of length $\ell := D_{\mathcal{T}}(x, x')$ in $\mathcal{T}$ from $w_0 := x$ to $w_\ell := x'$. Note that the fact that, by assumption, x and x' are members of distinct cherries of $\mathcal{T}$ implies that e must be the first and e' the last interior edge of that path, i.e., $e = \{w_1, w_2\}$ and $e' = \{w_{\ell-2}, w_{\ell-1}\}$ and, therefore, also $\ell \geq 4$, $w_1 = v$, $w_2 = u$, $w_{\ell-2} = u'$, $w_{\ell-1} = v'$ and, hence,

$$\ell = D_{\mathcal{T}}(x, x') = D_{\mathcal{T}}(x, u) + D_{\mathcal{T}}(u, u') + D_{\mathcal{T}}(u', x')$$

as well as

$$\ell - 2 = D_{\mathcal{T}}(x', u) = D_{\mathcal{T}}(x', u') + D_{\mathcal{T}}(u', u) = 2 + D_{\mathcal{T}}(u, u')$$

must hold.

The first assertion follows directly from the fact that e and e' share a vertex if and only if $u = u'$ and $\ell = 4$ holds in which case the support of every quartet that — like q' — distinguishes e' must contain some leaf that is separated from u by e and, hence, must be an element of $\{x, y\}$. Likewise, the support of every quartet that — like q — distinguishes e must contain some leaf that is separated from u by e' and, hence, must be an element of $\{x', y'\}$. For example, considering the tree $\mathcal{T}$ depicted in Figure 6.1, $6 \in \{a, b\}$ or $7 \in \{a, b\}$ must hold for every quartet of the form $12|ab$ that distinguishes the vertical edge $e_{\mathcal{T}}(1, 2)$ in that tree.

To illustrate the second assertion and its proof, consider for example the tree $\mathcal{T}$ depicted in Figure 6.1 and the quartets $q := 12|37$ and $q' := 34|57$: Clearly, putting $x := 1$, $y := 2$, $x' := 3$, $y' := 4$, $a := x' = 3$, $a' := 5$, and $b = b' := 7$, the assumptions of ($\mathcal{Q}$2-ii) are fulfilled, and $D(b, x) + D(a, a') = D(7, 1) + D(3, 5) < D(1, 3) + D(7, 5) = D(x, a) + D(b, a')$ indeed holds. Furthermore, the path from $b = 7$ to $a = 3$ considered below in the proof of ($\mathcal{Q}$2–ii) must first pass through the lower vertex u of the vertical edge $e_{\mathcal{T}}(1, 2)$ and then through $u' := v_{\mathcal{T}}(5)$ and $v' := v_{\mathcal{T}}(3)$, and $3 = D_{\mathcal{T}}(7, u') = D_{\mathcal{T}}(7, u) + D_{\mathcal{T}}(u, u') = 2 + 1$ holds.

Now, to prove ($\mathcal{Q}$2–ii) in general, we return to the notations introduced before and assume that $e \cap e' = \emptyset$ holds. Note first that this implies that $u \neq u'$ and, hence, $\ell \geq 5$ must hold.

Further, assuming that also $a = x'$ and $b = b'$ holds, we have

$$\begin{aligned} D_{\mathcal{T}}(b, a) &= D_{\mathcal{T}}(a, u) + D_{\mathcal{T}}(u, b) = D_{\mathcal{T}}(x', u) + D_{\mathcal{T}}(u, b) \\ &= D_{\mathcal{T}}(x', u') + D_{\mathcal{T}}(u', u) + D_{\mathcal{T}}(u, b) \\ &= D_{\mathcal{T}}(a, u') + D_{\mathcal{T}}(u', u) + D_{\mathcal{T}}(u, b) \end{aligned}$$

and, therefore (cf. the discussion following Equation (1.3) in Chapter 1) also

$$D_{\mathcal{T}}(b, u') = D_{\mathcal{T}}(b, u) + D_{\mathcal{T}}(u, u').$$

Thus, in view of

$$\begin{aligned} D_{\mathcal{T}}(b, x) &= D_{\mathcal{T}}(b, u) + D_{\mathcal{T}}(u, x), \\ D_{\mathcal{T}}(a, a') &= D_{\mathcal{T}}(x', a') = D_{\mathcal{T}}(x', u') + D_{\mathcal{T}}(u', a') \\ &= D_{\mathcal{T}}(a, u') + D_{\mathcal{T}}(u', a'), \\ D_{\mathcal{T}}(b, a') &= D_{\mathcal{T}}(b', a') = D_{\mathcal{T}}(b', u') + D_{\mathcal{T}}(u', a') \\ &= D_{\mathcal{T}}(b, u') + D_{\mathcal{T}}(u', a') \\ &= D_{\mathcal{T}}(b, u) + D_{\mathcal{T}}(u, u') + D_{\mathcal{T}}(u', a'), \quad \text{and} \\ D_{\mathcal{T}}(x, a) &= D_{\mathcal{T}}(x, x') = \ell = D_{\mathcal{T}}(x, u) + D_{\mathcal{T}}(u, u') + D_{\mathcal{T}}(u', x') \\ &= D_{\mathcal{T}}(x, u) + D_{\mathcal{T}}(u, u') + D_{\mathcal{T}}(u', a), \end{aligned}$$

we have indeed,

$$\begin{aligned}
&D_{\mathcal{T}}(b,x) + D_{\mathcal{T}}(a,a') \\
&\quad = D_{\mathcal{T}}(b,u) + D_{\mathcal{T}}(u,x) + D_{\mathcal{T}}(a,u') + D_{\mathcal{T}}(u',a') \\
&\quad < D_{\mathcal{T}}(b,u) + D_{\mathcal{T}}(u,x) + D_{\mathcal{T}}(a,u') + D_{\mathcal{T}}(u',a') + 2D_{\mathcal{T}}(u,u') \\
&\quad = D_{\mathcal{T}}(b,u) + D_{\mathcal{T}}(u,u') + D_{\mathcal{T}}(u',a') + D_{\mathcal{T}}(x,u) + D_{\mathcal{T}}(u,u') \\
&\qquad + D_{\mathcal{T}}(u',a) \\
&\quad = D_{\mathcal{T}}(b,a') + D_{\mathcal{T}}(x,a),
\end{aligned}$$

as claimed. □

($\mathcal{Q}$3) We will say that a subset $\mathcal{Q}$ of $\mathcal{Q}(X)$ distinguishes all interior edges of some phylogenetic X-tree $\mathcal{T}=(V,E,\varphi)$ if there exists, for every interior edge $e \in E_{int}(\mathcal{T})$ of $\mathcal{T}$, some quartet $q = q_e \in \mathcal{Q}$ that distinguishes e, and that $\mathcal{Q}$ is gap free relative to $\mathcal{T}$ if every $q \in \mathcal{Q}$ is gap free relative to $\mathcal{T}$.

For instance, the four interior edges of the tree $\mathcal{T}$ depicted in Figure 6.1 are distinguished, going more or less from top to bottom, by the — obviously gap free — quartets 12|37, 16|45, 15|34, 23|67, respectively. So, the quartet system $\mathcal{Q} := \{12|37, 16|45, 15|34, 23|67\}$ distinguishes all interior edges of that tree, and it is also gap free relative to $\mathcal{T}$.

($\mathcal{Q}$4) Next, given a phylogenetic X-tree $\mathcal{T}=(V,E,\varphi)$ and a subset $\mathcal{Q}$ of $\mathcal{Q}_{\mathcal{T}}$, let us consider the $\mathcal{Q}$-indexed family $\mathcal{E}_{int}(\mathcal{Q}) := \big(E_{int}(\mathcal{T}_q)\big)_{q\in\mathcal{Q}}$ of one-element subsets of $\binom{V}{2}$. Clearly, if a quartet system $\mathcal{Q}$ defines a (necessarily binary) phylogenetic X-tree $\mathcal{T}$, it must distinguish all interior edges of $\mathcal{T}$, that is, we must have $E_{int}(\mathcal{T}) \subseteq \bigcup_{q\in\mathcal{Q}} E_{int}(\mathcal{T}_q)$ and, therefore,

$$|\mathcal{Q}| \geq |E| = n - 3. \tag{6.1}$$

So, provided there exists — as will be demonstrated in Lemma 6.2 — at least one sparse definitive quartet system for $\mathcal{T}$, it is exactly the sparse definitive quartet systems that have minimal size among all definitive quartet systems $\mathcal{Q} \subseteq \mathcal{Q}_{\mathcal{T}}$ for $\mathcal{T}$.

To study such quartet systems $\mathcal{Q}$ — and, later on and much more generally, arbitrary systems of finite labeled trees — it is worthwhile to note that, quite generally, given any finite index set $\mathcal{I}$, any subset F of a finite set F', and any $\mathcal{I}$-indexed family $\mathcal{F} = (F_i)_{i\in\mathcal{I}}$ of non-empty subsets of F', the trivial fact that two finite sets A and B coincide if and only if $A \subseteq B$ and $|A| \geq |B|$ holds, implies that the following five assertions are all equivalent:

(i) The family $\mathcal{F}$ forms an $\mathcal{I}$-indexed partition of F.
(ii) Every member F_i of $\mathcal{F}$ is contained in F, one has $F_i \cap F_j = \emptyset$ for any two distinct indices $i, j \in \mathcal{I}$, and $\sum_{i\in\mathcal{I}} |F_i| \geq |F|$ holds.

(ii′) Every member F_i of $\mathcal{F}$ is contained in F, one has $F_i \cap F_j = \emptyset$ for any two distinct indices $i, j \in \mathcal{I}$, and $\sum_{i\in\mathcal{I}} |F_i| = |F|$ holds.
(iii) One has $F \subseteq \bigcup_{i\in\mathcal{I}} F_i$ and $\sum_{i\in\mathcal{I}} |F_i| \leq |F|$.
(iii′) One has $F \subseteq \bigcup_{i\in\mathcal{I}} F_i$ and $\sum_{i\in\mathcal{I}} |F_i| = |F|$.

In particular, denoting the *(formal) disjoint union*

$$\amalg_{i\in\mathcal{I}} F_i := \{(i, f) : i \in \mathcal{I}, f \in F_i\}$$

of the members of the family $\mathcal{F}$ by $\amalg\mathcal{F}$, the following five assertions all are equivalent:

(i) The canonical map $\psi_{\mathcal{F}} : \amalg\mathcal{F} \to F' : (i, f) \to f$ is a bijection from $\amalg\mathcal{F}$ onto F.
(ii) The map $\psi_{\mathcal{F}}$ is injective and one has $\sum_{i\in\mathcal{I}} |F_i| \geq |F|$.
(ii′) The map $\psi_{\mathcal{F}}$ is injective and one has $\sum_{i\in\mathcal{I}} |F_i| = |F|$.
(iii) The set F is contained in the image of $\psi_{\mathcal{F}}$ and one has $\sum_{i\in\mathcal{I}} |F_i| \leq |F|$.
(iii′) The set F is contained in the image of $\psi_{\mathcal{F}}$ and one has $\sum_{i\in\mathcal{I}} |F_i| = |F|$.

Thus, returning — with $\mathcal{T}$ and $\mathcal{Q}$ as above — to the $\mathcal{Q}$-indexed family $\mathcal{E}_{int}(\mathcal{Q}) := \big(E_{int}(\mathcal{T}_q)\big)_{q\in\mathcal{Q}}$ of one-element subsets of $\binom{V}{2}$ introduced above, the following five assertions are all equivalent:

($\mathcal{Q}$4–i) The family $\mathcal{E}_{int}(\mathcal{Q})$ forms a $\mathcal{Q}$-indexed partition of $E_{int}(\mathcal{T})$ (by one-element subsets — so, not the partition, but its indexing is of any interest here).
($\mathcal{Q}$4–ii) Every member $E_{int}(\mathcal{T}_q)$ of $\mathcal{E}_{int}(\mathcal{Q})$ is contained in $E_{int}(\mathcal{T})$ (i.e., $\mathcal{Q}$ is gap free relative to $\mathcal{T}$), one has $E_{int}(\mathcal{T}_q) \cap E_{int}(\mathcal{T}_{q'}) = \emptyset$ for any two distinct quartets $q, q' \in \mathcal{Q}$, and $|\mathcal{Q}| \geq E_{int}(\mathcal{T}) = n - 3$.
($\mathcal{Q}$4–ii′) $\mathcal{Q}$ is gap free relative to $\mathcal{T}$, one has $E_{int}(\mathcal{T}_q) \cap E_{int}(\mathcal{T}_{q'}) = \emptyset$ for any two distinct quartets $q, q' \in \mathcal{Q}$, and $|\mathcal{Q}| = n - 3$.
($\mathcal{Q}$4–iii) One has $E_{int}(\mathcal{T}) \subseteq \bigcup_{q\in\mathcal{Q}} E_{int}(\mathcal{T}_q)$ — i.e., $\mathcal{Q}$ distinguishes all interior edges of $\mathcal{T}$, and $|\mathcal{Q}| \leq n - 3$ holds.
($\mathcal{Q}$4–iii′) $\mathcal{Q}$ distinguishes all interior edges of $\mathcal{T}$, and $|\mathcal{Q}| = n - 3$ holds.

($\mathcal{Q}$5) Note also that $supp(\mathcal{Q}) = X$ must hold in case $\mathcal{T} = (V, E, \varphi)$ is a binary X-tree and $\mathcal{Q}$ is a subset of $\mathcal{Q}(X)$ that distinguishes all interior edges of $\mathcal{T}$: Indeed, consider an arbitrary leaf $a \in X$, the interior vertex $v_{\mathcal{T}}(a)$ contained in the pendant edge $e_{\mathcal{T}}(a)$ containing $\varphi(a)$, and an interior edge e containing $v_{\mathcal{T}}(a)$. Then $a \in supp(q)$ must hold for every quartet q that distinguishes e. Clearly, there are exactly two such interior edges if and only if a is not part of a cherry while, otherwise, there is exactly one such edge (Figure 6.2 illustrates these two cases).

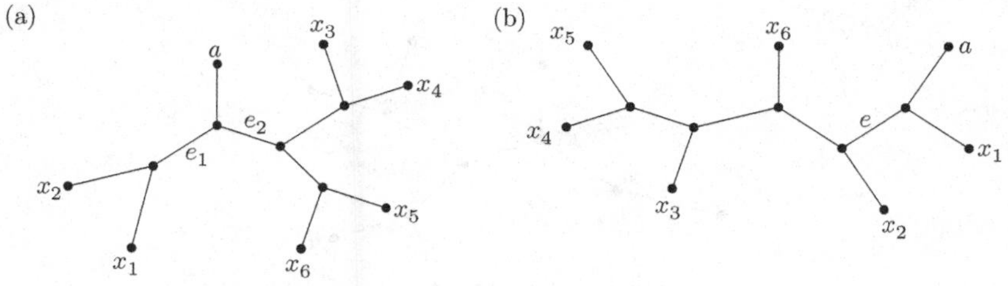

Figure 6.2 (a) If a quartet q distinguishes the edge e_1 or the edge e_2, then $a \in supp(q)$ holds. (b) If a quartet q distinguishes the edge e, then $a \in supp(q)$ holds.

So, we must also have $exc(\mathcal{Q}) = n - 3 - |\mathcal{Q}|$ in case $\mathcal{Q}$ distinguishes all interior edges of $\mathcal{T}$ and, therefore, also $exc(\mathcal{Q}) \leq 0$. Furthermore, denoting, for every $q \in \mathcal{Q}$ the unique 2-subset of V in $E_{int}(\mathcal{T}_q)$ by e_q and assuming that $\bigcup supp(\mathcal{Q}) = X$ holds (whether or not $\mathcal{Q}$ distinguishes all interior edges of $\mathcal{T}$), the following five assertions are all equivalent:

($\mathcal{Q}$5–i) The canonical map

$$\psi_{\mathcal{Q}} : \mathcal{Q} \to \binom{V}{2} : q \mapsto e_q$$

is a bijection from $\mathcal{Q}$ onto $E_{int}(\mathcal{T})$.

($\mathcal{Q}$5–ii) The map $\psi_{\mathcal{Q}}$ is an injective map from $\mathcal{Q}$ into $E_{int}(\mathcal{T})$ (i.e., $\mathcal{Q}$ is gap free relative to $\mathcal{T}$) and one has $exc(\mathcal{Q}) \leq 0$.

($\mathcal{Q}$5–ii′) The map $\psi_{\mathcal{Q}}$ is an injective map from $\mathcal{Q}$ onto $E_{int}(\mathcal{T})$ (i.e., $\mathcal{Q}$ is gap free relative to $\mathcal{T}$) and one has $exc(\mathcal{Q}) = 0$.

($\mathcal{Q}$5–iii) The set $E_{int}(\mathcal{T})$ is contained in the image of $\psi_{\mathcal{Q}}$ (i.e., $\mathcal{Q}$ distinguishes all interior edges of $\mathcal{T}$) and one has $exc(\mathcal{Q}) \geq 0$.

($\mathcal{Q}$5–iii′) $\mathcal{Q}$ distinguishes all interior edges of $\mathcal{T}$, and one has $exc(\mathcal{Q}) = 0$.

Finally, as mentioned before, we present a fundamental observation concerning the *existence* of sparse definitive quartet systems:

Lemma 6.2 *Let $\mathcal{T} = (V, E, \varphi)$ be a binary X-tree with $n \geq 4$. Then, there exists a sparse quartet system $\mathcal{Q} \subseteq \mathcal{Q}_{\mathcal{T}}$ that defines $\mathcal{T}$. More specifically, there exists an excess-free quartet system $\mathcal{Q} = \{q_1, q_2, \ldots, q_{n-3}\}$ that defines $\mathcal{T}$ and for which, in addition, $\left|\bigcup_{j=1}^{i} supp(q_j)\right| = i + 3$ holds for all $i = 1, 2, \ldots, n - 3$.*

Proof We prove the lemma by induction on the number n of elements in X. Assume, without loss of generality, that $\mathcal{T}$ is a simple binary X-tree.

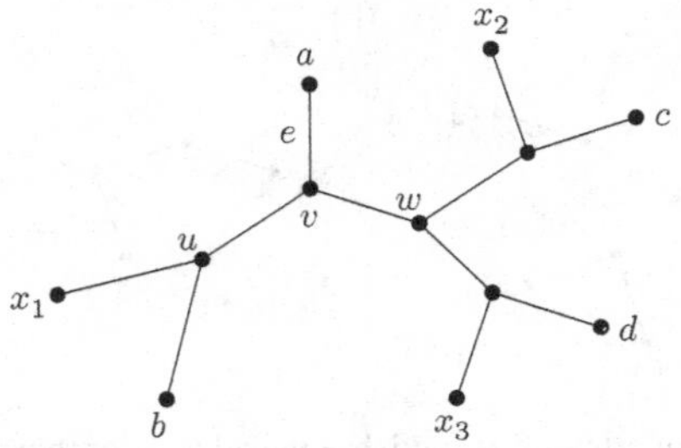

Figure 6.3 An example of the selection of a suitable quartet $q = ab|cd$ in the proof of Lemma 6.2.

If $n = 4$, then $\mathcal{Q} := \mathcal{Q}_\mathcal{T}$ contains a single quartet and, clearly, defines $\mathcal{T}$. If $n > 4$ holds, there must exist a quartet $q = ab|cd \in \mathcal{Q}_\mathcal{T}$ that distinguishes an edge of $\mathcal{T}$ such that the two paths in $\mathcal{T}$ from c and d to b both pass through the interior vertex $v := v_\mathcal{T}(a)$ in the pendant edge $e = e_\mathcal{T}(a) = \{v, a\} \in E$ with $a \in e$ and, therefore, also through the two other vertices $u, w \in V$ with $\{v, u\}, \{v, w\} \in E$ (one of which might coincide with b). So, these two paths diverge immediately after passing through these three vertices, and q distinguishes the first edge in the path from v to c or, as well, to d (see Figure 6.3). It follows that $\mathcal{T}' := (V', E', \varphi')$ defined by $V' := V - \{a, v\}$, $E' := (E - \{\{v, a\}, \{u, v\}, \{v, w\}\}) + \{u, w\}$, and $\varphi' := \varphi|_{X-a}$, is a binary X'-tree for $X' := X - a$. By induction, there exists some quartet system $\mathcal{Q}' = \{q_1, q_2, \ldots, q_{|X'|-3}\}$ of size $|X'| - 3 = n - 4$ that defines $\mathcal{T}'$ for which we may assume that also $\left|\bigcup_{j=1}^{i} supp(q_j)\right| = i + 3$ holds for all $i = 1, 2, \ldots, |X'| - 3$.

Now, put $\mathcal{Q} := \mathcal{Q}' + ab|cd$ and note that the quartet system $\mathcal{Q}$ has the correct size (and structure), is contained in $\mathcal{Q}_\mathcal{T}$, and defines $\mathcal{T}$: Indeed, if $\mathcal{T}^* = (V^*, E^*, \varphi^*)$ is any phylogenetic X-tree with $\mathcal{Q} \subseteq \mathcal{Q}_{\mathcal{T}^*}$, if $v^* \in V^*$ is the unique vertex in V^* that is adjacent to $\varphi^*(a)$, and if u^* and w^* are the other two vertices in V^* that are adjacent to v^*, then deleting $\varphi^*(a)$ and v^* as well as all edges in E^* that contain v^*, adding instead the edge $\{u^*, w^*\}$, and restricting φ^* to X', one obtains an X'-tree whose quartet system must contain $\mathcal{Q}'$. So, we can assume, without loss of generality, that this X'-tree actually coincides with $\mathcal{T}'$ which, in turn, then implies that also $\varphi^*(x) = x$ holds for every $x \in X'$, that u coincides with u^*, and w with w^*.

This, however, implies that $\mathcal{T}^*$ as well as $\mathcal{T}$ are both obtained, up to canonical isomorphism, from $\mathcal{T}'$ by (i) splitting the edge $\{u, w\}$ into two by inserting an additional vertex v between u and w, (ii) adding another new vertex dubbed a appended by the additional pendant edge $\{v, a\}$ to v, and (iii) extending the map $\varphi' : X' \to V'$ to a map φ from X into $V' \cup \{v, a\}$ by just putting $\varphi(a) := a$. Hence, also $\mathcal{T}$ and $\mathcal{T}^*$ must be canonically isomorphic. □

In view of this result, it is natural to ask which excess-free quartet systems define a binary X-tree $\mathcal{T}$. The example presented in Figure 6.1 shows that, at least, not all such quartet systems arise according to the construction used to establish Lemma 6.2 as the (joint) support of any three of its four quartets is all of $X = \langle 7 \rangle$.

The rest of this chapter is devoted to showing that, however, an appropriate variant of this procedure does indeed yield all excess-free quartet systems that define $\mathcal{T}$.

6.2 On set and tree systems

Suppose that $\mathcal{T}$ is a binary X-tree and that $\mathcal{Q} \subseteq \mathcal{Q}_\mathcal{T}$ is an excess-free quartet system that distinguishes all edges in $\mathcal{T}$. Then, how can we check whether $\mathcal{T}$ is defined by $\mathcal{Q}$?

One simple scheme would be to first consider pairs of quartets and to check whether any of these pairs define a phylogenetic tree on their (joint) support and then, if this is the case, to try to iteratively piece together further pairs of quartets or larger, already obtained trees — checking at each stage whether the relevant quartets define the resulting tree — until $\mathcal{T}$ is either obtained or it is found that it is not possible to obtain $\mathcal{T}$ in this way. We illustrate this process for the quartet system $\mathcal{Q} = \{15|34, 16|45, 12|37, 23|67\}$ depicted in Figure 6.4.

Clearly, at each stage in such a process, it is essential to be able to decide when two phylogenetic trees define another one. To deal with such a task, it is of advantage to first generalize the concepts that we introduced for quartet systems to more general set and tree systems. To this end, assume that, as above, $\mathcal{T} = (V, E, \varphi)$ is a fixed binary X-tree to which we will always refer in the rest of this chapter.

($\mathcal{X}$ 1) Recall first that, given a subset Y of X, the binary Y-tree obtained by restricting $\mathcal{T}$ to Y is denoted by $\mathcal{T}|_Y$. We will also denote by $V(Y) = V(Y|\mathcal{T})$ and $E(Y) = E(Y|\mathcal{T})$ the set of its vertices and edges, respectively, by $V_{int}(Y) = V_{int}(Y|\mathcal{T})$ the set of its interior vertices, and by $E_{int}(Y) = E_{int}(Y|\mathcal{T}) := E(Y) \cap \binom{V_{int}(Y)}{2} \subseteq \binom{V}{2}$ the set of its interior edges. Recall that $|V(Y)| = 2|Y| - 2$, $|E(Y)| = 2|Y| - 3$, $|V_{int}(Y)| = |Y| - 2$, and $|E_{int}(Y)| = |Y| - 3$ must hold in case $|Y| \geq 3$.

Further, we'll say that an edge $e \in E$ of $\mathcal{T}$ is *distinguished* by Y if $e \in E(Y)$ holds, and that Y is *gap free* relative to $\mathcal{T}$ if the set $V_{int}(Y)$ of interior vertices

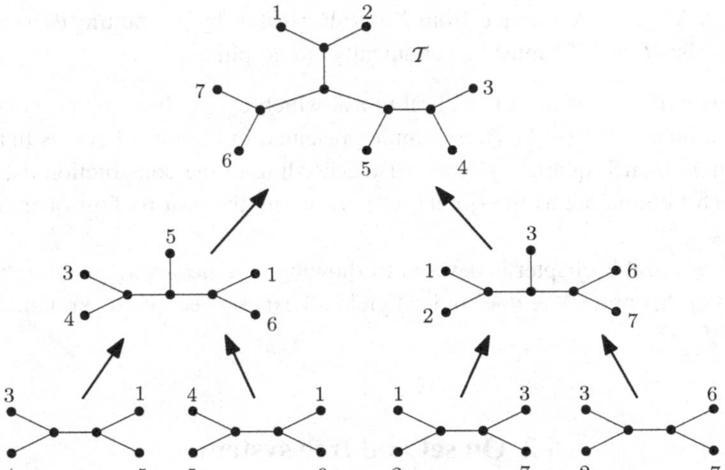

Figure 6.4 Building up the ⟨7⟩-tree $\mathcal{T}$ in Figure 6.1 from the quartet system $\mathcal{Q} = \{15|34, 16|45, 12|37, 32|67\}$ starting, for each quartet $q \in \mathcal{Q}$, with the phylogenetic tree defined by q.

of $\mathcal{T}|_Y$ is a connected or, equivalently — a convex subset of V in which case $E_{int}(Y)$ is the set of edges in the graph $\mathcal{T}[V_{int}(Y)]$ induced by $\mathcal{T}$ on $V_{int}(Y)$.

It is easy to see that Y is gap free if and only if $E_{int}(Y) \subseteq E$ — or, equivalently, $E_{int}(Y) \subseteq E_{int}(\mathcal{T})$ — holds.

($\mathcal{X}$ 2) Next, given a collection $\mathcal{X} \subseteq \mathbb{P}_{\geq 4}(X)$ of subsets Y of X of cardinality at least 4,

(i) we define the *tree system* associated to $\mathcal{X}$, denoted by $\mathcal{T}_{\mathcal{X}}$, to be the $\mathcal{X}$-indexed family $(\mathcal{T}|_Y)_{Y \in \mathcal{X}}$ of finite labeled trees $\mathcal{T}|_Y$ obtained by restricting the X-tree $\mathcal{T}$ to the subsets Y in $\mathcal{X}$,

(ii) we put $\bigcup \mathcal{X} := \bigcup_{Y \in \mathcal{X}} Y$,

(iii) we define the *excess* of $\mathcal{X}$, denoted by $exc(\mathcal{X})$, by $exc(\mathcal{X}) := |\bigcup \mathcal{X}| - 3 - \sum_{Y \in \mathcal{X}}(|Y| - 3)$, in case $\mathcal{X}$ is not empty and by $exc(\mathcal{X}) := 0$ else,

(iv) we define $\mathcal{X}$ to be *excess free* if $exc(\mathcal{X}) = 0$ holds,

(v) we define $\mathcal{X}$ to distinguish all interior edges of $\mathcal{T}$ if every interior edge of $\mathcal{T}$ is distinguished by some $Y \in \mathcal{X}$,

(vi) we define $\mathcal{X}$ to be *gap free* (relative to $\mathcal{T}$) if every $Y \in \mathcal{X}$ is gap free,

(vii) we define $\mathcal{X}$ to be *slim* if $exc(\mathcal{Y}) \geq 0$ holds for every subsystem $\mathcal{Y}$ of $\mathcal{X}$, and

(viii) we'll say that $\mathcal{X}$ *defines* $\mathcal{T}$ if every X-tree $\mathcal{T}'$ for which $\mathcal{T}'|_Y$ is isomorphic to $\mathcal{T}|_Y$ for all $Y \in \mathcal{X}$ must be isomorphic to $\mathcal{T}$.

Clearly, if $\mathcal{X}$ consists of exactly two distinct subsets Y_1 and Y_2, one has

$$exc(\mathcal{X}) = |Y_1 \cup Y_2| - 3 - (|Y_1| - 3) - (|Y_2| - 3) = 3 - |Y_1 \cap Y_2|, \quad (6.2)$$

so, $\mathcal{X}$ is excess free if and only if $|Y_1 \cap Y_2| = 3$ holds, and it has positive excess if and only if $|Y_1 \cap Y_2| \leq 2$ holds.

($\mathcal{X}$ 3) Note that, given a quartet system $\mathcal{Q} \subseteq \mathcal{Q}(X)$, the following hold for the associated collection $\mathcal{X}_{\mathcal{Q}} = \{supp(q) : q \in \mathcal{Q}\}$ of 4-subsets Y of X that are of the form $Y = supp(q)$ for some $q \in \mathcal{Q}$:

(1) One has $exc(\mathcal{X}_{\mathcal{Q}}) = exc(\mathcal{Q})$ for the collection $\mathcal{X}_{\mathcal{Q}} = \{supp(q) : q \in \mathcal{Q}\}$ of 4-subsets Y of X that are of the form $Y = supp(q)$ for some $q \in \mathcal{Q}$ provided $\mathcal{Q}$ contains, for every 4-subset Y of X, at most one quartet q with support Y (which holds in particular if $\mathcal{T}$ displays $\mathcal{Q}$).

(2) If $\mathcal{Q}$ is displayed by $\mathcal{T}$, then it distinguishes all interior edges of $\mathcal{T}$ if and only if $\mathcal{X}_{\mathcal{Q}}$ does.

(3) If $\mathcal{Q}$ is displayed by $\mathcal{T}$, then it is gap free if and only if so is $\mathcal{X}_{\mathcal{Q}}$.

(4) And if $\mathcal{Q}$ is displayed by $\mathcal{T}$, then it defines $\mathcal{T}$ if and only if $\mathcal{X}_{\mathcal{Q}}$ defines $\mathcal{T}$.

($\mathcal{X}$ 4) Let us now associate to $\mathcal{X}$ the $\mathcal{X}$-indexed family $\mathcal{E}_{int}(\mathcal{X}) := \big(E_{int}(Y)\big)_{Y \in \mathcal{X}}$. Then, the following five assertions are all equivalent:

($\mathcal{X}$ 4–i) The family $\mathcal{E}_{int}(\mathcal{X})$ forms an $\mathcal{X}$-indexed partition of $E_{int}(\mathcal{T})$.

($\mathcal{X}$ 4–ii) Every member $E_{int}(Y)$ of $\mathcal{E}_{int}(\mathcal{X})$ is contained in $E_{int}(\mathcal{T})$ (i.e., $\mathcal{X}$ is gap free relative to $\mathcal{T}$), one has $E_{int}(Y) \cap E_{int}(Y') = \emptyset$ for any two distinct subsets $Y, Y' \in \mathcal{T}$, and $\sum_{Y \in \mathcal{X}} |E_{int}(Y)| \geq E_{int}(\mathcal{T}) = n - 3$ holds.

($\mathcal{X}$ 4–ii′) $\mathcal{X}$ is gap free relative to $\mathcal{T}$, one has $E_{int}(Y) \cap E_{int}(Y') = \emptyset$ for any two distinct subsets $Y, Y' \in \mathcal{T}$, and $\sum_{Y \in \mathcal{X}} |E_{int}(Y)| = n - 3$.

($\mathcal{X}$ 4–iii) One has $E_{int}(\mathcal{T}) \subseteq \bigcup_{Y \in \mathcal{X}} E_{int}(Y)$ (i.e., $\mathcal{X}$ distinguishes all interior edges of $\mathcal{T}$) and $\sum_{Y \in \mathcal{X}} |E_{int}(Y)| \leq n - 3$ holds.

($\mathcal{X}$ 4–iii′) $\mathcal{X}$ distinguishes all interior edges of $\mathcal{T}$, and $\sum_{Y \in \mathcal{X}} |E_{int}(Y)| = n - 3$ holds.

In particular, $\mathcal{X}$ must be slim in this case in view of the fact that the family $\mathcal{E}_{int}(\mathcal{Y})$ must, for every subsystem $\mathcal{Y}$ of $\mathcal{X}$, form a $\mathcal{Y}$-indexed collection of pairwise disjoint subsets of $E_{int}(\mathcal{T}|_{\bigcup \mathcal{Y}})$ implying that $exc(\mathcal{Y}) = |\bigcup \mathcal{Y}| - 3 - \sum_{Y \in \mathcal{Y}} |E_{int}(Y)| = |E_{int}(\mathcal{T}|_{\bigcup \mathcal{Y}})| - \sum_{Y \in \mathcal{Y}} |E_{int}(Y)| \geq 0$ must hold.

($\mathcal{X}$5) Note also that the argument presented in ($\mathcal{Q}$5) yields also that $\bigcup \mathcal{X} = X$ must hold in case $\mathcal{X}$ distinguishes all interior edges of $\mathcal{T}$.

So, we must also have $n = |\bigcup \mathcal{X}|$ in case $\mathcal{X}$ distinguishes all interior edges of $\mathcal{T}$ and, therefore, also $exc(\mathcal{X}) \leq 0$ as this yields $|E_{int}(\mathcal{T})| \leq \sum_{Y \in \mathcal{X}} |E_{int}(Y)|$ and, therefore,

$$exc(\mathcal{X}) = \left|\bigcup \mathcal{X}\right| - 3 - \sum_{Y \in \mathcal{X}} (|Y| - 3) = |E_{int}(\mathcal{T})| - \sum_{Y \in \mathcal{X}} |E_{int}(Y)| \leq 0,$$

as claimed. Furthermore, denoting the (formal) disjoint union

$$\amalg_{Y \in \mathcal{X}} E_{int}(Y) := \{(Y, e) : Y \in \mathcal{X}, e \in E_{int}(Y)\}$$

of the members of the family $\mathcal{E}_{int}(\mathcal{X})$ by $\amalg \mathcal{E}_{int}(\mathcal{X})$ and assuming that $\bigcup \mathcal{X} = X$ holds (whether or not $\mathcal{Q}$ distinguishes all interior edges of $\mathcal{T}$), the following five assertions are all equivalent in this case:

($\mathcal{X}$5–i) The canonical map

$$\psi_{\mathcal{X}} : \amalg \mathcal{E}_{int}(\mathcal{X}) \to \binom{V}{2} : (Y, e) \mapsto e$$

is a bijection from $\amalg \mathcal{E}_{int}(\mathcal{X})$ onto $E_{int}(\mathcal{T})$.

($\mathcal{X}$5–ii) The map $\psi_{\mathcal{X}}$ is an injective map from $\amalg E_{int}(\mathcal{X})$ into $E_{int}(\mathcal{T})$ (i.e., $\mathcal{X}$ is gap free relative to $\mathcal{T}$) and one has $exc(\mathcal{X}) \leq 0$.

($\mathcal{X}$5–ii′) The map $\psi_{\mathcal{X}}$ is an injective map from $\amalg \mathcal{E}_{int}(\mathcal{X})$ onto $E_{int}(\mathcal{T})$ (i.e., $\mathcal{X}$ is gap free relative to $\mathcal{T}$) and one has $exc(\mathcal{X}) = 0$.

($\mathcal{X}$5–iii) The set $E_{int}(\mathcal{T})$ is contained in the image of $\psi_{\mathcal{X}}$ (i.e., $\mathcal{X}$ distinguishes all interior edges of $\mathcal{T}$) and one has $exc(\mathcal{T}) \geq 0$.

($\mathcal{X}$5–iii′) $\mathcal{X}$ distinguishes all interior edges of $\mathcal{T}$, and one has $exc(\mathcal{T}) = 0$.

If all of this holds, we will also say that $\mathcal{X}$ *partitions* $E_{int}(\mathcal{T})$.

($\mathcal{X}$6) Note that if a set system $\mathcal{X} \subseteq \mathbb{P}_{\geq 4}(X)$ consisting of exactly two subsets Y_1 and Y_2 partitions $E_{int}(\mathcal{T})$, one must have

($\mathcal{X}$6–i) $|Y_1 \cap Y_2| = 3$ in view of (6.2),

($\mathcal{X}$6–ii) the median v of the three elements in $Y_1 \cap Y_2$ is the only vertex in the intersection $V_{int}(Y_1) \cap V_{int}(Y_2)$, no two of these three elements can form a cherry in $\mathcal{T}$, and

($\mathcal{X}$6–iii) an edge $e \in E_v$ that is incident to v is contained in $E_{int}(Y_i)$ for $i = 1$ or 2 if and only if all elements $x \in X$ that are separated from v by e are contained in Y_i.

In particular, both subsets Y_1 and Y_2 must contain a cherry of $\mathcal{T}$ as, for both indices $i = 1, 2$, any element $x \in Y_i$ whose distance to v is as large as possible must be part of a cherry in Y_i.

Finally, we define a collection $\mathcal{P}$ of labeled trees to be a *binary tree system* if it is finite and all labeled trees in $\mathcal{P}$ are finite and binary, have at least one internal edge (and, therefore, a support of cardinality at least 4), and are pairwise non-isomorphic. In analogy to the concepts we have defined above for quartet and set systems, given a binary tree system $\mathcal{P}$,

(i) we denote by $supp(\mathcal{P})$, the $\mathcal{P}$-indexed family $supp(\mathcal{P}) := (supp(\mathcal{T}'))_{\mathcal{T}' \in \mathcal{P}}$ of finite sets,

(ii) we put

$$exc(\mathcal{P}) := \left|\bigcup supp(\mathcal{P})\right| - 3 - \sum_{\mathcal{T}' \in \mathcal{P}} (|supp(\mathcal{T}')| - 3)$$

in case $\mathcal{P}$ is non-empty, and $exc(\mathcal{P}) := 0$ else,

(iii) we define $\mathcal{P}$ to be *excess free* if $exc(\mathcal{P}) = 0$ holds, and

(iv) we define $\mathcal{P}$ to be *slim* if $exc(\mathcal{P}') \geq 0$ holds for every subsystem $\mathcal{P}'$ of $\mathcal{P}$ or, equivalently, if the collection

$$\mathcal{X}_{\mathcal{P}} := \{supp(\mathcal{T}') : \mathcal{T}' \in \mathcal{P}\}$$

of sets that are of the form $Y = supp(\mathcal{T}')$ for some $\mathcal{T}' \in \mathcal{P}$ is slim and $supp(\mathcal{T}') = supp(\mathcal{T}'')$ implies $\mathcal{T}' = \mathcal{T}''$ for all $\mathcal{T}', \mathcal{T}'' \in \mathcal{P}$.

Furthermore, a given binary tree system $\mathcal{P}$ is said

($\mathcal{P}$-i) to be *displayed* by a binary labeled tree if every tree in $\mathcal{P}$ is displayed by that tree,

($\mathcal{P}$-ii) to be *compatible* if there exists a binary labeled tree displaying $\mathcal{P}$,

($\mathcal{P}$-iii) to *define* a binary X-tree $\mathcal{T}$ if $\mathcal{T}$ displays $\mathcal{P}$ and is defined by the set system $\mathcal{X}_{\mathcal{P}}$ and

($\mathcal{P}$-iv) to be *definitive* if it defines some binary labeled tree.

Furthermore, referring to some given binary X-tree $\mathcal{T}$, we define a binary labeled tree $\mathcal{T}'$ to be gap free relative to $\mathcal{T}$, if $\mathcal{T}$ displays $\mathcal{T}'$ and $supp(\mathcal{T}')$ is gap free relative to $\mathcal{T}$. Moreover, we define a binary tree system $\mathcal{P}$ to be gap free relative to $\mathcal{T}$ if this holds for all of its members, and we define $\mathcal{P}$ to distinguish all interior edges of $\mathcal{T}$ if $\mathcal{T}$ displays $\mathcal{P}$ and $\mathcal{X}_{\mathcal{P}}$ distinguishes all interior edges of $\mathcal{T}$.

Clearly, an excess-free binary tree system $\mathcal{P}$ with $\bigcup supp(\mathcal{P}) = X$ that is displayed by $\mathcal{T}$ is gap free if and only if it distinguishes all edges of $\mathcal{T}$ in which case it must also be slim.

As announced already above, we will use these definitions in the following sections to characterize sparse definitive quartet systems and analogously defined tree and set systems.

6.3 Constructing trees from quartet, tree, and set systems

To show how trees can be constructed from quartet, tree, and set systems, we begin with a result that characterizes compatible as well as definitive pairs of binary phylogenetic trees:

Lemma 6.3 *Consider three subsets X_1, X_2, X' of a finite set X with $X = X_1 \cup X_2$ and $X' = X_1 \cap X_2$ and, for each $\nu = 1, 2$, a binary X_ν-tree $\mathcal{T}_\nu = (V_\nu, E_\nu, \varphi_\nu)$. Then, $\mathcal{P} := \{\mathcal{T}_1, \mathcal{T}_2\}$ is compatible if and only if there exists a binary X'-tree $\mathcal{T}'$ that is isomorphic to the two X'-trees $\mathcal{T}_1|_{X'}$ and $\mathcal{T}_2|_{X'}$ obtained by restricting $\mathcal{T}_1$ and $\mathcal{T}_2$ to X' — and, therefore, in particular whenever $|X'| \le 3$ holds.*

Moreover, if such an X'-tree $\mathcal{T}' = (V', E', \varphi')$ exists, then $\mathcal{P}$ is definitive if and only if the canonical embeddings $V' \to V_1$ and $V' \to V_2$ induced by the isomorphisms $\mathcal{T}' \to \mathcal{T}_1|_{X'}$ and $\mathcal{T}' \to \mathcal{T}_2|_{X'}$ map every edge $\{u', v'\} \in E'$ either onto an edge in E_1 or an edge in E_2.

Proof If $\mathcal{P}$ is compatible and $\mathcal{T} = (V, E, \varphi)$ is a phylogenetic X-tree that displays $\mathcal{P}$, then the binary X'-tree $\mathcal{T}' := \mathcal{T}|_{X'}$ is isomorphic to both, $\mathcal{T}_1|_{X'}$ and $\mathcal{T}_2|_{X'}$, implying that $\mathcal{T}'$ is displayed by both, $\mathcal{T}_1$ and $\mathcal{T}_2$, as required.

Now, suppose that, conversely, there exists a binary X'-tree $\mathcal{T}' = (V', E', \varphi')$ that is displayed by both, $\mathcal{T}_1$ and $\mathcal{T}_2$. Without loss of generality, we may assume that $V' = V_1 \cap V_2$ and $\varphi' = \varphi_1|_{X'} = \varphi_2|_{X'}$ holds.

We will construct a phylogenetic X-tree $\mathcal{T} = (V, E, \varphi)$ with $V := V_1 \cup V_2$ that displays $\mathcal{P}$ by inserting, for each edge $e' = \{u', v'\} \in E'$ in $\mathcal{T}'$, the subtrees of $\mathcal{T}_1$ and $\mathcal{T}_2$ emanating from the vertices between u' to v' in $\mathcal{T}_1$ and $\mathcal{T}_2$ into $\mathcal{T}'$ along that edge in any order compatible with their order in $\mathcal{T}_1$ and $\mathcal{T}_2$, respectively.

More specifically, consider, for every edge $e' = \{u', v'\} \in E'$ and each $\nu = 1, 2$, the unique path $\mathbf{p}_\nu(u', v') := \mathbf{p}_{\mathcal{T}_\nu}(u', v')$ in $\mathcal{T}_\nu$ joining u' and v', let $\ell_\nu := \ell_\nu(e')$ denote the (necessarily positive) length of $\mathbf{p}_\nu(u', v')$, and let $u_0^{(\nu)} := u', u_1^{(\nu)}, \dots, u_{\ell_\nu}^{(\nu)} := v'$ denote the sequence of vertices of $\mathbf{p}_\nu(u', v')$. To define the edge set E of $\mathcal{T}$, choose, for every edge $e' = \{u', v'\} \in E'$, a linear order "$\preceq_{e'}$" of the vertices in the subset $V(e') := \{u_i^\nu : \nu = 1, 2;\ i = 0, \dots, \ell_\nu\}$ of $V = V_1 \cup V_2$ of cardinality $\ell_1 + \ell_2$ such that $u_i^{(\nu)} \preceq_{e'} u_j^{(\nu)}$ holds, for $\nu = 1, 2$, for all $i, j \in \{0, \dots, \ell_\nu\}$ with $i \le j$. Now, we define two distinct vertices $a, b \in V$ to be adjacent in $\mathcal{T}$ if there exists either an edge $e' \in E'$ with

$a, b \in V(e')$ and there is no $c \in V(e') - \{a, b\}$ with either $a \preceq_{e'} c \preceq_{e'} b$ or $b \preceq_{e'} c \preceq_{e'} a$, or there is no such edge $e' \in V'$ and $\{a, b\}$ is either an edge in $\mathcal{T}_1$ or in $\mathcal{T}_2$.

Finally, we define the map $\varphi : X \to V$ by

$$\varphi(x) := \begin{cases} \varphi_1(x) & \text{if } x \in X_1, \\ \varphi_2(x) & \text{if } x \in X_2, \end{cases}$$

so that — in view of $\varphi' = \varphi_1|_{X'} = \varphi_2|_{X'}$ — we have $\varphi(x) = \varphi'(x')$ for all $x' \in X'$. It is not hard to check that $\mathcal{T}$ displays $\mathcal{T}_1$ and $\mathcal{T}_2$, as required.

It remains to note that, given any edge $e' \in E'$, one has $\ell_1(e') = 1$ or $\ell_2(e') = 1$ — or, equivalently, $e' \in E_1$ or $e' \in E_2$ — if and only if there is exactly one ordering "$\preceq_{e'}$" of $V(e')$ that satisfies the requirements above, and that, in consequence, there is exactly one X-tree $\mathcal{T}$ that displays $\mathcal{T}_1$ and $\mathcal{T}_2$ if and only if this holds for all $e' \in E'$. □

With this result in hand, we can now, given a binary X-tree $\mathcal{T}$ and a sparse quartet system $\mathcal{Q} \subseteq \mathcal{Q}_\mathcal{T}$, present a sufficient condition for $\mathcal{Q}$ to define $\mathcal{T}$. To illustrate the basic idea, consider the scheme presented in Figure 6.4: For each phylogenetic tree $\mathcal{T}'$ in that figure, let $\mathcal{Q}^{\mathcal{T}'} := \mathcal{Q}_{\mathcal{T}'} \cap \mathcal{Q}$ denote the set of quartets in $\mathcal{Q}$ that are displayed by $\mathcal{T}'$. This yields the collection

$$\mathcal{H} := \big\{\{15|34\}, \{16|45\}, \{12|37\}, \{32|67\}, \{15|34, 16|45\}, \{12|37, 32|67\}, \mathcal{Q}\big\}$$

of seven subsets of $\mathcal{Q}$. Note that this collection has the following special properties:

($\mathcal{H}$ 1) It is a hierarchy over $\mathcal{Q}$.

($\mathcal{H}$ 2) It is, in fact, a maximal hierarchy over $\mathcal{Q}$.

($\mathcal{H}$ 3) Every cluster $\mathcal{Q}'$ in $\mathcal{H}$ is excess free (which, in case $\mathcal{Q} = \mathcal{Q}'$, is equivalent to the assertion $|\mathcal{Q}| = n - 3 = |E_{int}(\mathcal{T})| = 4$).

And this is by no means an accident — indeed, our main result states:

Theorem 6.4 *Suppose that $\mathcal{T}$ is a binary X-tree, and that $\mathcal{Q}$ is an excess- and gap-free subset of $\mathcal{Q}_\mathcal{T}$. Then, $\mathcal{Q}$ defines $\mathcal{T}$ if and only if the collection*

$$Exc_0(\mathcal{Q}) := \{\mathcal{Q}' : \mathcal{Q}' \subseteq \mathcal{Q} \text{ and } exc(\mathcal{Q}') = 0\}$$

contains a maximal hierarchy over $\mathcal{Q}$.

More generally, an excess- and gap-free system $\mathcal{X} \subseteq \mathbb{P}_{\geq 4}(X)$ of subsets of X of cardinality at least 4 defines $\mathcal{T}$ if and only if the collection

$$Exc_0(\mathcal{X}) := \{\mathcal{Y} \subseteq \mathcal{X} : \mathcal{Y} \neq \emptyset, exc(\mathcal{Y}) = 0\}$$

contains a maximal hierarchy over $\mathcal{X}$.

Remark 6.5 *More specifically, $Exc_0(\mathcal{X})$ — being a collection of set systems — could be called a* hyper set system. *However, trying to avoid such bloated notation, we will call such systems of set systems just a collection of set systems — and still may sometimes also call a set system just a collection of sets.*

Remark 6.6 *This theorem was first established for quartet systems in [26]. The proof that we shall give here will be based on an ingenious and much simpler (yet still rather involved) approach developed by Stefan Grünewald (see [84]).*

Proof To show that a subset $\mathcal{X}$ of $\mathbb{P}_{\geq 4}(X)$ defines $\mathcal{T}$ if it is excess and gap free and $Exc_0(\mathcal{X})$ contains a maximal hierarchy over $\mathcal{X}$, we will use induction on $|\mathcal{X}|$.

If $|\mathcal{X}| = 1$, our claim holds obviously as $\mathcal{X} = \{X\}$ and, hence, $\mathcal{T}_{\mathcal{X}} = \{\mathcal{T}|_X\} = \{\mathcal{T}\}$ must hold in this case. Otherwise, fix some maximal hierarchy $\mathcal{H}$ over $\mathcal{X}$ that is contained in $Exc_0(\mathcal{X})$ and let $\mathcal{X}_1$ and $\mathcal{X}_2$ denote the two children of $\mathcal{X}$ in $\mathcal{H}$. For $\nu = 1, 2$, put $X_\nu := supp(\mathcal{T}_{\mathcal{X}_\nu}) = \bigcup_{Y \in \mathcal{X}_\nu} Y$ and let $\mathcal{T}_\nu = (V_\nu, E_\nu, \varphi_\nu) := \mathcal{T}|_{X_\nu}$ denote the binary X_ν-tree obtained by restricting $\mathcal{T}$ to X_ν.

We claim first that $\mathcal{X}_\nu$ defines $\mathcal{T}_\nu$ for both indices $\nu \in \{1, 2\}$: Clearly, as $A \subseteq B$ implies $\mathcal{T}|_A = (\mathcal{T}|_B)|_A$ for all subsets A, B of X, $\mathcal{T}_\nu$ displays $\mathcal{T}_{\mathcal{X}_\nu}$ and $Exc_0(\mathcal{X}_\nu) = \{\mathcal{X}' \in Exc_0(\mathcal{X}) : \mathcal{X}' \subseteq \mathcal{X}_\nu\}$ contains a maximal hierarchy over $\mathcal{X}_\nu$. So, by induction, it remains to note that this tree system also distinguishes every interior edge of $\mathcal{T}_\nu$, that is, that $E_{int}(\mathcal{T}_\nu) \subseteq \bigcup_{Y \in \mathcal{X}_\nu} E_{int}(Y)$ holds.

Yet, as our assumptions imply that $E_{int}(\mathcal{T})$ is the disjoint union of the sets $E_{int}(Y)$ $(Y \in \mathcal{X})$ and as each such set $E_{int}(Y)$ with $Y \in \mathcal{X}_\nu$ is, therefore, a subset of $E_{int}(\mathcal{T}) \cap \binom{V_\nu}{2} \subseteq E_{int}(\mathcal{T}_\nu)$, our assumption $\mathcal{X}_\nu \in Exc_0(\mathcal{X})$ implies $|\bigcup_{Y \in \mathcal{X}_\nu} E_{int}(Y)| = \sum_{Y \in \mathcal{X}_\nu} |E_{int}(Y)| = |supp(\mathcal{T}_{\mathcal{X}_\nu})| - 3 = |X_\nu| - 3 = |E_{int}(\mathcal{T}_\nu)|$. So, $E_{int}(\mathcal{T}_\nu)$ must actually coincide with $\bigcup_{Y \in \mathcal{X}_\nu} E_{int}(Y)$, establishing (by induction) that, as claimed, $\mathcal{X}_\nu$ indeed defines $\mathcal{T}_\nu$.

It remains to show that the pair $\{\mathcal{T}_1, \mathcal{T}_2\}$ is definitive (as this clearly yields that $\mathcal{X}$ defines $\mathcal{T} = (V, E, \varphi)$ as every phylogenetic X-tree $\mathcal{T}'$ that displays $\mathcal{T}_{\mathcal{X}}$ must also display $\mathcal{T}_{\mathcal{X}_1}$ and $\mathcal{T}_{\mathcal{X}_2}$ and, hence, also $\mathcal{T}_1$ and $\mathcal{T}_2$ and must, therefore, be canonically isomorphic to $\mathcal{T}$). To this end, we note first that $X = X_1 \cup X_2$ and $V = V_1 \cup V_2$ must hold, the latter as every vertex $v \in V - X$ must be contained in one interior edge e of $\mathcal{T}$ which, in turn, must be contained in either V_1 or V_2. In consequence, $X' := X_1 \cap X_2$ must have cardinality 3 in view of

$$\begin{aligned}|X'| &= |X_1| + |X_2| - |X_1 \cup X_2| = (|X_1| - 3) + (|X_2| - 3) - (n-3) + 3\\ &= |E_{int}(\mathcal{T}_1)| + |E_{int}(\mathcal{T}_2)| - |E_{int}(\mathcal{T})| + 3\\ &= \sum_{Y \in \mathcal{X}_1} |E_{int}(Y)| + \sum_{Y \in \mathcal{X}_2} |E_{int}(Y)| - \sum_{Y \in \mathcal{X}} |E_{int}(Y)| + 3 = 3,\end{aligned}$$

and the vertex set V' of the subtree $\mathcal{T}' := \mathcal{T}|_{X'}$ obtained by restricting $\mathcal{T}$ to the set X' must coincide with its superset $V_1 \cap V_2$ in view of $|V'| = 2|X'| - 2 = 4$ and

$$\begin{aligned}|V'| \le |V_1 \cap V_2| &= |V_1| + |V_2| - |V_1 \cup V_2| = |V_1| + |V_2| - |V|\\ &= (2|X_1| - 2) + (2|X_2| - 2) - (2n - 2)\\ &= 2(|X_1| + |X_2| - |X_1 \cup X_2|) - 2 = 2 \cdot 3 - 2 = 4.\end{aligned}$$

Now, to show that the pair $\mathcal{T}_1, \mathcal{T}_2$ is definitive, let y_1, y_2, y_3 denote the three elements in X with $X' = \{y_1, y_2, y_3\}$, and let $v \in V$ denote the median of y_1, y_2, and y_3 in $\mathcal{T}$, i.e., the unique element in V for which $V' = \{y_1, y_2, y_3, v\}$ holds. In view of Lemma 6.3, it suffices to note that none of the three edges $e_1 := \{v, y_1\}, e_2 := \{v, y_2\}$, and $e_3 := \{v, y_3\}$ of $\mathcal{T}'$ is subdivided by both, $\mathcal{T}_1$ and $\mathcal{T}_2$. Otherwise, however, there would exist two adjacent vertices $v_1 \in V_1 - V'$ and $v_2 \in V_2 - V'$ with $e := \{v_1, v_2\} \in E_{int}(\mathcal{T})$ which is impossible as every interior edge of $E_{int}(\mathcal{T})$ is either an interior edge of $E_{int}(\mathcal{T}_1)$ or of $E_{int}(\mathcal{T}_2)$ and, thus, contained — as a subset — in either V_1 or V_2.

This establishes the easier direction of Theorem 6.4. The other direction will be established in the last section of this chapter while a crucial result from [84] that allowed Stefan Grünewald to substantially simplify the original proof will be established in the next section. □

6.4 Slim tree systems

In this section, we deal with the problem of deciding whether a given tree system $\mathcal{P}$ is compatible, which is commonly known as the *Supertree Problem.* It has received considerable attention in the literature (cf. for example [24]). The remarkably simple sufficient (but, of course, not necessary) compatibility criterion from [84] can be viewed as a partial converse of the fact implied by

($\mathcal{X}$ 4) that every excess-free tree system $\mathcal{P}$ must be slim which is displayed by a binary phylogenetic tree $\mathcal{T}$ and distinguishes all edges of $\mathcal{T}$:

Theorem 6.7 [84] *Every slim binary tree system $\mathcal{P}$ is compatible.*

Proof We will follow Stefan Grünewald's rather ingenious and intricate proof proceeding by induction relative to $|\mathcal{P}| + |\bigcup \mathcal{X}_{\mathcal{P}}|$. So, suppose that $\mathcal{P}$ is a slim binary tree system and that the theorem holds for all such tree systems $\mathcal{P}'$ with $|\mathcal{P}'| + |\bigcup \mathcal{X}_{\mathcal{P}'}| < |\mathcal{P}| + |\bigcup \mathcal{X}_{\mathcal{P}}|$.

Recall that, by definition of a binary tree system, $\mathcal{P}$ is finite and every tree $\mathcal{T}'$ in $\mathcal{P}$ is finite and binary, and has at least four leaves. Let us further assume that every tree $\mathcal{T}' \in \mathcal{P}$ is simple, and let $V(\mathcal{T}')$ and $E(\mathcal{T}')$ denote its vertex and its edge set, respectively. The proof proceeds in altogether three steps:

Step 1: Next note that $\mathcal{P}$ must also be compatible if there exists a proper subset $\mathcal{P}_0 \subset \mathcal{P}$ with $exc(\mathcal{P}_0) = 0$ and $1 < |\mathcal{P}_0|$: Indeed, any subset $\mathcal{P}_0$ of $\mathcal{P}$ must be slim and, therefore, also compatible if it is a proper subset. So, there must exist a binary $\bigcup supp(\mathcal{P}_0)$-tree $\mathcal{T}_0$ that displays $\mathcal{P}_0$.

Further, denoting by $\mathcal{P}/\mathcal{P}_0$ the tree system $(\mathcal{P} - \mathcal{P}_0) + \mathcal{T}_0$, it is clear that

- $\mathcal{P}$ must be compatible if $\mathcal{P}/\mathcal{P}_0$ is (as any tree displaying $\mathcal{P}/\mathcal{P}_0$ will also display $\mathcal{P}$),
- we have $\bigcup supp(\mathcal{P}/\mathcal{P}_0) = \bigcup supp(\mathcal{P})$,
- $|\mathcal{P}/\mathcal{P}_0| < |\mathcal{P}|$ holds in case $1 < |\mathcal{P}_0|$, and
- $\mathcal{P}/\mathcal{P}_0$ must also be slim in case $exc(\mathcal{P}_0) = 0$ as this implies

$$|E_{int}(\mathcal{T}_0)| = |supp(\mathcal{T}_0)| - 3 = \left|\bigcup supp(\mathcal{P}_0)\right| - 3 = \sum_{\mathcal{T}' \in \mathcal{P}_0} (|supp(\mathcal{T}')| - 3).$$

So, by induction, $\mathcal{P}/\mathcal{P}_0$ and, hence, also $\mathcal{P}$ must be compatible for any such subset $\mathcal{P}_0$ of $\mathcal{P}$.

Step 2: We will now show that there must exist two distinct elements $x, y \in X := \bigcup supp(\mathcal{P})$ such that (i) the set $\mathcal{P}_{\{x,y\}} := \{\mathcal{T}' \in \mathcal{P} : x, y \in supp(\mathcal{T}')\}$ is non-empty and (ii) the pair x, y forms a cherry in every tree $\mathcal{T}'$ in $\mathcal{P}_{\{x,y\}}$.

Note that this claim is quite plausible as any cherry x, y in any tree that displays $\mathcal{P}$ must be a cherry in any tree $\mathcal{T}' \in \mathcal{P}_{\{x,y\}}$, and it must be contained in any tree $\mathcal{T}' \in \mathcal{P}$ that distinguishes the interior edge leading to that cherry.

To establish our claim, we consider, for any two distinct elements $x, y \in X$, also the two subsets

$$\mathcal{P}_{x \vee y} := \{\mathcal{T}' \in \mathcal{P}_{\{x,y\}} : \text{the two elements } x, y \text{ form a cherry in } \mathcal{T}'\}$$

and $\mathcal{P}_{x|y} := \mathcal{P}_{\{x,y\}} - \mathcal{P}_{x\vee y}$, and the directed graph $G_{\{x,y\}} = (\mathcal{P}, A_{\{x,y\}})$ with vertex set $\mathcal{P}$ and edge set all pairs $(\mathcal{T}', \mathcal{T}'') \in \mathcal{P}_{x\vee y} \times \mathcal{P}_{x|y}$. And we consider the arc-labeled directed graph $G_{\mathcal{P}} = \big((\mathcal{P}, A_{\mathcal{P}}), \lambda_{\mathcal{P}}\big)$ with vertex set $\mathcal{P}$, edge set the "formal" disjoint union

$$\begin{aligned} A_{\mathcal{P}} &:= \coprod_{\{x,y\}\in\binom{X}{2}} A_{\{x,y\}} \\ &= \left\{(\{x,y\},(\mathcal{T}',\mathcal{T}'')) : \{x,y\} \in \binom{X}{2}, (\mathcal{T}',\mathcal{T}'') \in A_{\{x,y\}}\right\} \\ &= \left\{(\{x,y\},(\mathcal{T}',\mathcal{T}'')) : \{x,y\} \in \binom{X}{2}, \mathcal{T}' \in \mathcal{P}_{x\vee y}, \mathcal{T}'' \in \mathcal{P}_{x|y}\right\} \end{aligned}$$

of the edge sets $A_{\{x,y\}}$ of the various graphs $G_{\{x,y\}}$, and labeling map

$$\lambda = \lambda_{\mathcal{P}} : A_{\mathcal{P}} \to \binom{X}{2} : (\{x,y\}; (\mathcal{T}',\mathcal{T}'')) \mapsto \{x,y\}$$

that labels, for each $\{x,y\} \in \binom{X}{2}$, each arc $\overrightarrow{a} = \big(\{x,y\}; (\mathcal{T}',\mathcal{T}'')\big)$ from $G_{\{x,y\}}$ with the pair $\{x,y\}$.

Clearly, our claim above is equivalent to the assertion that there exists some $\{x,y\} \in \binom{X}{2}$ with $\mathcal{P}_{x\vee y} = \mathcal{P}_{\{x,y\}} \neq \emptyset$. Yet, otherwise, taking into account that the vertex set of every finite binary tree with at least four distinct leaves contains at least two distinct — and, hence, disjoint — cherries, there would at least exist two distinct arcs emanating from every vertex of $G_{\mathcal{P}}$ all of which, in addition, must have disjoint labels. Hence, our claim will follow from the actually even slightly stronger assertion that there exists some vertex $\mathcal{T}'$ of $G_{\mathcal{P}}$ from which at most one arc emanates and, hence, from the even still slightly stronger assertion that there is no non-empty subdigraph of $G_{\mathcal{P}}$ containing exactly two distinct arcs emanating from all of its vertices.

So, let us assume that, to the contrary, there exist such subgraphs (as presented for instance in Figure 6.5) and that $G' = (\mathcal{P}', A')$ is one such subgraph for which $|\mathcal{P}'|$ is as small as possible. The remainder of the proof is concerned with showing that $exc(\mathcal{P}') < 0$ must hold for any such subgraph of $G_{\mathcal{P}}$ which, once established, will yield the desired contradiction as $\mathcal{P}$ was supposed to be slim.

To this end, we first note that G' must be connected, and that $|A'| = 2|\mathcal{P}'|$ must hold. Further, we put $X' := \bigcup supp(\mathcal{P}')$ and we define a cycle C in G' to be *monochromatic* if there exists some (necessarily unique !) element $x = x_C \in X'$ for which $x \in \lambda(\overrightarrow{a})$ holds for every arc $\overrightarrow{a}$ in C in which case we will also say that C is *x-colored*.

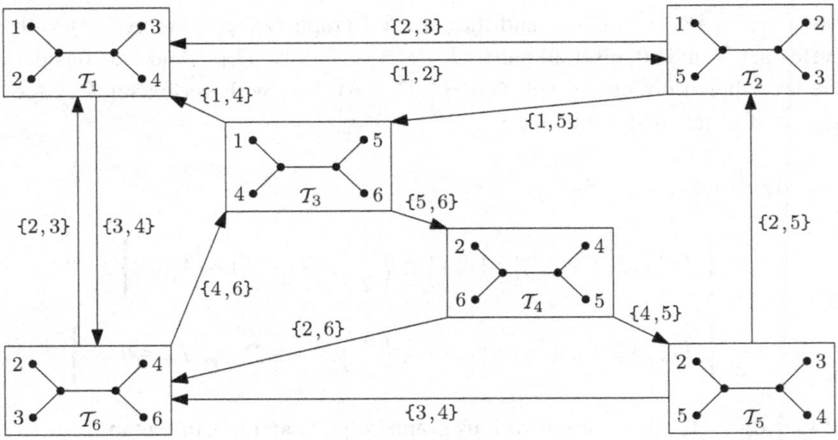

Figure 6.5 An example of a directed graph G' as considered in the proof of Theorem 6.7 with vertex set $\mathcal{P}' = \{\mathcal{T}_1, \ldots, \mathcal{T}_6\}$.

Clearly, as $\lambda(\overrightarrow{a'}) \cap \lambda'(\overrightarrow{a''}) = \emptyset$ holds, by assumption, for any two arcs $\overrightarrow{a'}$ and $\overrightarrow{a''}$ in A' emanating from the same vertex in $\mathcal{P}'$, every monochromatic cycle C in G' is a directed cycle in G', and every arc $\overrightarrow{a} \in A'$ is contained in at most two monochromatic cycles.

Let us now define, for every $x \in X'$, the subdigraph $G_x = (\mathcal{P}_x, A_x)$ of G' with vertex set $\mathcal{P}_x := \{\mathcal{T}' \in \mathcal{P}' : x \in supp(\mathcal{T}')\}$ and edge set $A_x := \{\overrightarrow{a} \in A' : x \in \lambda'(\overrightarrow{a})\}$. Note that G_x is well defined, since, by construction, the starting and the end point of every arc $\overrightarrow{a} \in A_x$ must be contained in $\mathcal{P}_x$. Note also that $\sum_{\mathcal{T}' \in \mathcal{P}'} |supp(\mathcal{T}')| = \sum_{x \in X'} |\mathcal{P}_x|$ must hold as both numbers coincide with the cardinality of the set $\{(x, \mathcal{T}') \in X' \times \mathcal{P}' : x \in supp(\mathcal{T}')\}$. Moreover, if we remove, for every (necessarily x-colored and, hence, directed) cycle C in G_x, an arbitrary arc of this cycle from G_x, then the resulting subdigraph of G_x is a forest. Therefore — in view of Lemma 1.2 — we have

$$|\mathcal{P}_x| \geq |A_x| - |\mathcal{C}_x| + 1, \tag{6.3}$$

where $\mathcal{C}_x$ is the set of x-colored cycles in G_x. Thus, denoting by $\mathcal{C}' := \bigcup_{x \in X'} \mathcal{C}_x$ the set of all monochromatic cycles, we have

$$\begin{aligned} \sum_{x \in X'} |\mathcal{P}_x| &\geq \sum_{x \in X'} (|A_x| - |\mathcal{C}_x| + 1) \\ &= |X'| + 2|A'| - |\mathcal{C}'| \end{aligned}$$

and, therefore,

$$\begin{aligned} exc(\mathcal{P}') &= |X'| - 3 - \sum_{T' \in \mathcal{P}'} (|supp(T')| - 3) \\ &= |X'| - 3 + 3|\mathcal{P}'| - \sum_{T' \in \mathcal{P}'} |supp(T')| \\ &= |X'| - 3 + 3|\mathcal{P}'| - \sum_{x \in X'} |\mathcal{P}_x| \\ &\le |X'| - 3 + 3|\mathcal{P}'| - (|X'| + 4|\mathcal{P}'| - |\mathcal{C}'|) \\ &= |\mathcal{C}'| - |\mathcal{P}'| - 3. \end{aligned}$$

In consequence, it suffices to show that the inequality $|\mathcal{C}'| \le |\mathcal{P}'| + 2$ must hold. However, the cyclomatic number $c(G')$ of G' coincides, in view of Theorem 1.3, with $|A'| - |\mathcal{P}'| + 1 = |\mathcal{P}'| + 1$ implying that $|\mathcal{C}'| \le |\mathcal{P}'| + 2$ holds if and only if

$$|\mathcal{C}'| \le c(G') + 1 \tag{6.4}$$

holds.

This, however, is obvious if the cycles in $\mathcal{C}'$, considered as vectors in the cycle space $\mathcal{C}(G')$ of G', are linearly independent as, then, even $|\mathcal{C}'| \le \dim \mathcal{C}(G') = c(G')$ must hold. Otherwise, let $\mathcal{C}^*$ denote an inclusion-minimal subset of $\mathcal{C}'$ that, considered as a set of vectors in the cycle space $\mathcal{C}(G')$, is linearly dependent. Clearly, $|\mathcal{C}^*| \le c(G') + 1$ must hold for any such subset $\mathcal{C}^*$ of $\mathcal{C}'$ while the sum of the vectors in $\mathcal{C}^*$ must vanish. Thus, if $G^* = (\mathcal{P}^*, A^*)$ denotes the subdigraph of G' with vertex set $\mathcal{P}^*$ and edge set A^* the union of the vertices and edges, respectively, of all cycles in $\mathcal{C}^*$, every arc $\vec{a} = (\{x, y\}; (T', T''))$ in A^* must be contained in an even and, by construction, positive number of cycles in $\mathcal{C}^*$ and, hence, in at least two such cycles implying that at least — and, hence, exactly — two distinct arcs must emanate in G^* from the endpoint T' of $\vec{a}$, *viz.* the two distinct arcs emanating from T'' in G'. However, as (again by construction) every vertex in G^* is an endpoint of some arc in G^*, this must therefore hold for every vertex in G^*. In consequence, we must have $G^* = G'$ and, hence, also $\mathcal{C}' = \mathcal{C}^*$ in view of our particular choice of G'. Thus, also $|\mathcal{C}'| = |\mathcal{C}^*| \le c(G') + 1$ and, therefore, $exc(\mathcal{P}') < 0$ must hold.

Step 3: It follows that, as already mentioned above, there must exist two elements $x, y \in X := supp(T')$ with $\mathcal{P}_{\{x,y\}} = \mathcal{P}_{x \vee y} \neq \emptyset$.

We shall construct a new tree system $\mathcal{P}_{y \to x}$ of simple phylogenetic trees: To this end, we define, for every $T' \in \mathcal{P}$, a phylogenetic tree $T'_{y \to x}$ as follows: If $y \notin supp(T')$ holds, we put $T'_{y \to x} := T'$. If $y \in supp(T')$, but

$x \notin supp(\mathcal{T}')$ holds, we just replace y by x in $\mathcal{T}'$, i.e., we replace the vertex set $V(\mathcal{T}')$ of $\mathcal{T}'$ by $V\left(\mathcal{T}'_{y\to x}\right) := V(\mathcal{T}')\Delta\{y, x\}$ and its edge set $E(\mathcal{T}')$ by $E\left(\mathcal{T}'_{y\to x}\right) := E(\mathcal{T}')\Delta\{\{y, v_{\mathcal{T}'}(y)\}, \{x, v_{\mathcal{T}'}(y)\}\}$, and note that the support (or leaf set) of $\mathcal{T}'_{y\to x}$ then coincides with $supp(\mathcal{T}')\Delta\{y, x\}$. Finally, if $x, y \in supp(\mathcal{T}')$ holds, we put $\mathcal{T}'_{y\to x} := \mathcal{T}'|_{supp(\mathcal{T}')-y}$ (i.e., we replace $V(\mathcal{T}')$ by $V\left(\mathcal{T}'_{y\to x}\right) := V(\mathcal{T}') - \{y, v_{\mathcal{T}'}(y)\}$ and $E(\mathcal{T}')$ by $E\left(\mathcal{T}'_{y\to x}\right) = (E(\mathcal{T}') - \{e \in E(\mathcal{T}') : v_{\mathcal{T}'}(y) \in e\}) + \{x, w\}$ where w is the unique interior vertex of $V(\mathcal{T}')$ that is adjacent to $v_{\mathcal{T}'}(y)$).

Clearly, the excess of the resulting tree system $\mathcal{P}_{y\to x} := \left\{\mathcal{T}'_{y\to x} : \mathcal{T}' \in \mathcal{P}\right\}$ must be non-negative: Indeed, $\bigcup supp\left(\mathcal{P}_{y\to x}\right) = X - y$ holds by construction. Furthermore, the cardinality $m = m_{x\vee y}$ of $\mathcal{P}_{x\vee y} = \mathcal{P}_{\{x,y\}}$ is, by our choice of x and y, positive while $\left|E_{int}\left(\mathcal{T}'_{y\to x}\right)\right| = |E_{int}(\mathcal{T}')| - 1$ holds, by construction, for every tree $\mathcal{T}' \in \mathcal{P}_{x\vee y}$ and $\left|E_{int}\left(\mathcal{T}'_{y\to x}\right)\right| = |E_{int}(\mathcal{T}')|$ for every other tree. So, we have

$$\sum_{\mathcal{T}'\in\mathcal{P}_{y\to x}} (|supp(\mathcal{T}')| - 3) = \left(\sum_{\mathcal{T}'\in\mathcal{P}} (|supp(\mathcal{T}')| - 3)\right) - m$$

and, therefore,

$$\begin{aligned} exc(\mathcal{P}_{y\to x}) &= \left|\bigcup supp(\mathcal{P}_{y\to x})\right| - 3 - \sum_{\mathcal{T}'\in\mathcal{P}_{y\to x}} (|supp(\mathcal{T}')| - 3) \\ &= \left(\left|\bigcup supp(\mathcal{P})\right| - 1\right) - 3 - \left(\left(\sum_{\mathcal{T}'\in\mathcal{P}} (|supp(\mathcal{T}')| - 3)\right) - m\right) \\ &= exc(\mathcal{P}) + (m-1) \geq 0. \end{aligned}$$

Furthermore, if $\mathcal{P}_{y\to x}$ is compatible, then so is $\mathcal{P}$: Indeed, assume that $\mathcal{T}'' = (V'', E'')$ is a simple phylogenetic $\bigcup supp(\mathcal{P}_{y\to x})$-tree that displays $\mathcal{P}_{y\to x}$. Assume also, that y is not contained in V'', let u denote an additional element not contained in V'', and consider the simple phylogenetic X-tree

$$\mathcal{T}''' := \left(V'' \cup \{u, y\}, (E'' - e_{\mathcal{T}''}(x)) \cup \left\{\{u, v_{\mathcal{T}''}(x)\}, \{u, y\}, \{u, x\}\right\}\right).$$

Then, it is easy to check that $\mathcal{T}'''$ displays $\mathcal{P}$.

And finally, if $\mathcal{P}_{y\to x}$ is not compatible, it can, by induction, not be slim, implying that there exists a subset $\mathcal{P}' \subseteq \mathcal{P}_{y\to x}$ with $exc(\mathcal{P}') < 0$ and, therefore, also $1 < |\mathcal{P}'| < |\mathcal{P}_{y\to x}|$. Define $\mathcal{P}'' := \left\{\mathcal{T}' \in \mathcal{P} : \mathcal{T}'_{y\to x} \in \mathcal{P}'\right\}$. Then, we must have $x, y \in \bigcup supp(\mathcal{P}'')$ as well as $\sum_{\mathcal{T}' \in \mathcal{P}''}(|supp(\mathcal{T}')| - 3) \geq \sum_{\mathcal{T}' \in \mathcal{P}'}(|supp(\mathcal{T}')| - 3)$ and, therefore, also

$$\begin{aligned} 0 \leq exc(\mathcal{P}'') &= \left| \bigcup_{\mathcal{T}'' \in \mathcal{P}''} supp(\mathcal{T}'') \right| - 3 - \sum_{\mathcal{T}' \in \mathcal{P}''} (|supp(\mathcal{T}')| - 3) \\ &\leq \left| \bigcup_{\mathcal{T}' \in \mathcal{P}'} supp(\mathcal{T}') \right| + 1 - 3 - \sum_{\mathcal{T}' \in \mathcal{P}'} (|supp(\mathcal{T}')| - 3) \\ &= exc(\mathcal{P}') + 1 \leq 0. \end{aligned}$$

Thus, $exc(\mathcal{P}'') = 0$ must hold, implying that $\mathcal{P}$ must be compatible also in this case in view of $1 < |\mathcal{P}'| = |\mathcal{P}''| < |\mathcal{P}_{y \to x}| = |\mathcal{P}|$ and the observation presented in *Step 2*. □

Corollary 6.8 *Let $\mathcal{P}$ be an excess-free tree system that defines a binary X-tree $\mathcal{T} = (V, E, \varphi)$. Then $\mathcal{P}$ is slim, and every non-empty excess-free subsystem $\mathcal{P}' \subseteq \mathcal{P}$ of $\mathcal{P}$ is definitive.*

Proof Indeed, $\mathcal{P}$ must be slim in view of ($\mathcal{X}$4). To establish also the second assertion, consider an arbitrary non-empty subset $\mathcal{P}_0 \subseteq \mathcal{P}$ with $exc(\mathcal{P}_0) = 0$, and put $X_0 := \bigcup supp(\mathcal{P}_0)$ and $\mathcal{T}_0 := \mathcal{T}|_{X_0}$. We have to show that $\mathcal{P}_0$ defines $\mathcal{T}_0$. So, assume that there exists a second binary X_0-tree $\mathcal{T}_*$ that displays $\mathcal{P}_0$, but is not isomorphic to $\mathcal{T}_0$, and consider the tree system $\mathcal{P}^{(\mathcal{P}_0 \to \mathcal{T}_*)} := (\mathcal{P} - \mathcal{P}_0) + \mathcal{T}_*$. Note first that $\mathcal{P}^{(\mathcal{P}_0 \to \mathcal{T}_*)}$ is slim. Indeed, consider an arbitrary subset $\mathcal{P}'$ of $\mathcal{P}^{(\mathcal{P}_0 \to \mathcal{T}_*)}$ and put $X' := \bigcup supp(\mathcal{P}')$. If $\mathcal{P}'$ does not contain $\mathcal{T}_*$, we have $exc(\mathcal{P}') \geq 0$ since $\mathcal{P}'$ is then also a subset of $\mathcal{P}$. If $\mathcal{T}_* \in \mathcal{P}'$ holds, consider the subset $\mathcal{P}'_{(\mathcal{T}_* \to \mathcal{P}_0)} := (\mathcal{P}' - \mathcal{T}_*) \cup \mathcal{P}_0$ of $\mathcal{P}$ and note that $\bigcup supp\left(\mathcal{P}'_{(\mathcal{T}_* \to \mathcal{P}_0)}\right) = \bigcup supp(\mathcal{P}' - \mathcal{T}_*) \cup \bigcup supp(\mathcal{P}_0) = \bigcup supp(\mathcal{P}' - \mathcal{T}_*) \cup supp(\mathcal{T}_*) = \bigcup supp(\mathcal{P}') = X'$ holds. Combining this with the fact that also $|E_{int}(\mathcal{T}_*)| = |X_0| - 3 = \sum_{\mathcal{T}' \in \mathcal{P}_0}(|supp(\mathcal{T}')| - 3)$ holds in view of $exc(\mathcal{P}_0) = 0$, we see that

$$\begin{aligned} exc(\mathcal{P}') &= |X'| - 3 - \sum_{\mathcal{T}' \in \mathcal{P}'} (|supp(\mathcal{T}')| - 3) \\ &= |X'| - 3 - |E_{int}(\mathcal{T}_*)| - \sum_{\mathcal{T}' \in (\mathcal{P}' - \mathcal{T}_*)} (|supp(\mathcal{T}')| - 3) \\ &= |X'| - 3 - \sum_{\mathcal{T}' \in \mathcal{P}_0} (|supp(\mathcal{T}')| - 3) - \sum_{\mathcal{T}' \in (\mathcal{P}' - \mathcal{T}_*)} (|supp(\mathcal{T}')| - 3) \end{aligned}$$

$$= \left|\bigcup supp\left(\mathcal{P}'_{(\mathcal{T}_*\to\mathcal{P}_0)}\right)\right| - 3 - \sum_{\mathcal{T}'\in\mathcal{P}'_{(\mathcal{T}_*\to\mathcal{P}_0)}} (|supp(\mathcal{T}')| - 3)$$

$$= exc\left(\mathcal{P}'_{(\mathcal{T}_*\to\mathcal{P}_0)}\right) \geq 0$$

must also hold in this case.

So, there exists, by Theorem 6.7, a binary X-tree $\tilde{\mathcal{T}} = (\tilde{V}, \tilde{E}, \tilde{\varphi})$ that displays $\mathcal{P}^{(\mathcal{P}_0\to\mathcal{T}_*)}$ and, therefore, also $\mathcal{P}_0$. But $\tilde{\mathcal{T}}$ and $\mathcal{T}$ are not isomorphic since their restrictions $\mathcal{T}_0 = \mathcal{T}|_{X_0}$ and $\mathcal{T}_* = \tilde{\mathcal{T}}|_{X_0}$ to X_0 are not isomorphic, contradicting our assumption that $\mathcal{P}$ defines $\mathcal{T}$. □

6.5 Definitive set systems

This last section is devoted to establishing the other direction of Theorem 6.4. So, assume that $\mathcal{X}$ is an excess- and gap-free subset of $\mathbb{P}_{\geq 4}(X)$ that defines a simple binary tree $\mathcal{T}$. We have to show that $Exc_0(\mathcal{X})$ contains a maximal hierarchy over $\mathcal{X}$.

Note first that, in view of ($\mathcal{X}$5), we must have $\bigcup \mathcal{X} = X$ and that, if a 2-subset $p = \{x, y\} \subseteq X$ forms a cherry in $\mathcal{T}$, it must form a cherry in the tree $\mathcal{T}|_Y$ obtained by restricting $\mathcal{T}$ to any subset Y of X of cardinality at least 4 that contains p, and there must exist exactly one such subset $Y = Y_p \in \mathcal{X}$, *viz.* the unique subset Y in $\mathcal{X}$ for which the unique interior edge incident with $v_{\mathcal{T}}(x) = v_{\mathcal{T}}(y)$ is contained in $E_{int}(Y)$.

We may now assume for the sake of contradiction that, with $\mathcal{T}$ and $\mathcal{X}$ as above, $Exc_0(\mathcal{X})$ does not contain a maximal hierarchy over $\mathcal{X}$ and that $\mathcal{T}$ and $\mathcal{X}$ are chosen so that $n + |\mathcal{X}|$ is as small as possible.

As before, we proceed in a number of steps to derive a contradiction:

Step 1: We must have $|\mathcal{X}| \geq 3$ and, therefore, also $n = |E_{int}(\mathcal{T})| + 3 \geq |\mathcal{X}| + 3 \geq 6$ as well as, according to Corollary 6.8, $exc(\mathcal{Y}) > 0$ for every subset $\mathcal{Y}$ of $\mathcal{X}$ with $1 < |\mathcal{Y}| < |\mathcal{X}|$: Indeed, if $exc(\mathcal{Y}_0) = 0$ were to hold for some subset $\mathcal{Y}_0$ of $\mathcal{X}$ with $1 < |\mathcal{Y}_0| < |\mathcal{X}|$, the set system $\mathcal{X}' := (\mathcal{X} - \mathcal{Y}_0) + Y_0$ with $Y_0 := \{\bigcup \mathcal{Y}_0\}$ would also be an excess-free set system that defines $\mathcal{T}$ while $\mathcal{Y}_0$ would be an excess-free set system that defines $\mathcal{T}|_{\bigcup \mathcal{Y}_0}$. Thus, by induction, there must exist a maximal hierarchy $\mathcal{H}' \subseteq Exc_0(\mathcal{X}')$ as well as a maximal hierarchy $\mathcal{H}'' \subseteq Exc_0(\mathcal{Y}_0)$ which together would give rise to a maximal hierarchy

$$\mathcal{H} := \mathcal{H}'' \cup \{\mathcal{Y}' \in \mathcal{H}' : Y_0 \notin \mathcal{Y}'\} \cup \{(\mathcal{Y}' - \{Y_0\}) \cup \mathcal{Y}_0 : \mathcal{Y}' \in \mathcal{H}', Y_0 \in \mathcal{Y}'\}$$

contained in $Exc_0(\mathcal{X})$.

So, in view of the last observation in ($\mathcal{X}4$), we must have

$$\mathcal{Y} \subseteq \mathcal{X} \quad \text{and} \quad 1 < |\mathcal{Y}| < |\mathcal{X}| \Rightarrow exc(\mathcal{Y}) \geq 1. \tag{6.5}$$

In particular, given any two distinct subsets Y_1 and Y_2 in $\mathcal{X}$, we must have

$$|Y_1 \cap Y_2| = |Y_1| + |Y_2| - |Y_1 \cup Y_2| = 3 - exc(\{Y_1, Y_2\}) \leq 2.$$

Step 2: Next note that this, in turn, implies that, given any element $x_0 \in X$, we do not only necessarily have $x_0 \in \bigcup \mathcal{X}$ in view of ($\mathcal{X}5$): There must exist at least two distinct subsets $Y \in \mathcal{X}$ with $x_0 \in Y$ or, equivalently

$$Y \subseteq \bigcup_{Y' \in \mathcal{X} - \{Y\}} Y' \tag{6.6}$$

must hold for every subset Y in $\mathcal{Y}$. Indeed, if there were at most and, hence, exactly one such subset $Y_0 \in \mathcal{X}$ and if $|Y_0| = 4$ were to hold for this subset, we may put $\mathcal{Y} := \mathcal{X} - \{Y_0\}$ and note that, in view of the last observation in ($\mathcal{X}4$), one would have

$$\begin{aligned} 0 \leq exc(\mathcal{Y}) &= \left|\bigcup \mathcal{Y}\right| - 3 - \sum_{Y \in \mathcal{X} - \{Y_0\}} (|Y| - 3) \\ &\leq \left(\left|\bigcup \mathcal{X}\right| - 1\right) - 3 - \sum_{Y \in \mathcal{X} - \{Y_0\}} (|Y| - 3) \\ &= \left|\bigcup \mathcal{X}\right| - 3 - \sum_{Y \in \mathcal{X}} (|Y| - 3) = exc(\mathcal{X}) = 0 \end{aligned}$$

in contradiction to (6.5). Otherwise (cf. Figure 6.6), one may put

$$X' := X - x_0, \; Y_0' := Y_0 - x_0, \; \text{and } \mathcal{X}' := (\mathcal{X} - Y_0) + Y_0'$$

and note that $X' = \bigcup \mathcal{X}'$ must hold and that, given any subset $\mathcal{Y}$ of $\mathcal{X}$, one would have $exc(\mathcal{Y}') = exc(\mathcal{Y})$ for the subset $\mathcal{Y}' := \{Y - x_0 : Y \in \mathcal{Y}\}$ of $\mathcal{X}'$: Indeed, this would hold obviously in case $Y_0 \notin \mathcal{Y}$ and, therefore, $\mathcal{Y}' = \mathcal{Y}$, but also in case $Y_0 \in \mathcal{Y}$ as this would imply

$$\begin{aligned} exc(\mathcal{Y}') &= \left|\bigcup \mathcal{Y}'\right| - 3 - \sum_{Y' \in \mathcal{Y}'} (|Y'| - 3) \\ &= \left(\left|\bigcup \mathcal{Y}\right| - 1\right) - 3 - \sum_{Y \in \mathcal{Y}} (|Y| - 3 - \delta_{Y, Y_0}) \\ &= \left|\bigcup \mathcal{Y}\right| - 3 - \sum_{Y \in \mathcal{Y}} (|Y| - 3) = exc(\mathcal{Y}). \end{aligned}$$

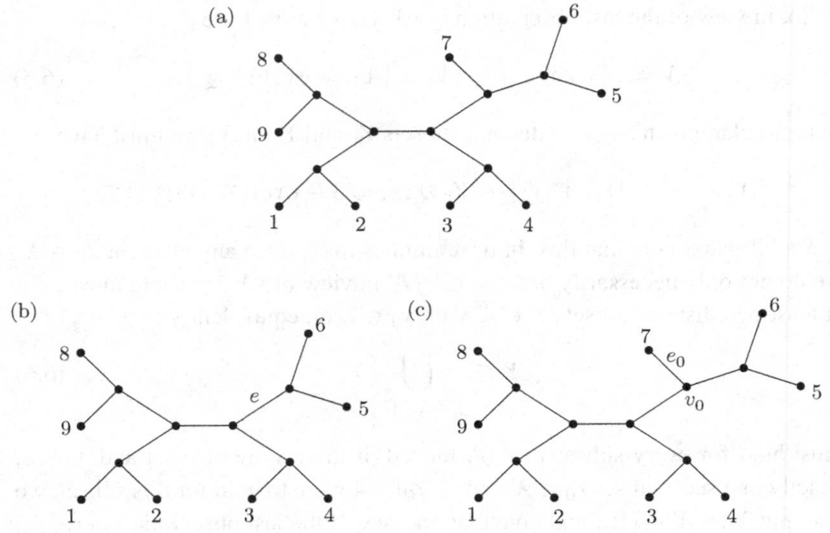

Figure 6.6 For the $\langle 9\rangle$-tree $\mathcal{T}$ depicted in (a), consider the set system $\mathcal{X}$ consisting of the three subsets $Y_0 = \{2, 3, 4, 5, 6, 7\}$, $Y_1 = \{1, 4, 6, 9\}$, $Y_2 = \{1, 2, 5, 8, 9\}$ of $\langle 9\rangle$. The element $x_0 = 7$ occurs only in Y_0. In (b), the tree $\mathcal{T}' := \mathcal{T}|_{\langle 9\rangle - 7}$ is depicted, with e denoting the edge obtained after eliminating the element $x_0 = 7$, and in (c) the tree $\mathcal{T}_{new}$ obtained from $\mathcal{T}'$ by adding a pendant edge to a vertex v_0 inserted into the edge e leading to a new leaf labeled by the element $x_0 = 7$ is depicted. Clearly, $\mathcal{T}_{new}$ is isomorphic to $\mathcal{T}$.

Furthermore, any X'-tree $\mathcal{T}'$ that, like the X'-tree $\mathcal{T}|_{X'}$, were to display all the trees $\mathcal{T}|_{Y'}$ with $Y' \in \mathcal{X}'$ could be extended to an X-tree $\mathcal{T}_{new}$ that would display all the trees $\mathcal{T}|_Y$ with $Y \in \mathcal{X}$ by adding a pendant edge $e_0 = \{x_0, v_0\}$ that is attached to a vertex v_0 that is inserted into that edge e of $\mathcal{T}|_{Y_0'}$ that is split into two in $\mathcal{T}|_{Y_0}$.

So, $\mathcal{T}_{new}$ must be isomorphic to $\mathcal{T}$ implying that $\mathcal{T}|_{X'}$ is defined by $\mathcal{X}'$. Thus, using induction, $Exc_0(\mathcal{X}')$ would contain a maximal hierarchy over $\mathcal{X}'$ implying that also $Exc_0(\mathcal{X})$ would contain a maximal hierarchy over $\mathcal{X}$ in contradiction to our assumption.

Step 3: Now denote, for any given pair x, y of distinct elements x, y in X, by $X_{y\to x}$ the set $X - y$, by $h_{y\to x}$ the map

$$h_{y\to x} : X \to X_{y\to x} : z \to \begin{cases} x & \text{if } z = y, \\ z & \text{otherwise,} \end{cases}$$

and put $\mathcal{X}_{y\to x} := \{h_{y\to x}(Y) \;:\; Y \in \mathcal{X}, |h_{y\to x}(Y)| \geq 4\}$ where $h_{y\to x}(Y)$ of course denotes, for every subset Y of X, the set $\{h_{y\to x}(z) : z \in Y\}$.

Clearly, $Y, Y' \subseteq X$ and $h_{y\to x}(Y) = h_{y\to x}(Y')$ implies $Y - \{x, y\} = h_{y\to x}(Y) - x = h_{y\to x}(Y') - x = Y' - \{x, y\} \subseteq Y \cap Y'$. So, $Y, Y' \in \mathcal{X}$ and $h_{y\to x}(Y) = h_{y\to x}(Y')$ implies $Y = Y'$: Indeed, if $|Y \cap Y'| \leq 2$ were to hold in case $Y \neq Y'$, $h_{y\to x}(Y) = h_{y\to x}(Y')$ would imply that $Y - \{x, y\}$ and $Y' - \{x, y\}$ both must contain exactly two elements in case $Y \neq Y'$ which, in turn, would imply $\{x, y\} \subseteq Y \cap Y'$ and, hence, $Y = Y'$.

Step 4: Next, assume that we are given two distinct elements $x, y \in X$ such that (i) there exist subsets in $\mathcal{X}$ that contain both elements x, y and that (ii) the two elements x, y form a cherry in $\mathcal{T}|_Y$ for every such subset $Y \in \mathcal{X}$ (as would surely be the case if the two elements x, y were to form a cherry in $\mathcal{T}$). Then, $X_{y\to x} = \bigcup \mathcal{X}_{y\to x}$ must hold: Indeed, assume that there exists some $x_0 \in X_{y\to x} - \bigcup \mathcal{X}_{y\to x}$ and note that $\{x, y\} \subseteq Y$ and $|Y| = 4$ would then necessarily hold for every $Y \in \mathcal{X}$ with $x_0 \in Y$. If $x_0 \neq x$, this would imply that there is only one $Y \in \mathcal{X}$ with $x_0 \in Y$ which, in view of what we learned in *Step 2*, is impossible. However, if $x_0 = x$ were to hold, we would have $\{x, y\} \subseteq Y$ and $|Y| = 4$ for every $Y \in \mathcal{X}$ with $x_0 \in Y$ (as was already mentioned above), x, y would form a cherry in $\mathcal{T}|_Y$ for every such subset Y, and there would be at least two distinct such subsets in $\mathcal{X}$ — so, x, y cannot form a cherry in $\mathcal{T}$. Then, however, $x_0 \in Y'$ and, therefore, $\{x, y\} \subseteq Y'$ would hold for that subset $Y' \in \mathcal{X}$ for which $E_{int}(Y')$ contains the first interior edge on the path from $x = x_0$ to y and x, y could not form a cherry in $\mathcal{T}|_{Y'}$ for this subset Y'.

Step 5: Continuing with the assumption that the two elements x, y form a cherry in $\mathcal{T}|_Y$ for every such subset $Y \in \mathcal{X}$ and that such subsets Y exist, we now claim that the two elements x, y must form a cherry in $\mathcal{T}$ in this case and that the set system $\mathcal{X}_{y\to x}$ is slim and defines the tree $\mathcal{T}_{y\to x} := \mathcal{T}|_{X_{y\to x}}$: Indeed, given any subset $\mathcal{Y}$ of $\mathcal{X}$, put $\delta_{\mathcal{Y},\{x,y\}} := 1$ in case $\{x, y\} \subseteq \bigcup \mathcal{Y}$ and $\delta_{\mathcal{Y},\{x,y\}} := 0$ else, and put

$$\mathcal{Y}_{\supseteq x,y} := \big\{Y \in \mathcal{Y} : Y \supseteq \{x, y\}\big\}.$$

Then, given any pair of non-empty subsets $\mathcal{Y} \subseteq \mathcal{X}$ and $\mathcal{Y}' \subseteq \mathcal{X}_{y\to x}$ with

$$\mathcal{Y} = h_{y\to x}^{-1}(\mathcal{Y}') := \big\{Y \in \mathcal{X} : h_{y\to x}(Y) \in \mathcal{Y}'\big\}$$

of $\mathcal{X}$, we have

$$\begin{aligned} exc(\mathcal{Y}') &= \left|\bigcup \mathcal{Y}'\right| - 3 - \sum_{Y' \in \mathcal{Y}'} (|Y'| - 3) \\ &= \left(\left|\bigcup \mathcal{Y}\right| - \delta_{\mathcal{Y},\{x,y\}}\right) - 3 - \sum_{Y \in \mathcal{Y}_{\supseteq x,y}} (|Y| - 4) - \sum_{Y \in \mathcal{Y} - \mathcal{Y}_{\supseteq x,y}} (|Y| - 3) \\ &= exc(\mathcal{Y}) + |\mathcal{Y}_{\supseteq x,y}| - \delta_{\mathcal{Y},\{x,y\}} \end{aligned}$$

and, therefore,

$$|\mathcal{Y}_{\supseteq x,y}| = \delta_{\mathcal{Y},\{x,y\}} \Rightarrow exc(\mathcal{Y}) = exc(\mathcal{Y}'). \tag{6.7}$$

So, $exc(\mathcal{Y}) = exc(\mathcal{Y}')$ holds in particular if $\mathcal{Y}$ contains exactly one subset Y with $\{x, y\} \subseteq Y$. It follows also that $\mathcal{X}_{y \to x}$ must be slim as $exc(\mathcal{Y}') = exc(\mathcal{Y}) + |\mathcal{Y}_{\supseteq x,y}| - \delta_{\mathcal{Y},\{x,y\}} \geq 1 + |\mathcal{Y}_{\supseteq x,y}| - 1 \geq 0$ must surely hold in case $\mathcal{Y} \neq \mathcal{X}$ while $\mathcal{Y} = \mathcal{X}$ implies $exc(\mathcal{Y}) = 0$ and $\mathcal{Y}_{\supseteq x,y} \neq \emptyset$ and, therefore, $exc(\mathcal{Y}') = exc(\mathcal{Y}) + |\mathcal{Y}_{\supseteq x,y}| - \delta_{\mathcal{Y},\{x,y\}} = |\mathcal{Y}_{\supseteq x,y}| - \delta_{\mathcal{Y},\{x,y\}} \geq 1 - 1 = 0$.

So, there must exist at least one $X_{y \to x}$-tree that displays the tree system $\mathcal{T}_{\mathcal{X}_{y \to x}}$. Furthermore, given any such $X_{y \to x}$-tree $\mathcal{T}' = (V', E', \varphi')$, the tree $\mathcal{T}_{new} := (V_{new}, E_{new}, \varphi_{new})$ defined by $V_{new} := V' \cup \{u, w\}$ for some two distinct elements u, w not yet involved in our set-up, $E_{new} := E' \cup \big\{\{u, \varphi'(x)\}, \{w, \varphi'(x)\}\big\}$, and φ_{new} defined by $\varphi_{new}(z) := \varphi'(z)$ in case $z \in X - \{x, y\}$, $\varphi_{new}(x) := u$, and $\varphi_{new}(y) := w$ displays the tree system $\mathcal{T}_{\mathcal{X}}$ and must therefore, by assumption, be canonically isomorphic to $\mathcal{T}$. Thus, as $u = \varphi_{new}(x)$ and $w = \varphi_{new}(y)$ form, by construction, a cherry in $\mathcal{T}_{new}$, x, and y must form a cherry in $\mathcal{T}$, as claimed. In addition, there must be exactly one subset $Y = Y_p \in \mathcal{X}$ that contains the 2-set $p = \{x, y\}$, and the set system $\mathcal{X}_{y \to x}$ must define any $X_{y \to x}$-tree that displays $\mathcal{X}_{y \to x}$ and, therefore, in particular, the tree $\mathcal{T}_{y \to x}$.

Step 6: It follows that, given any pair of subsets $\mathcal{Y} \subseteq \mathcal{X}$ and $\mathcal{Y}' \subseteq \mathcal{X}_{y \to x}$ as specified in *Step 5* with $Y_p \notin \mathcal{Y}$, $1 < |\mathcal{Y}| < |\mathcal{X}|$, and $exc(\mathcal{Y}') = 0$, we must have $p \subseteq \bigcup \mathcal{Y}$ as these assumptions imply $1 \leq exc(\mathcal{Y}) = exc(\mathcal{Y}') - |\mathcal{Y}_{\supseteq x,y}| + \delta_{\mathcal{Y},\{x,y\}} = \delta_{\mathcal{Y},\{x,y\}}$. We claim that also $|Y_p| = 4$ must hold in this case: By induction, there must exist a bipartition $\mathcal{X}^1_{y \to x}$ and $\mathcal{X}^2_{y \to x}$ of $\mathcal{X}_{y \to x}$ into two non-empty disjoint excess-free subsets of $\mathcal{X}_{y \to x}$ for which, furthermore,

$$\left|\bigcup \mathcal{X}^1_{y \to x} \cap \bigcup \mathcal{X}^2_{y \to x}\right| = 3 \tag{6.8}$$

must hold as the two subsets $\bigcup \mathcal{X}^1_{y \to x}$ and $\bigcup \mathcal{X}^2_{y \to x}$ together must form an excess-free set system (of cardinality 2) implying also that both, $\bigcup \mathcal{X}^1_{y \to x}$ and $\bigcup \mathcal{X}^2_{y \to x}$, must contain a cherry of $\mathcal{T}_{y \to x}$.

For $i=1,2$, let $\mathcal{X}^i := \left\{Y \in \mathcal{X} : h_{y\to x}(Y) \in \mathcal{X}^i_{y\to x}\right\}$ denote the "pre-image" of $\mathcal{X}^i_{y\to x}$ in $\mathcal{X}$. If $|Y_p| > 4$ were to hold, we may assume — without loss of generality — that, say, $h_{y\to x}(Y_p) = Y_p - y \in \mathcal{X}^1_{y\to x}$ holds which in turn, in view of (6.7), would imply that also $\mathcal{X}^1$ must be excess free. So, as $|\mathcal{X}^1| < |\mathcal{X}|$ must hold, we must have $|\mathcal{X}^1| = 1$ and, hence, $\mathcal{X}^1 = \{Y_p\}$ as well as $\mathcal{X}^2_{y\to x} = \mathcal{X}_{y\to x} - \{h_{y\to x}(Y_p)\}$ and $|h_{y\to x}(Y_p) \cap \bigcup(\mathcal{X}_{y\to x} - \{h_{y\to x}(Y_p)\})| = 3$. In view of (6.6), this would, in turn, imply that $4 < |Y_p| = |Y_p \cap \bigcup(\mathcal{X} - \{Y_p\})| \le 4$ must hold, clearly a contradiction.

So, $|Y_p| = 4$ must hold for every 2-subset p of X that forms a cherry in $\mathcal{T}$.

Step 7: In consequence, we must have $Y_p \notin \bigcup \mathcal{X}^i$ for $i=1,2$ and, therefore, $\{x,y\} \subseteq \bigcup \mathcal{X}^i$ for each $i \in \{1,2\}$ with $|\mathcal{X}^i| > 1$. That is, we must have $\mathcal{X}^i_{\supseteq x,y} = \emptyset$, $\delta_{\mathcal{X}^i_{y\to x},\{x,y\}} = 1$, and

$$exc(\mathcal{X}^i) = exc\left(\mathcal{X}^i_{y\to x}\right) - |\mathcal{Y}_{\supseteq x,y}| + \delta_{\mathcal{Y},\{x,y\}} = 0 - 0 + 1 = 1$$

for each such i.

Furthermore, denoting the set $Y_p - p$ by $\overline{p}$, it is obvious that this 2-subset forms a cherry in the Y_p-tree $\mathcal{T}|_{Y_p}$, but not in $\mathcal{T}$ (as $X = Y_p$ would otherwise hold) implying that there must exist some subset $Y \in \mathcal{X} - \{Y_p\} = \mathcal{X}^1 \dot\cup \mathcal{X}^2$ with $\overline{p} \subseteq Y$ for which, moreover, $\overline{p}$ does not form a cherry in $\mathcal{T}|_Y$. Clearly, we must have $Y_p \cap Y = \overline{p}$ or, equivalently, $x, y \notin Y$ for every subset $Y \in \mathcal{X} - \{Y_p\}$ that contains $\overline{p}$ as $|Y_1 \cap Y_2| \le 2$ must hold for any two distinct subsets $Y_1, Y_2 \in \mathcal{X}$.

Let us now fix some subset Y_0 in $\mathcal{X} - \{Y_p\}$ with $\overline{p} \subseteq Y_0$ and, therefore, $p \cap Y_0 = \emptyset$. We claim:

(7.1) $\mathcal{X}_{y\to x} - \{Y_0\}$ is an excess-free set system containing at least two distinct sets. In particular, $|Y_0 \cap \bigcup(\mathcal{X} - \{Y_p, Y_0\})| = 3$ and $p \subseteq \bigcup(\mathcal{X} - \{Y_p, Y_0\})$ must hold.

(7.2) There exists a unique 2-subset p' of Y_0 that forms a cherry in $\mathcal{T}$ implying that, in particular, $p' \ne \overline{p}$, $Y_0 = Y_{p'}$ and $|Y_0| = |Y_{p'}| = 4$ must hold.

(7.3) $\overline{p}$ and p' have a unique element in common that we will denote by $a(p)$ while we will denote the unique element in $\overline{p} \cap \overline{p'} = \overline{p} \cap (Y_{p'} - p') = \overline{p} - p' \subseteq Y_{p'} - p' = \overline{p'}$ by $b(p)$. Clearly, $a(p)$ is also the unique element in $Y_0 - \bigcup \mathcal{X}^2$.

(7.4) And there is only one subset Y in $\mathcal{X} - \{Y_p\}$ that contains $\overline{p}$ which we will, henceforth, denote by Y^p. Clearly, the correspondingly defined unique subset $Y^{p'}$ in $\mathcal{X}$ that is distinct from $Y_{p'}$ and contains $\overline{p'} = Y_{p'} - p'$ must be distinct from Y_p as $p \cap \overline{p'} = p \cap (Y^p - p') \subseteq p \cap Y^p = \emptyset$ holds. The unique cherry of $\mathcal{T}$ contained in $Y^{p'}$ must have a non-empty intersection with $\overline{p'}$ that coincides with $\{a(p')\}$. We must have $b(p)=b(p')$. And the quartets $xb(p)|a(p)a(p')$ and $yb(p)|a(p)a(p')$ must be contained in $\mathcal{Q}_{\mathcal{T}}$.

Proof (7.1) Assume that, say, $Y_0 \in \mathcal{X}^1$ holds. If $|\mathcal{X}^1| > 1$ were to hold, we would also have $\{x, y\} \subseteq \bigcup \mathcal{X}^1$ and, therefore, $Y_p \subseteq \{x, y\} \cup Y_0 \subseteq \bigcup \mathcal{X}^1$ implying that

$$exc(\mathcal{X}^1 + Y_p) = \left|\bigcup \mathcal{X}^1 \cup Y_p\right| - 3 - \sum_{Y \in \mathcal{X}^1} (|Y| - 3) - (|Y_p| - 3)$$

$$= \left|\bigcup \mathcal{X}^1\right| - 3 - \sum_{Y \in \mathcal{X}^1} (|Y| - 3) - 1$$

$$= exc(\mathcal{X}^1) - 1 = 0$$

would hold in contradiction to $1 < |\mathcal{X}^1 + Y_p| < |\mathcal{X}|$.

So, we must have $|\mathcal{X}^1| = 1$ and, hence, $\mathcal{X}^1 = \{Y_0\}$, $\mathcal{X}^2 = \mathcal{X} - \{Y_p, Y_0\}$, $\mathcal{X}^2_{y\to x} = \mathcal{X}_{y\to x} - \{Y_0\}$ and, therefore, also $exc(\mathcal{X}_{y\to x} - \{Y_0\}) = exc\left(\mathcal{X}^2_{y\to x}\right) = 0$, and $Y_0 \cap \bigcup \mathcal{X}^2_{y\to x} = Y_0 \cap \bigcup \mathcal{X}^2$ (as $\bigcup \mathcal{X}^2_{y\to x} - p = \bigcup \mathcal{X}^2 - p$ and $p \cap Y_0 = \emptyset$ holds) and, therefore, also $3 = \left|Y_0 \cap \bigcup \mathcal{X}^2_{y\to x}\right| = |Y_0 \cap \bigcup \mathcal{X}^2| = |Y_0 \cap \bigcup (\mathcal{X} - \{Y_p, Y_0\})|$.

Furthermore, we must have $p \subseteq \bigcup \mathcal{X}^2$ as, otherwise, one of the two elements x and y would be contained in Y_p, only, in contradiction to what we have learned in *Step 2*. So, we must also have $\left|\mathcal{X}^2_{y\to x}\right| = |\mathcal{X}^2| \geq 2$ because, if $\mathcal{X}^2$ were to contain a single set Y', only, $\bigcup \mathcal{X}^2$ could not contain p (as this set Y' would need to be distinct from Y_p, the only set in $\mathcal{X}$ that contains p). So, $p \subseteq \bigcup \mathcal{X}^2$ and $|\mathcal{X}^2| \geq 2$ must hold.

(7.2) Next, note that $Y_0 = \bigcup \mathcal{X}^1$ must contain a cherry $p' = \{x', y'\}$ of $\mathcal{T}_{y\to x}$ which must actually form a cherry of $\mathcal{T}$, too (as Y_0 contains neither x nor y). Furthermore, it can neither be contained in $\bigcup \mathcal{X}^2_{y\to x}$ (as every cherry of $\mathcal{T}_{y\to x}$ that is contained in Y_0 must contain at least one element that is not contained in $\bigcup \mathcal{X}^2_{y\to x}$) nor, in consequence, in $\bigcup \mathcal{X}^2$. And it must be distinct from $\overline{p}$ (as $\overline{p}$ does not form a cherry in $\mathcal{T}$). So, we must have $Y_0 = Y_{p'}$ as well as $|Y_0| = |Y_{p'}| = 4$.

(7.3) Thus, there must exist a unique element $a \in Y_0$ that is not contained in $\bigcup \mathcal{X}^2$. This element $a = a(p) = a(p, Y_0) \in Y_0$ must also be contained in p', again as every cherry of $\mathcal{T}_{y\to x}$ that is contained in Y_0 must contain at least one element that is not contained in $\bigcup \mathcal{X}^2_{y\to x}$ and, hence, not in $\bigcup \mathcal{X}^2$.

And in view of (6.6) applied to $Y := Y_0$, a must also be contained in Y_p and, hence, in $\overline{p} = Y_0 \cap Y_p$. So, we must have $\{a\} = p' \cap \overline{p}$ in view of $p' \neq \overline{p}$, and we can denote the unique element in $\overline{p}$ that is distinct from a by $b = b(p) = b(p, Y_0)$.

(7.4) It now follows also that there can be no other subset $Y_0' \in \mathcal{X} - \{Y_p\}$ with $\overline{p} \subseteq Y_0'$ as this would yield a subset $Y = Y_0' \in \mathcal{X}^2$ with $a \in Y_0'$ in contradiction to the fact that $a \notin \bigcup \mathcal{X}^2$ must hold.

So, we can now also write Y^p for that unique subset $Y \in \mathcal{X} - \{Y_p\}$ with $\overline{p} \subseteq Y$, and we see that the two elements $a = a(p, Y_0)$ and $b = b(p, Y_0)$ do, in fact, depend only on p and not on the subset Y_0 (as, given p, there is only one choice for this subset).

Furthermore, we must have $b(p) = b(p')$ as $b(p)$ cannot be contained in p' and $b(p) \neq b(p')$ would, therefore, imply $b(p) = a(p')$ in contradiction to $b(p) \in Y_p \cap Y_{p'}$ and the fact that $a(p')$ (just like $a(p)$) can only be contained in one subset $Y \in \mathcal{X}$ in which it is not part of a cherry of $\mathcal{T}$ that is contained in Y.

Finally, Lemma 6.1 applied to $q := xy|a(p)b(p)$ and $q' := a(p)y'|a(p')$ $b(p)$, implies that the two quartets $xb(p)|a(p)a(p')$ and $yb(p)|a(p)a(p')$ must both be contained in $\mathcal{Q}_{\mathcal{T}}$. □

Step 8: One can now derive the required contradiction as follows: Starting with one cherry p_0 of $\mathcal{T}$, one can consider the sequence $p_1, p_2, p_3, \ldots$ of cherries defined recursively by $p_{k+1} := p_k'$ for all $k \in \mathbb{N}_{>0}$ and the associated elements $a_k := a(p_k)$ and $b_k := b(p_k)$ in X. In view of (7.4), $a_0 := b(p_0)$ must coincide with $b(p_k)$ and $a_0 a_k | a_{k+1} a_{k+2}$ must be contained in $\mathcal{Q}_{\mathcal{T}}$ for all $k \in \mathbb{N}_{>0}$. Thus, invoking Corollary 3.8, $|\{a_0, a_1, \ldots, a_\ell\}| = \ell + 1$ must hold for all $\ell \geq 3$. This, however, is impossible in view of the finiteness of X.

Note that another instructive, yet slightly more laborious way to derive a final contradiction would be to note that, also in view of the finiteness of X, there must exist just some cherry p_1 (not necessarily the cherry $p_1 = p_0'$ introduced above) for which $p_{\ell+1} = p_1$ holds for some $\ell > 1$ for the recursively defined sequence of cherries $p_1, p_2, p_3, \ldots$ with $p_{k+1} := p_k'$. Consequently, the set system $\mathcal{Y}_0$ consisting of all $Y \in \mathcal{X}$ that are not of the form $Y = Y_{p_i}$ for some such sequence $p_1, p_2, \ldots, p_{\ell+1}$ with $p_{\ell+1} = p_1$ must be excess free as

$$0 = exc(\mathcal{X}) = \left|\bigcup \mathcal{X}\right| - 3 - \sum_{Y \in \mathcal{X}} (|Y| - 3)$$

$$= \left|\bigcup \mathcal{Y}_0\right| + \ell - 3 - \sum_{y \in \mathcal{Y}_0} (|Y| - 3) - \ell$$

$$= \left|\bigcup \mathcal{Y}_0\right| - 3 - \sum_{y \in \mathcal{Y}_0} (|Y| - 3) = exc(\mathcal{Y}_0)$$

must hold. So, $|\mathcal{Y}_0| \leq 1$ must hold which could also be used to derive a contradiction.

We finally note that, as a consequence of Theorem 6.4, it can be decided in polynomial time whether an excess-free quartet or tree system is definitive in which case the binary X-tree defined by it can be reconstructed in polynomial time (see [27] for more on this).

7
From metrics to split systems and back

In previous chapters, we have seen how to relate split systems and metrics to trees and networks. Clearly, this also provides — though only indirectly — a way to relate split systems and metrics. Here, we consider constructions that *directly* derive split systems from metrics. More specifically, we present two constructions in the first two sections that partly invert the map $\mathbb{R}^{\Sigma(X)} \to \mathbb{R}^{X \times X} : \nu \mapsto D_\nu$ considered already in Chapter 3 by associating weighted split systems to a metric, we discuss the formal algebraic properties of the map from weighted split systems to metrics in Section 7.3, and, in the last section, we discuss a surprising relationship between the Buneman graph and the tight span.

In what follows, X will always be a finite set of cardinality n, D a metric defined on X, ν a weighted split system $\nu : \Sigma(X) \to \mathbb{R}_{\geq 0}$ defined on X, and Σ a collection of X-splits.

7.1 Buneman splits

For D as above, we put

$$D(aa' : bb') := \frac{1}{2}\big(D(a,b) + D(a',b') - D(a,a') - D(b,b')\big)$$

and

$$D_+(aa' : bb') := \max\{D(aa' : bb'), 0\}$$

for all $a, a', b, b' \in X$, and we put

$$D(A : B) := \min_{\substack{a,a' \in A \\ b,b' \in B}} D(aa' : bb')$$

and

$$D_+(A:B) := \max\{D(A:B), 0\} = \min_{\substack{a,a' \in A \\ b,b' \in B}} D_+(aa':bb')$$

for any pair A, B of subsets of X.

Note that

$$D(ax:bb') + D(aa':xb') = D(aa':bb') \tag{7.1}$$

holds for all $a, a', b, b', x \in X$ in view of $2\,D(ax:bb') + 2\,D(aa':xb') = D(a,b) + D(x,b') - D(a,x) - D(b,b') + D(a,x) + D(a',b') - D(a,a') - D(x,b') = D(a,b) - D(b,b') + D(a',b') - D(a,a') = 2\,D(aa':bb')$.

Note also that, as $D(aa':bb')$ may differ from $D(a'a:bb')$, so it may differ from $D(\{a,a'\}:\{b,b'\})$, and $D_+(aa':b'b)$ from $D_+(\{a,a'\}:\{b,b'\})$. However, we have $D(\{a,a'\}:\{b,b'\}) = \min\{D(aa':bb'), D(aa':b'b)\}$ and $D_+(\{a,a'\}:\{b,b'\}) = \min\{D_+(aa':bb'), D_+(aa':b'b)\}$ for all $a, a', b, b' \in X$ as the triangle inequality implies that

$$\begin{aligned} 2\,D(aa':bb') &= D(a,b) + D(a',b') - D(a,a') - D(b,b') \\ &\leq 2\,D(aa:bb') = D(a,b) + D(a,b') - D(b,b') \end{aligned} \tag{7.2}$$

and more such inequalities always hold.

Further, $D(aa:bb) = D(a,b)$ holds for all $a, b \in X$, we have $D(A:B) = \infty$ in case $A = \emptyset$ or $B = \emptyset$, we have $D(A:B) \leq 0$ in case $A \cap B \neq \emptyset$, and we have $0 < D(A:B) < \infty$ if and only if (i) A, B is a *partial split* of X, i.e., one has $A \cap B = \emptyset \neq A, B$, and (ii) $D(a,b) + D(a',b') > D(a,a') + D(b,b')$ holds for all $a, a' \in A$ and all $b, b' \in B$, in which case the pair A, B will also be called a *strong* partial D-split — or just a *strong* D-split if, in addition, A, B is a split, i.e., $A \cup B = X$ holds.

Clearly, a (partial) split A, B of X is a strong (partial) D-split if and only if

$$D(a,a') + D(b,b') < \min\{D(a,b) + D(a',b'), D(a,b') + D(a',b)\}$$

holds for all $a, a' \in A$ and all $b, b' \in B$ implying that there can be no strong partial D-split A', B' with $a, b \in A'$ and $a', b' \in B'$ or $a, b' \in A'$ and $a, b' \in B'$ in case A, B is a strong partial D-split and $a, a' \in A$ and $b, b' \in B$ holds. In particular, any two strong D-splits must be compatible — a fact that was noted already by Peter Buneman who also suggested these concepts (cf. [36]).

In case $S = A|B$ is a split of X, we also write $\nu_D(S) = \nu_D(A|B)$ instead of $D_+(A:B)$ and refer to this number as the *Buneman index* of S relative to D, thus obtaining a weighted split system $\nu_D : \Sigma(X) \to \mathbb{R}_{\geq 0}$ associated to D. By definition, this index is never negative, and it is positive whenever S is a strong D-split, that is, the support $supp(\nu_D) = \{S \in \Sigma(X) : \nu_D(S) \neq 0\}$ of ν_D

coincides with the collection, denoted also by Σ_D, of all strong D-splits and is, therefore, always compatible. In view of Theorem 3.4, the weighted split system ν_D thus corresponds to an edge-weighted X-tree that is determined by ν_D — and, hence, by D — up to canonical isomorphism, which we refer to as the *Buneman tree* $\mathcal{T}_D$ associated to D.

We now show that Buneman's construction is *consistent*, that is, if $\mathcal{T} = (V, E, \omega, \varphi)$ is an edge-weighted X-tree and D coincides with the induced metric $D_{\mathcal{T}}$, then ν_D coincides with the weighted split system $\nu_{\mathcal{T}}$ associated to $\mathcal{T}$. To this end, we define, for every split $S \in \Sigma(X)$, the associated *split metric* D_S by putting $D_S(x, y) := 1$ if $S(x) \neq S(y)$ holds, and $D(x, y) : = 0$ else — which is clearly a metric, though a highly degenerate one. More generally, every metric that is a non-zero scalar multiple of a split metric of the form D_S for some split $S \in \Sigma(X)$ will also be called a split metric. So, a metric D is a split metric if and only if the partition Π_D of X into the equivalence classes of the equivalence relation "$\sim_D$" associated to D is a split of X (that is, the equivalence relation "$\sim_D$" defined by putting, for all $x, y \in X$, $x \sim_D y \iff \forall_{z \in X} D(x, z) = D(y, z)$ or, equivalently, $x \sim_D y \iff D(x, y) = 0$). Clearly, given a weighted split system $\nu : \Sigma(X) \to \mathbb{R}_{\geq 0}$, the associated map D_ν defined in Chapter 3 can now also be written as the sum $\sum_{S \in \Sigma(X)} \nu(S) D_S$. Recall also that $D_\nu = D_{\mathcal{T}}$ holds in case ν coincides with the weighted split system $\nu_{\mathcal{T}}$ associated to an edge-weighted X-tree $\mathcal{T}$. So, to establish the consistency claim, it suffices to show that $\nu_D = \nu$ holds for $D := D_\nu$ for every weighted split system ν with compatible support.

Further, we will use the symbol $A|B$ also for a pair of two subsets A, B of X that, together, form a partial split of X; we denote the set of all partial splits of X by $\Sigma_{\text{part}}(X)$; we will say that the partial split $A'|B'$ *extends* the partial split $A|B$ if either $A \subseteq A'$ and $B \subseteq B'$ or $A \subseteq B'$ and $B \subseteq A'$ holds in which case we'll also write $A|B \preceq A'|B'$; and we denote, for any pair A, B of non-empty subsets of X, by $\Sigma(A|B)$ the set

$$\Sigma(A|B) = \Sigma_X(A|B) := \{S \in \Sigma(X) : \forall_{a \in A, b \in B} S(a) \neq S(b)\}$$

and, for any weighted split system ν and A, B as above, by $\nu(A|B)$ the sum

$$\nu(A|B) = \nu\big(\Sigma(A|B)\big) := \sum_{S \in \Sigma(A|B)} \nu(S)$$

over the values of ν on all splits S in $\Sigma(A|B)$. In case $A = \{a_1, a_2, \dots, a_i\}$ and $B = \{b_1, b_2, \dots, b_j\}$, we will also write $a_1 a_2 \dots a_i | b_1 b_2 \dots b_j$ or $a_1 a_2 \dots a_i | B$ or $A | b_1 b_2 \dots b_j$ for $A|B$ and, correspondingly, $\Sigma(a_1 a_2 \dots a_i | b_1 b_2 \dots b_j)$ for $\Sigma(A|B)$, and $\nu(a_1 a_2 \dots a_i | b_1 b_2 \dots b_j)$ for $\nu(A|B)$.

Clearly, $\Sigma(A|B)$ is non-empty if and only if $A|B$ is a partial X-split in which case $\Sigma(A|B) = \{S \in \Sigma(X) : A|B \preceq S\}$ holds.

Further, we have

$$\Sigma(A|B) = \Sigma(A + x|B) \,\dot{\cup}\, \Sigma(A|B + x)$$

and, therefore, also

$$\nu(A|B) = \nu(A + x|B) + \nu(A|B + x) \tag{7.3}$$

as well as

$$\nu(A|B) \geq \nu(A'|B')$$

for every weighted split system ν, every element $x \in X$, and all pairs A, B and A', B' of non-empty subsets of X with $A \subseteq A'$ and $B \subseteq B'$.

Also, if Σ is a compatible subset of $\Sigma(X)$ and $\mathcal{T} = (V, E, \varphi)$ is an X-tree associated with Σ (i.e., with $\Sigma = \Sigma_{\mathcal{T}}$), then $\Sigma \cap \Sigma(A|B)$ consists of all splits S in Σ (if any) that correspond to edges in the intersection $\bigcap_{a \in A, b \in B} E_T[a, b]$ with $T := (V, E)$. It is easy to see (but not required here) that, if such edges exist, they form a path, stretching from the unique (!) vertex u in $\bigcap_{a \in A, b \in B} [\varphi(a), \varphi(b)]_{D_T} \cap \bigcup_{a, a' \in A} [\varphi(a), \varphi(a')]_{D_T}$ to the equally unique vertex v in $\bigcap_{a \in A, b \in B} [\varphi(a), \varphi(b)]_{D_T} \cap \bigcup_{b, b' \in B} [\varphi(b), \varphi(b')]_{D_T}$.

We also have $D_\nu(a, b) = \nu\big(\Sigma(a|b)\big) = \nu(a|b)$ for any two elements $a, b \in X$ and, therefore, also

$$\begin{aligned} D_\nu(a, b) &= \nu(a|b) = \nu(aa'|b) + \nu(a|ba') \\ &= \nu(aa'b'|b) + \nu(aa'|bb') + \nu(ab'|ba') + \nu(a|ba'b') \end{aligned}$$

for all $a, b, a', b' \in X$. In turn, this implies that also

$$\begin{aligned} 2\, D_\nu(aa' : bb') &= D_\nu(a, b) + D_\nu(a', b') - D_\nu(a, a') - D_\nu(b, b') \\ &= \nu(aa'b'|b) + \nu(aa'|bb') + \nu(ab'|ba') + \nu(a|ba'b') \\ &\quad + \nu(a'ab|b') + \nu(a'a|b'b) + \nu(a'b|b'a) + \nu(a'|b'ab) \\ &\quad - \nu(abb'|a') - \nu(ab|a'b') - \nu(ab'|a'b) - \nu(a|a'bb') \\ &\quad - \nu(baa'|b') - \nu(ba|b'a') - \nu(ba'|b'a) - \nu(b|b'aa') \\ &= 2\, \nu(aa'|bb') - 2\, \nu(ab|a'b') \end{aligned}$$

and, therefore,

$$D_\nu(aa' : bb') = \nu(aa'|bb') - \nu(ab|a'b') \tag{7.4}$$

holds for every weighted split system ν and all $a, b, a', b' \in X$. We will use this formula now to establish that Buneman's construction is indeed consistent, that is, that ν_D coincides with the weighted split system $\nu_{\mathcal{T}}$ associated to an edge-weighted X-tree $\mathcal{T} = (V, E, \omega, \varphi)$ in case D coincides with the induced metric $D_{\mathcal{T}}$.

Theorem 7.1 *Given a metric D and a weighted split system ν, the following two assertions are equivalent:*

(i) *The support $supp(\nu)$ of ν is compatible, i.e., there exists some edge-weighted X-tree $\mathcal{T} = (V, E, \omega, \varphi)$ with $\nu = \nu_{\mathcal{T}}$, and one has $D = D_\nu$.*

(ii) *D is treelike, i.e., there exists some edge-weighted X-tree $\mathcal{T} = (V, E, \omega, \varphi)$ with $D = D_{\mathcal{T}}$, and $\nu = \nu_D$ holds.*

In other words,

- *The inverse of the canonical one-to-one correspondence between treelike metrics D and weighted split systems ν with compatible support implied by Chapter 3 (given by associating the metric D_ν to any such split system ν) is given by associating the split system ν_D to any treelike metric D.*
- *If $\mathcal{T} = (V, E, \omega, \varphi)$ is an edge-weighted X-tree and D coincides with the induced metric $D_{\mathcal{T}}$, then ν_D coincides with the weighted split system $\nu_{\mathcal{T}}$ associated to $\mathcal{T}$.*
- *A metric D is treelike if and only if $D_{\nu_D} = D$ holds in which case ν_D is the only weighted split system ν with compatible support for which $D = D_\nu$ holds.*
- *And conversely, a weighted system ν of X-splits has a compatible support if and only if it is of the form $\nu = \nu_D$ for some metric D in which case this holds in particular for the metric $D := D_\nu$ (but also, in case $n \geq 4$, for many other metrics).*
- *In particular, associating the metric D_{ν_D} to any metric D, defines an idempotent operator $\mathbf{B}_X : D \mapsto D_{\nu_D}$ — which we also call the* Buneman operator *for X — from the space $\mathbb{M}(X)$ of all metrics defined on X into itself whose image consists exactly of the space $\mathbb{T}(X)$ of all treelike metrics defined on X.*

Proof In view of our previous remarks, it suffices to show that, given a weighted split system ν with compatible support and a split $S = A|B$ of X, one has $\nu(S) = \nu_D(S)$ for the induced metric $D := D_\nu$. To this end, recall first that we have $\nu_{D_\nu}(S) = \max\{D_\nu(A : B), 0\}$ and

$$D_\nu(A : B) = \min_{\substack{a,a' \in A \\ b,b' \in B}} D_\nu(aa' : bb') = \min_{\substack{a,a' \in A \\ b,b' \in B}} \big(\nu(aa'|bb') - \nu(ab|a'b')\big)$$

as well as either $\nu(aa'|bb') = \nu(\Sigma(aa'|bb')) = 0$ or $\nu(ab|a'b') = \nu(\Sigma(ab|a'b')) = 0$ for all $a, a', b, b' \in X$ as our assumption that $supp(\nu)$ is compatible implies that $\Sigma(aa'|bb') \cap \Sigma(ab|a'b') \cap supp(\nu)$ must be empty. Thus,

$$D_\nu(A : B) = \min_{\substack{a,a' \in A \\ b,b' \in B}} D_\nu(aa' : bb')$$

is positive if and only if there exists, for all $a, a' \in A$ and $b, b' \in B$, some $S' = A'|B' \in supp(\nu)$ with $a, a' \in A'$ and $b, b' \in B'$ in which case $D_\nu(A : B)$ coincides with $\min_{a,a' \in A, b,b' \in B} \sum_{S' \in \Sigma(aa'|bb')} \nu(S')$.

Thus, we have $\nu_D(S) \geq \nu(S)$ for every $S = A|B \in supp(\nu)$ as $a, a' \in A$ and $b, b' \in B$ implies $S \in \Sigma(aa'|bb')$ and, therefore,

$$\nu(aa'|bb') = \sum_{S' \in \Sigma(aa'|bb')} \nu(S') \geq \nu(S) > 0.$$

This, in turn, implies that $D_\nu(aa' : bb') = \nu(aa'|bb') \geq \nu(S) > 0$ and, therefore, also $\nu(ab|a'b') = \nu(\Sigma(ab|a'b')) = 0$ must hold for all $a, a' \in A$ and $b, b' \in B$ and, hence,

$$D_\nu(A : B) = \min_{a,a' \in A, b,b' \in B} D_\nu(aa' : bb') \geq \nu(S) > 0$$

and

$$\nu_D(S) = \max\{D_\nu(A : B), 0\} = D_\nu(A : B) \geq \nu(S).$$

Furthermore, $\nu_D(A|B) = \max\{D_\nu(A : B), 0\} \geq \nu(S) = 0$ holds, essentially by definition, for every split $S = A|B \in \Sigma(X) - supp(\nu)$. So, our claim that equality holds for every split $S \in \Sigma(X)$ follows immediately from the following characterization of splits in compatible split systems:

Lemma 7.2 *Given a compatible split system $\Sigma \subseteq \Sigma(X)$, a split $S = A|B$ in $\Sigma(X)$, and some elements $a \in A$ and $b \in B$, the following three assertions are equivalent:*

(i) $S \in \Sigma$,

(ii) *there exists some $a' \in A$ and $b' \in B$ with $\Sigma(aa'|bb') \cap \Sigma = \{S\}$,*

(iii) *there exists, for every $a' \in A$ and $b' \in B$, some split $S' \in \Sigma$ with $S'(a) = S'(a') \neq S'(b) = S'(b')$.*

Indeed, applying this lemma to $\Sigma := supp(\nu)$, we see that also $\nu_D(S) \leq \nu(S)$ and, therefore, $\nu_D(S) = \nu(S)$ must hold for every $S \in \Sigma(X)$: The implication "(iii)$\Rightarrow$ (i)" implies that $supp(\nu_D)$ must coincide with $supp(\nu)$ — so, we have $\nu_D(S) = \nu(S) = 0$ for every split $S \in \Sigma(X) - supp(\nu)$. And the implication "(i)$\Rightarrow$ (ii)" implies that

$$\nu_D(S) = \min_{\substack{a,a' \in A \\ b,b' \in B}} \nu\big(\Sigma(aa'|bb')\big) = \min_{\substack{a,a' \in A \\ b,b' \in B}} \nu\big(\Sigma(aa'|bb') \cap supp(\nu)\big) \leq \nu(S)$$

must also hold for every split $S = A|B \in supp(\nu)$ as there will always exist some $a, a' \in A$ and $b, b' \in B$ with $\Sigma(aa'|bb') \cap supp(\nu) = \{S\}$.

Proof of Lemma 7.2 It is obvious that "(ii)$\Rightarrow$(i)$\Rightarrow$(iii)" holds. Conversely, if there exists, for every $a' \in A$ and $b' \in B$, some split $S' \in \Sigma$ with $S'(a) = S'(a') \neq S'(b) = S'(b')$, we must have $S \in \Sigma$, i.e., (iii) implies (i): Indeed, otherwise, we would have $\Sigma \cap \Sigma(A|B) = \Sigma \cap \{S\} = \emptyset$, and there would exist a "minimal" partial X-split $A'|B'$ with $a \in A' \subseteq A$ and $b \in B' \subseteq B$ such that $\Sigma \cap \Sigma(A'|B') = \emptyset$ holds. So, assuming that (iii) holds, either $|A'| \geq 3$ or $|B'| \geq 3$ must hold, and assuming, without loss of generality, that $|A'| \geq 3$ holds, there would exist two distinct elements $a_1, a_2 \in A' - a$ and two splits $S_1, S_2 \in \Sigma$ with $S_1 \in \Sigma(A' - a_1|B') - \Sigma(A'|B')$ and $S_2 \in \Sigma(A' - a_2|B') - \Sigma(A'|B')$ implying that $S_1 \in \Sigma(aa_2|ba_1)$ and $S_2 \in \Sigma(aa_1|ba_2)$ must hold in contradiction to the fact that the two splits $S_1, S_2 \in \Sigma$ must be compatible.

Finally, also "(i)$\Rightarrow$ (ii)" must hold as, applying "(i) $\Longleftrightarrow$ (iii)" for $\Sigma' := \Sigma - S$, we see that there must exist some $a' \in A$ and $b' \in B$ with $\Sigma(aa'|bb') \cap \Sigma' = \emptyset$ and, hence, with $\Sigma(aa'|bb') \cap \Sigma = \{S\}$. □

Remark 7.3 *In this context, it is worth noting that, given a treelike metric D as above, two elements $a, b \in X$, and a non-negative real number ρ, the binary relation "$\overset{ab|\rho}{\underset{D}{\sim}}$" defined on X by putting*

$$x \overset{ab|\rho}{\underset{D}{\sim}} y \iff D(a,x) + D(b,y) = D(a,y) + D(b,x)$$
$$\geq D(a,b) + D(x,y) + \rho$$

for all $x, y \in X$, yields an equivalence relation on the set $\{x \in X : ab + \rho \leq ax + bx\}$: Indeed, if $x \overset{ab|\rho}{\underset{D}{\sim}} y$ and $y \overset{ab|\rho}{\underset{D}{\sim}} z$ holds for some $x, y, z \in X$, we have

$$\begin{aligned} & D(a,x) + D(b,z) \\ &= \big(D(a,x) + D(b,y)\big) - D(b,y) - D(a,y) + \big(D(a,y) + D(b,z)\big) \\ &= \big(D(a,y) + D(b,x)\big) - D(b,y) - D(a,y) + \big(D(a,z) + D(b,y)\big) \\ &= D(a,z) + D(b,x) \end{aligned}$$

and, therefore, also

$$\begin{aligned}
D(a,b)+D(x,z)+D(b,y) &\le D(a,b)+D(x,z)+\rho+D(b,y)\\
&= D(a,b)+\big(D(x,z)+D(b,y)\big)+\rho\\
&\le D(a,b)+\max\{D(b,x)+D(y,z),D(b,z)+D(x,y)\}+\rho\\
&= \max\begin{Bmatrix} D(b,x)+(D(a,b)+D(z,y)+\rho)\\ D(b,z)+(D(a,b)+D(x,y)+\rho)\end{Bmatrix}\\
&\le \max\{D(b,x)+D(a,z)+D(b,y),D(b,z)+D(a,x)+D(b,y)\}\\
&= \max\{D(a,z)+D(b,x),D(a,x)+D(b,z)\}+D(b,y)\\
&= D(a,x)+D(b,z)+D(b,y)=D(a,z)+D(b,x)+D(b,y),
\end{aligned}$$

which immediately implies our claim.

Remark 7.4 *There is also a simple "geometric" way of establishing that, given a split $S=A|B$ in a compatible split system Σ and some elements $a\in A$ and $b\in B$, there exist some elements $a'\in A$ and $b'\in B$ with $\Sigma(aa'|bb')\cap\Sigma=\{S\}$ by referring to any X-tree $\mathcal{T}=\mathcal{T}_\Sigma=(V,E,\varphi)$ associated with Σ: Indeed, let $e=e_S=\{u,v\}$ denote the edge in E corresponding to S, and assume that, say, $A=\varphi^{-1}\big(T^{(e)}(v)\big)$ — and, hence, also $B=\varphi^{-1}\big(T^{(e)}(u)\big)$ — holds for the tree $T:=(V,E)$. Then, the fact that the degree of every unlabeled vertex in V is at least 3 implies that there must exist some $a'\in A$ with $v=med_T(a,a',u)$ and some $b'\in B$ with $u=med_T(b,b',v)$ which in turn implies easily that e must be the only edge in $\mathcal{T}$ that separates a and a' from b and b'.*

An interesting variant of Buneman's approach, going back to the Russian linguist Juri Derenick Apresjan [4], is to consider, for any fixed element $a\in X$, the weighted split system $\nu_D^{(a)}$ defined by putting, for every split S in $\Sigma(X)$,

$$\nu_D^{(a)}(S):=\max\left\{0,\ \min_{\substack{x,y\in\overline{S}(a)\\ z\in S(a)}} D(xy:az)\right\}$$

and its support, denoted also by $\Sigma_D^{(a)}$, that contains exactly those splits S in $\Sigma(X)$ for which

$$D(x,y)+D(a,z)<D(a,x)+D(y,z) \tag{7.5}$$

and, therefore, also

$$D(x,y)+D(a,z)<\min\{D(a,x)+D(y,z),D(a,y)+D(x,z)\}$$

holds for all $x,y\in\overline{S}(a)$ and $z\in S(a)$.

Equivalently, subtracting $D(a,x)+D(a,y)+D(a,z)$ from both sides of (7.5) implies that one may alternatively define $\Sigma_D^{(a)}$ by requiring that it consists of exactly those splits S in $\Sigma(X)$ for which

$$D(x,y)-D(a,x)-D(a,y)<D(x,z)-D(a,x)-D(a,z)$$

or, again equivalently,

$$D(x,y)-D(a,x)-D(a,y)<\min\left\{\begin{matrix}D(x,z)-D(a,x)-D(a,z)\\ D(y,z)-D(a,y)-D(a,z)\end{matrix}\right\} \tag{7.6}$$

holds, for all $x,y\in\overline{S}(a)$ and $z\in S(a)$.

Referring to Inequality (7.6) may sometimes be advantageous because each of the terms that are to be compared now depends only on three of the four elements a,x,y,z under consideration (cf. also our discussion of the "Farris transform" in Chapter 9 below).

Clearly, we have

$$\Sigma_D=\bigcap_{a\in X}\Sigma_D^{(a)}.$$

Moreover, generalizing the fact that the collection Σ_D consisting of all strong D-splits is compatible, the following holds:

Lemma 7.5 (Apresjan's Lemma [4]) *Given any metric D defined on a finite set X, the split system $\Sigma_D^{(a)}$ is compatible for every element $a\in X$.*

Proof Suppose that S_1, S_2 are two splits in $\Sigma_D^{(a)}$ and assume that there were elements x, x_1, x_2 in X with

$$x\in\overline{S_1}(a)\cap\overline{S_2}(a),\quad x_1\in\overline{S_1}(a)\cap S_2(a),\quad\text{and}\quad x_2\in S_1(a)\cap\overline{S_2}(a).$$

Then,

$$D(x,x_1)+D(a,x_2)<D(x,x_2)+D(a,x_1)$$

would hold in view of $x,x_1\in\overline{S_1}(a)$ and $x_2\in S_1(a)$, and

$$D(x,x_2)+D(a,x_1)<D(x,x_1)+D(a,x_2)$$

in view of $x,x_2\in\overline{S_2}(a)$ and $x_1\in S_2(a)$ which is clearly impossible. □

For treelike metrics, the following also holds:

Lemma 7.6 *If D is a treelike metric, then $\Sigma_D=\Sigma_D^{(a)}$ holds for every $a\in X$.*

Proof Clearly, $\Sigma_D \subseteq \Sigma_D^{(a)}$ holds. Conversely, assume $S = A|B \in \Sigma_D^{(a)}$, assume that $\mathcal{T} = (V, E, \omega, \varphi)$ is an edge-weighted X-tree with $D = D_{\mathcal{T}}$, and note that $\mathcal{Q}_{\mathcal{T}}$ coincides with the set of all quartets $xx'|yy' \in \mathcal{Q}(X)$ for which $D(x, x') + D(y, y') < D(x, y) + D(x', y') = D(x, y') + D(x', y)$ holds. Thus, $S \in \Sigma_D^{(a)}$ implies that $ax|yy', ax'|yy' \in \mathcal{Q}_{\mathcal{T}}$ holds for all $x, x' \in S(a)$ and $y, y' \in \overline{S}(a)$ and hence, by transitivity, that also $xx'|yy' \in \mathcal{Q}_{\mathcal{T}}$ or, equivalently, $D(x, x') + D(y, y') < D(x, y) + D(x', y') = D(x, y') + D(x', y)$ holds implying in turn that $S \in \Sigma_D$ must hold, as claimed. □

For metrics D that result from biological data, the split system Σ_D generally tends to contain only very few non-trivial splits. Therefore, quite some effort has been spent to find variants of Buneman's construction that can result in larger collections of compatible splits and still are "canonical" and not, like, e.g., *Neighbor-Joining* [120], "approximative". This led, for instance, to the "refined Buneman trees" considered in [114].

Finally, it is worth noting that, from a computational point of view, the sets Σ_D and $\Sigma_D^{(a)}$, and other related collections of splits associated to D like, e.g., the split systems of the refined Buneman trees can always be computed in polynomial time (see e.g., [32]).

7.2 Weakly compatible split systems

In the last section, we presented some ways to associate a compatible split system to a metric D. We now explore some options for generating more general split systems which, although they will not necessarily correspond to a tree, can provide a more flexible way to represent data. While they may contain (and — in case they are not compatible — even must contain) some "false positives", they may yield some previously missed (or "false negative") splits and, thus, gather some more "true positives" — as well as hint at problems inherent in the data.

To this end, we define the *isolation index* $\alpha^D(aa' : bb')$ of two elements $a, a' \in X$ versus two further elements $b, b' \in X$ relative to a metric D by

$$\alpha^D(aa' : bb') := \max\{D(aa' : bb'), D(aa' : b'b), 0\}$$

$$= \frac{1}{2}\left(\max\begin{Bmatrix} D(a, b) + D(a', b') \\ D(a, b') + D(a', b) \\ D(a, a') + D(b, b') \end{Bmatrix} - D(a, a') - D(b, b')\right),$$

and we define the *isolation index* $\alpha^D(A:B)$ of any two subsets A, B of X relative to D by

$$\alpha^D(A:B) := \min_{\substack{a,a' \in A \\ b,b' \in B}} \alpha^D(aa':bb').$$

The following assertions follow immediately from our definitions:

(S1) $\alpha^D(aa:bb) = D(a,b)$ holds for all $a, b \in X$,
(S2) $\alpha^D(A:B) = \min\{\alpha^D(A-a_1:B), \alpha^D(A-a_2:B), \alpha^D(A-a_3:B)\}$ holds for any two subsets A, B of X and any three distinct elements a_1, a_2, a_3,
(S3) $\alpha^D(A':B') \geq \alpha^D(A'':B'')$ holds for any two pairs A', B' and A'', B'' of subsets of X with $A' \subseteq A''$ and $B' \subseteq B''$,
(S4) and, hence, in particular $D(a,b) \geq \alpha^D(A:B)$ for all $A, B \subseteq X$, $a \in A$, and $b \in B$, and
(S5) $D_+(A:B) \leq \alpha^D(A:B)$ also holds for all $A, B \subseteq X$.

Furthermore, we have $\alpha^D(A:B) = \infty$ in case $A = \emptyset$ or $B = \emptyset$, and we have

$$\alpha^D(\{a,a'\}:\{b,b'\}) = \alpha^D(aa':b,b')$$

for all $a, a', b, b' \in X$ in view of (7.2) and similar inequalities. So, we may also safely write $\alpha^D(a_1a_2 \ldots a_i : b_1b_2 \ldots b_j)$ for $\alpha^D(A:B)$ in case $A = \{a_1, a_2, \ldots, a_i\}$ and $B = \{b_1, b_2, \ldots, b_j\}$.

Using this notation, we begin by establishing the following crucial fact:

Lemma 7.7 *For every metric D and all $a_1, a_2, b_1, b_2, x \in X$, one has*

$$\alpha^D(a_1a_2x : b_1b_2) + \alpha^D(a_1a_2 : b_1b_2x) \leq \alpha^D(a_1a_2 : b_1b_2).$$

More generally, given any two subsets A, B of X, one has

$$\alpha^D(A+x:B) + \alpha^D(A:B+x) \leq \alpha^D(A:B)$$

for all $x \in X$ and, therefore, also

$$\sum_{A'|B' \in \Sigma_X(A|B)} \alpha^D(A':B') \leq \alpha^D(A:B).$$

Proof Assume, for a contradiction, that there exist $a_1, a_2, b_1, b_2, x \in X$ such that $\alpha^D(a_1a_2x : b_1b_2) + \alpha^D(a_1a_2 : b_1b_2x) > \alpha^D(a_1a_2 : b_1b_2)$ holds which — in view of (S3) — would imply that

$$\min\{\alpha^D(a_1a_2x : b_1b_2), \alpha^D(a_1a_2 : b_1b_2x), \alpha^D(a_1a_2 : b_1b_2)\} > 0$$

and also

$$\min \begin{Bmatrix} \alpha^D(a_1x : b_1b_2) + \alpha^D(a_1a_2 : xb_1b_2) \\ \alpha^D(a_2x : b_1b_2) + \alpha^D(a_1a_2 : xb_1b_2) \end{Bmatrix} > \alpha^D(a_1a_2 : b_1b_2)$$

must hold. Without loss of generality, we may therefore also assume that

$$2\alpha^D(a_1x : b_1b_2) = D(a_1, b_1) + D(x, b_2) - D(a_1, x) - D(b_1, b_2)$$

holds. Then, adding $D(a_1, a_2) + D(b_1, b_2)$ to both sides of the inequality

$$2\alpha^D(a_1x : b_1b_2) + 2\alpha^D(a_1a_2 : xb_2) > 2\alpha^D(a_1a_2 : b_1b_2)$$

and "clearing" the terms $D(x, b_2)$ on its left-hand side, we see that

$$D(a_1, b_1) - D(a_1, x) + \max \begin{Bmatrix} D(a_1, x) + D(a_2, b_2) \\ D(a_1, b_2) + D(a_2, x) \end{Bmatrix}$$
$$> \max \begin{Bmatrix} D(a_1, b_1) + D(a_2, b_2) \\ D(a_1, b_2) + D(a_2, b_1) \end{Bmatrix}$$

must hold. Hence, $D(a_1, x) + D(a_2, b_2) < D(a_1, b_2) + D(a_2, x)$ and, therefore,

$$D(a_1, b_1) + D(a_1, b_2) - D(a_1, x) + D(a_2, x)$$
$$> \max \begin{Bmatrix} D(a_1, b_1) + D(a_2, b_2) \\ D(a_1, b_2) + D(a_2, b_1) \end{Bmatrix}$$

follows which yields $D(a_1, b_2) + D(a_2, x) > D(a_1, x) + D(a_2, b_2)$. Similarly, the inequality

$$2\alpha^D(a_2x : b_1b_2) + 2\alpha^D(a_1a_2 : xb_2) > 2\alpha^D(a_1a_2 : b_1b_2)$$

implies that also $D(a_2, b_2) + D(a_1, x) > D(a_2, x) + D(a_1, b_2)$ must hold. As, together, this is impossible, the first assertion of Lemma 7.7 must hold.

It follows that, for any two subsets A, B of X, one has

$$\alpha^D(A + x : B) + \alpha^D(A : B + x) \le \alpha^D(aa'x : bb') + \alpha^D(aa' : bb'x)$$
$$\le \alpha^D(aa' : bb')$$

for all $a, a' \in A$ and $b, b' \in B$ and, therefore, also

$$\alpha^D(A + x : B) + \alpha^D(A : B + x) \le \min_{\substack{a,a' \in A \\ b,b' \in B}} \alpha^D(aa' : bb') = \alpha^D(A : B),$$

as claimed. So, a simple induction with respect to $|X - (A \cup B)|$ yields that also $\sum_{A'|B' \in \Sigma_X(A|B)} \alpha^D(A' : B') \le \alpha^D(A : B)$ must hold. □

Next, we denote the weighted split system that assigns the isolation index $\alpha^D(A:B)$ to any split $S=A|B$ in $\Sigma(X)$ by ν^D, i.e., we put $\nu^D(S)=\nu^D(A|B):=\alpha^D(A:B)$ for every split $S=A|B\in\Sigma(X)$.

Clearly, the above lemma implies that

$$\alpha^D(A:B)\geq\nu^D(A|B)=\nu^D\big(\Sigma(A|B)\big) \tag{7.7}$$

holds for any two subsets A,B of X. Furthermore, it was one of the starting observations in [10] that the split system $\Sigma^D:=supp(\nu^D)$ whose elements will also be called *weak D-splits* or just as well simply *D-splits*, has the following remarkable property:

(WeaC) For any three splits $A_1|B_1, A_2|B_2, A_3|B_3\in\Sigma^D$, at least one of the four intersections

$$A_1\cap A_2\cap A_3,\ A_1\cap B_2\cap B_3,\ B_1\cap A_2\cap B_3,\ B_1\cap B_2\cap A_3$$

is empty.

Proof Indeed, if there were splits $S_1=A_1|B_1$, $S_2=A_2|B_2$, and $S_3=A_3|B_3$ in Σ^D and elements $a_1,a_2,a_3,a_4\in X$ with $a_1\in A_1\cap A_2\cap A_3$, $a_2\in A_1\cap B_2\cap B_3$, $a_3\in B_1\cap A_2\cap B_3$, and $a_4\in B_1\cap B_2\cap A_3$, we would get

$$\max\{D(a_1,a_3)+D(a_2,a_4), D(a_1,a_4)+D(a_2,a_3)\}$$
$$> D(a_1,a_2)+D(a_3,a_4)$$

in view of $\nu^D(S_1)>0$ as well as

$$\max\{D(a_1,a_2)+D(a_3,a_4), D(a_1,a_4)+D(a_2,a_3)\}$$
$$> D(a_1,a_3)+D(a_2,a_4),$$

in view of $\nu^D(S_2)>0$, and

$$\max\{D(a_1,a_2)+D(a_3,a_4), D(a_1,a_3)+D(a_2,a_4)\}$$
$$> D(a_1,a_4)+D(a_2,a_3),$$

in view of $\nu^D(S_3)>0$ implying that none of the three sums $D(a_1,a_2)+D(a_3,a_4)$, $D(a_1,a_3)+D(a_2,a_4)$, and $D(a_1,a_4)+D(a_2,a_3)$ can coincide with the maximum of all three of them which is clearly impossible. □

Any split system $\Sigma\subseteq\Sigma(X)$ that satisfies the condition (WeaC) is called *weakly compatible*.

It is worth noting that, defining the weighted split system $\nu^D_{x|y}$, for any given metric D and any two elements $x, y \in X$, by putting

$$\nu^D_{x|y}(S) = \nu^D_{x|y}(A|B) := \min \left\{ \min_{\substack{a \in S(x) \\ b,b' \in S(y)}} \alpha^D(xa : bb'), \min_{\substack{a,a' \in S(x) \\ b \in S(y)}} \alpha^D(aa' : yb) \right\}$$

$$= \min \{\alpha^D(aa' : bb') : a, a' \in S(x)\,;\, b, b' \in S(y)\,;\, \{a, a', b, b'\} \cap \{x, y\} \neq \emptyset\}$$

for any $S = A|B \in \Sigma(X)$, the support $supp\left(\nu^D_{x|y}\right)$ of $\nu^D_{x|y}$ (which is clearly contained in $\Sigma_X(x|y)$) must also be weakly compatible: For any three splits S_1, S_2, S_3 in $supp\left(\nu^D_{x|y}\right)$, one just uses exactly the same inequalities as above, yet chooses a_1 to be x or a_4 to be y.

It is also worth noting that, given a metric D of the form $D := D_\nu$ for some weighted split system ν, (7.4) implies that

$$\begin{aligned}\alpha^D(aa' : bb') &= \max\{D(ab : a'b'), D(ab' : a'b), 0\} \\ &= \max\{\nu(aa'|bb') - \nu(ab|a'b'), \nu(aa'|bb') - \nu(ab'|a'b), 0\} \\ &= \nu(aa'|bb') - \min\{\nu(aa'|bb'), \nu(ab'|a'b), \nu(ab|a'b')\}\end{aligned}$$

holds for all $a, a', b, b' \in X$ implying that

$$\alpha^D(aa' : bb') = \nu(aa'|bb') \tag{7.8}$$

holds for all $a, a', b, b' \in X$ if and only if the support $supp(\nu)$ of ν is weakly compatible as $\min\{\nu(aa'|bb'), \nu(ab'|a'b), \nu(ab|a'b')\} = 0$ holds for some elements a, a', b, b' in X if and only if $supp(\nu)$ has an empty intersection with at least one of the three sets $\Sigma(aa'|bb'), \Sigma(ab'|a'b), \Sigma(ab|a'b')$.

Any metric D that is of the form $D = D_\nu$ for a weighted split system ν with weakly compatible support will also be called *totally split decomposable*. It is obvious that every compatible split system is weakly compatible, so every treelike metric is totally split decomposable.

In analogy with Theorem 7.1, we will now show that also the construction of ν^D is consistent provided it is restricted to the class of totally split decomposable metrics. That is, for any weighted split system ν with weakly compatible support, one has $\nu = \nu^D$ for the totally split decomposable metric $D := D_\nu$. In particular, if D is totally split decomposable, then $\nu = \nu^D$ is the only weighted split system with weakly compatible support for which $D = D_\nu$ holds. To establish this fact, we first prove the following result:

Theorem 7.8 *Given any metric D, one has $D(x, y) \geq D_\nu(x, y)$, for all $x, y \in X$, for the split system $\nu := \nu^D$. More specifically, given any weighted split system $\nu' : \Sigma(X) \to \mathbb{R}_{\geq 0}$ with $\nu'(S) \leq \nu(S)$ for every split S of X, the difference $D' := D - D_{\nu'} : X' \times X' \to \mathbb{R} : (x, y) \mapsto D(x, y) - D_{\nu'}(x, y)$ is a metric, and one has $\nu^{D'}(S) = \nu^D(S) - \nu'(S)$ for every split $S \in \Sigma(X)$.*

Proof We use induction on the size of $supp(\nu')$. Clearly, our claim is obvious in case $supp(\nu') = \emptyset$. Next, let us assume $|supp(\nu')| = 1$ holds, let S denote the unique split in $supp(\nu')$, and put $\rho := \nu'(S)$ so that $D' = D - \rho D_S$ holds.

Note first that D' is a metric: Indeed, given any two metrics D_1 and D_2 defined on X, there is obviously a unique largest real number $\mu = \mu(D_1|D_2) \geq 0$, the *metric index of D_1 relative to D_2*, such that $D_1 - \mu D_2$ is a metric, *viz.* the number

$$\mu(D_1|D_2) := \inf_{\substack{x,y,z \in X \\ D_2(x,z)+D_2(z,y)>D_2(x,y)}} \frac{D_1(x, z) + D_1(z, y) - D_1(x, y)}{D_2(x, z) + D_2(z, y) - D_2(x, y)}$$

(as, given any elements $x, y, z \in X$, one has $\mu \leq \frac{D_1(x,z)+D_1(z,y)-D_1(x,y)}{D_2(x,z)+D_2(z,y)-D_2(x,y)}$ for some $\mu \in \mathbb{R}_{\geq 0}$ if and only if one has $(D_1 - \mu D_2)(x, z) + (D_1 - \mu D_2)(z, y) \geq (D_1 - \mu D_2)(x, y)$).

Thus, our claim follows from observing that

$$\begin{aligned}\mu(D|D_S) &= \inf_{\substack{x,y,z \in X \\ S(x)=S(y)\neq S(z)}} \frac{D(x, z) + D(z, y) - D(x, y)}{2} \\ &\geq \inf_{\substack{a,a' \in A \\ b,b' \in B}} \frac{D(a, b) + D(a', b') - D(a, a') - D(b, b')}{2} \\ &= \alpha^D(A : B) = \nu_D(S) \geq \rho\end{aligned}$$

holds for every split $S = A|B$ of X.

Now, put $\nu' := \nu^{D'}$. We have to show that $\nu'(S') = \nu(S') - \rho\, \delta_{S,S'}$ holds for all $S' = A'|B' \in \Sigma(X)$. So, assume $a, a', b, b' \in X$ and consider the term

$$2\alpha^{D'}(aa' : bb') = \max \begin{Bmatrix} D'(a, b) + D'(a', b') \\ D'(a, b') + D'(a', b) \\ D'(a, a') + D'(b, b') \end{Bmatrix} - D'(a, a') - D'(b, b').$$

We claim that, for all $a, a', b, b' \in X$, we have

$$\alpha^{D'}(aa' : bb') = \begin{cases} \alpha^D(aa' : bb') - \rho & \text{if } S(a) = S(a') \neq S(b) = S(b'), \\ \alpha^D(aa' : bb') & \text{otherwise.} \end{cases} \tag{7.9}$$

Indeed, if $S(a) = S(a') \neq S(b) = S(b')$ holds, we have $D'(a, a') + D'(b, b') = D(a, a') + D(b, b') \leq \max\{D(a, b) + D(a', b'), D(a, b') + D(a', b)\} - 2\rho = \max\{D'(a, b) + D'(a', b'), D'(a, b') + D'(a', b)\}$ and, therefore,

$$\begin{aligned}
2\alpha^{D'}&(aa' : bb') \\
&= \max\{D'(a, b) + D'(a', b'), D'(a, b') + D'(a', b)\} \\
&\quad - D'(a, a') - D'(b, b') \\
&= \max\{D(a, b) + D(a', b'), D(a, b') + D(a', b)\} \\
&\quad - 2\rho - D(a, a') - D(b, b') \\
&= 2\alpha^{D}(aa' : bb') - 2\rho,
\end{aligned}$$

as claimed. Otherwise, we may assume, by symmetry, that either $S(a) = S(a') = S(b) = S(b')$ or $S(a) = S(a') = S(b) \neq S(b')$ or $S(a) = S(b) \neq S(a') = S(b')$ holds.

In the first case, we have $D'(x, y) = D(x, y)$ for all $x, y \in \{a, a', b, b'\}$. So, our claim holds for obvious reasons in this case.

In the second case, we have $D'(a, a') + D'(b, b') = D(a, a') + D(b, b') - \rho$, $D'(a, b) + D'(a', b') = D(a, b) + D(a', b') - \rho$, and $D'(a, b') + D'(a', b) = D(a, b') + D(a', b) - \rho$, and our claim holds again for obvious reasons.

In the third case, we have $D'(a, a') + D'(b, b') = D(a, a') + D(b, b') - 2\rho$, $D'(a, b) + D'(a', b') = D(a, b) + D(a', b')$, and $D'(a, b') + D'(a', b) = D(a, b') + D(a', b) - 2\rho$ as well as

$$D(a, b) + D(a', b') \leq \max\{D(a, a') + D(b, b'), D(a, b') + D(a', b)\} - 2\rho$$

and, therefore, also

$$D'(a, b) + D'(a', b') \leq \max\{D'(a, a') + D'(b, b'), D'(a, b') + D'(a', b)\}.$$

If $\max\{D(a, a') + D(b, b'), D(a, b') + D(a', b)\} = D(a, b') + D(a', b)$ and, therefore, also $\max\{D'(a, a') + D'(b, b'), D'(a, b') + D'(a', b)\} = D'(a, b') + D'(a', b)$ as well as $D'(a, b) + D'(a', b') = D(a, b) + D(a', b') \leq D(a, b') + D(a', b) - 2\rho = D'(a, b') + D'(a', b)$ holds, we get

$$\begin{aligned}
2\alpha^{D'}&(aa' : bb') \\
&= \max\left\{\begin{matrix} D'(a, b) + D'(a', b') \\ D'(a, b') + D'(a', b) \\ D'(a, a') + D'(b, b') \end{matrix}\right\} - D'(a, a') - D'(b, b') \\
&= \max\{D'(a, b') + D'(a', b) - D'(a, a') - D'(b, b'), 0\} \\
&= \max\{D(a, b') + D(a', b) - D(a, a') - D(b, b'), 0\}
\end{aligned}$$

$$= \max \begin{Bmatrix} D(a,b) + D(a',b') \\ D(a,b') + D(a',b) \\ D(a,a') + D(b,b') \end{Bmatrix} - D(a,a') - D(b,b')$$
$$= 2\alpha^D(aa' : bb'),$$

as claimed.

Otherwise, we must have $D(a,a') + D(b,b') > D(a,b') + D(a',b)$ implying, in view of $D(a,b') + D(a',b) \geq D(a,b) + D(a',b') + 2\rho$, that $\alpha^D(aa' : bb') = 0$ must hold. And we must have $D'(a,a') + D'(b,b') > D'(a,b') + D'(a',b)$ and, therefore, in view of

$$D'(a,b) + D'(a',b') \leq \max\{D'(a,a') + D'(b,b'), D'(a,b') + D'(a',b)\}$$

also

$$D'(a,b) + D'(a',b') \leq D'(a,a') + D'(b,b')$$

and, hence, $D'(a,a') + D'(b,b') \geq D'(a,b') + D'(a',b), D'(a,b) + D'(a',b')$ implying that $\alpha^{D'}(aa' : bb') = 0$ and, hence, $\alpha^{D'}(aa' : bb') = \alpha^D(aa' : bb')$ must hold also in this last case.

Hence, $\nu^{D'}(S') \leq \nu^D(S')$ must hold for every split $S' \in \Sigma(X)$, and equality must hold in case $S' = A'|B'$ is distinct from S: Indeed, note first that, in view of Lemma 7.7, we have $\alpha^D(aa' : bb') \geq \nu^D(S) + \nu^D(S') > \nu^D(S')$ for all $a, a' \in A'$ and $b, b' \in B'$ with $S(a) = S(a') \neq S(b) = S(b')$. Thus, choosing $a_1, a_2 \in A'$ and $b_1, b_2 \in B'$ with $\nu^D(S') = \alpha^D(a_1a_2 : b_1b_2)$, we cannot have $S(a_1) = S(a_2) \neq S(b_1) = S(b_2)$. So, we must have

$$\alpha^{D'}(a_1a_2 : b_1b_2) = \alpha^D(a_1a_2 : b_1b_2) \leq \alpha^D(aa' : bb') = \alpha^{D'}(aa' : bb')$$

for all $a, a' \in A'$ and $b, b' \in B'$ for which $S(a) = S(a') \neq S(b) = S(b')$ does not hold. Furthermore, $\alpha^{D'}(a_1a_2 : b_1b_2) = \alpha^D(a_1a_2 : b_1b_2) = \nu^D(S')$ also cannot exceed $\alpha^{D'}(aa' : bb') = \alpha^D(aa' : bb') - \rho$ in the remaining case $S(a) = S(a') \neq S(b) = S(b')$ as $\alpha^D(aa' : bb') - \rho \geq \nu^D(S) + \nu^D(S') - \rho \geq \nu^D(S') = \alpha^D(a_1a_2 : b_1b_2)$ must hold for all $a, a' \in A'$ and $b, b' \in B'$ with $S(a) = S(a') \neq S(b) = S(b')$.

Together, this implies that

$$\nu^D(S') = \alpha^D(a_1a_2 : b_1b_2) = \min_{\substack{a,a' \in A' \\ b,b' \in B'}} \alpha^{D'}(aa' : bb') = \nu^{D'}(S')$$

holds, as claimed.

The theorem now follows easily by induction with respect to $|supp(\nu')|$: Just choose any split $S \in supp(\nu')$, and apply the induction hypothesis to

$D' := D - \nu'(S)D_S$ and the weighted split system ν'' defined by $\nu''(S) := 0$, and $\nu''(S') := \nu'(S')$ for all splits S' distinct from S. □

Using this result, it is now fairly straightforward to show that ν^D is consistent:

Theorem 7.9 *Given a metric D and a weighted split system ν, the following two assertions are equivalent:*

(i) *The support* $\mathrm{supp}(\nu)$ *of ν is weakly compatible and one has $D = D_\nu$.*
(ii) *D is totally split decomposable and $\nu = \nu^D$ holds.*

In other words,

- *There is a canonical one-to-one correspondence between totally split-decomposable metrics D and weighted split systems ν with weakly compatible support given by associating, to any such split system ν, the metric D_ν and, in the other direction, the split system ν^D to any totally split-decomposable metric D.*
- *A metric D is totally split decomposable if and only if $D_{\nu^D} = D$ holds in which case ν^D is the only weighted split system ν with weakly compatible support for which $D = D_\nu$ holds.*
- *And conversely, a weighted system ν of X-splits has a weakly compatible support if and only if it is of the form $\nu = \nu^D$ for some metric D in which case this holds in particular for the metric $D := D_\nu$ (but also, in case $n \geq 5$, for many other metrics).*
- *In particular, associating the metric D_{ν^D} to any metric D, defines an idempotent operator $\mathbf{S}_X : D \mapsto D_{\nu^D}$ — which we also call the* split operator *for X — from the space $\mathbb{M}(X)$ into itself whose image consists exactly of the space $\mathbb{TSD}(X)$ consisting of all totally split-decomposable metrics defined on X.*
- *Moreover, defining a metric D to be* split prime *if $\nu^D(S) = 0$ holds for every split S of X, then associating, to any metric D, the metric $D_{\mathrm{sp}} := D - D_{\nu^D}$ defines an idempotent operator $\mathbf{SP} = \mathbf{SP}_X : D \mapsto D_{\mathrm{sp}}$ from $\mathbb{M}(X)$ onto the space $\mathbb{SP}(X)$ consisting of all split-prime metrics defined on X.*

Proof In view of our previous remarks, it suffices to show that $\nu = \nu^D$ holds, for every weighted split system ν with weakly compatible support, for the associated metric $D := D_\nu$. However, given any split $S = A|B \in \Sigma(X)$, we have

$$\nu^D(S) = \alpha^D(A|B) = \min_{\substack{a,a' \in A \\ b,b' \in B}} \alpha^D(aa' : bb') = \min_{\substack{a,a' \in A \\ b,b' \in B}} \nu(aa'|bb') \tag{7.10}$$

in view of (7.8). So, $\nu^D(S) > 0$ holds if and only if there exists, for all $a, a' \in A$ and $b, b' \in B$, some split $S' \in supp(\nu) \cap \Sigma(aa'|bb')$, and $\nu^D(S) \geq \nu(S)$ must hold for every split S in $\Sigma(X)$. Thus, equality follows from the following variant of Lemma 7.2 that characterizes splits in weakly compatible split systems:

Lemma 7.10 *Given a weakly compatible split system $\Sigma \subseteq \Sigma(X)$ and a split $S = A|B \in \Sigma(X)$, the following three assertions are equivalent:*

(i) $S \in \Sigma$,

(ii) *there exist some $a, a' \in A$ and $b, b' \in B$ with $\Sigma(aa'|bb') \cap \Sigma = \{S\}$,*

(iii) *there exists, for all $a, a' \in A$ and $b, b' \in B$, some split S' in $\Sigma(aa'|bb') \cap \Sigma$.*

Indeed, applying this lemma to $\Sigma := supp(\nu)$, we see that also $\nu^D(S) \leq \nu(S)$ and, therefore, $\nu^D(S) = \nu(S)$ must hold for every $S \in \Sigma(X)$: The implication "(iii)⇒ (i)" implies that $supp(\nu^D)$ must coincide with $supp(\nu)$ — so, we have $\nu^D(S) = \nu(S) = 0$ for every split $S \in \Sigma(X) - supp(\nu)$. And the implication "(i)⇒(ii)" implies that

$$\nu^D(S) = \min_{\substack{a,a' \in A \\ b,b' \in B}} \nu\big(\Sigma(aa'|bb')\big) = \min_{\substack{a,a' \in A \\ b,b' \in B}} \nu\big(\Sigma(aa'|bb') \cap supp(\nu)\big) \leq \nu(S)$$

must also hold for every split $S = A|B \in supp(\nu)$ as there will always exist some $a, a' \in A$ and $b, b' \in B$ with $\Sigma(aa'|bb') \cap supp(\nu) = \{S\}$.

Proof of Lemma 7.10 Again, it is obvious that "(ii)⇒(i)⇒(iii)" holds. Conversely, if there exists, for all $a, a' \in A$ and $b, b' \in B$, some split S' in $\Sigma(aa'|bb') \cap \Sigma$, we must have $S \in \Sigma$: Indeed, otherwise, we would have $\Sigma \cap \Sigma(A|B) = \Sigma \cap \{S\} = \emptyset$, and there would exist a "minimal" partial X-split $A'|B'$ with $A' \subseteq A$ and $B' \subseteq B$ such that $\Sigma \cap \Sigma(A'|B') = \emptyset$ holds. So, assuming that (iii) holds, either $|A'| \geq 3$ or $|B'| \geq 3$ must hold, and assuming, without loss of generality, that $|A'| \geq 3$ holds, there would exist three distinct elements $a_1, a_2, a_3 \in A'$ and three splits $S_1, S_2, S_3 \in \Sigma$ with $S_i \in \Sigma(A' - a_i|B') - \Sigma(A'|B')$ for all $i = 1, 2, 3$ implying that $S_1 \in \Sigma(a_2a_3|ba_1)$, $S_2 \in \Sigma(a_1a_3|ba_2)$, and $S_3 \in \Sigma(a_1a_2|ba_3)$ must hold for any $b \in B'$ in contradiction to the fact that the three splits $S_1, S_2, S_3 \in \Sigma$ must be weakly compatible.

Finally, also "(i)⇒ (ii)" must hold as, since "(i) ⟺ (iii)" holds for $\Sigma' := \Sigma - S$, we see that there must exist some $a, a' \in A$ and $b, b' \in B$ with $\Sigma(aa'|bb') \cap \Sigma' = \emptyset$ and, hence, with $\Sigma(aa'|bb') \cap \Sigma = \{S\}$. □

It follows that the results in Lemma 7.7 can be augmented for totally split-decomposable metrics as follows (cf. Theorem 6, [10]):

Theorem 7.11 *Given a metric D, the following assertions are equivalent:*

(TSD1) *D is totally split decomposable,*
(TSD2) $\alpha^D(a_1a_2x : b_1b_2) + \alpha^D(a_1a_2 : b_1b_2x) = \alpha^D(a_1a_2 : b_1b_2)$ *holds for all* $a_1, a_2, b_1, b_2, x \in X$,
(TSD3) $\alpha^D(A+x : B) + \alpha^D(A : B+x) = \alpha^D(A : B)$ *holds for all subsets A, B of X and all elements $x \in X$,*
(TSD4) $\sum_{A'|B' \in \Sigma_X(A|B)} \alpha^D(A' : B') = \alpha^D(A : B)$ *holds for all* $A, B \subseteq X$.

Furthermore, if ν is a weighted split system with weakly compatible support such that $D = D_\nu$ holds, we have

$$\alpha^D(A : B) = \nu(A|B) = \nu\big(\Sigma(A|B)\big) \tag{7.11}$$

for all non-empty subsets A, B of X implying that also $\alpha^D(A : B) = \nu^D(A|B)$ holds, for all A, B as above, in case D is totally split decomposable.

Sketch of proof That $\alpha^{D_\nu}(A : B) = \nu(A|B) = \nu\big(\Sigma(A|B)\big)$ holds for all non-empty subsets A, B of X in case ν is a weighted split system with weakly compatible support, follows immediately from Theorem 7.9 applied to the set $X' := A \cup B$, the weighted split system $\nu' := \nu|_{X'}$ defined on $\Sigma(X')$ by putting $\nu'(A'|B') := \nu(A'|B')$ for every split $A'|B'$ in $\Sigma(X')$, and the metric $D' := D_{\nu'} \in \mathbb{M}(X')$, noting that the support of ν' must be weakly compatible as every split in $supp(\nu')$ can be extended to one in $supp(\nu)$, and that $D' = D|_{X'}$ holds as

$$\begin{aligned} D'(a', b') = D_{\nu'}(a', b') &= \sum_{A'|B' \in \Sigma_{X'}(a'|b')} \nu'(A'|B') \\ &= \sum_{A'|B' \in \Sigma_{X'}(a'|b')} \left(\sum_{S \in \Sigma_X(A'|B')} \nu(S) \right) = \sum_{S \in \Sigma_X(a'|b')} \nu(S) \\ &= \nu\big(\Sigma_X(a'|b')\big) = D_\nu(a', b') = D(a', b') \end{aligned}$$

holds for all $a', b' \in X'$. Thus, we have $\alpha^{D'}(A' : B') = \alpha^D(A' : B')$ for any two subsets A', B' of X' as well as $\nu' = \nu^{D_{\nu'}} = \nu^{D'}$ implying that $\nu(A'|B') = \nu'(A' : B') = \nu^{D'}(A' : B') = \alpha^{D'}(A' : B') = \alpha^D(A' : B')$ must hold for every split $A'|B'$ of X' while $\nu(A'|B') = \alpha^D(A' : B') = 0$ holds for every pair A', B' of subsets of X' with $A' \cap B' \neq \emptyset$.

It follows immediately that (TSD1) implies (TSD4). So, as it is obvious that "(TSD4) $\Rightarrow$ (TSD3) $\Rightarrow$ (TSD2)" always holds, it remains to show that also "(TSD2) $\Rightarrow$ (TSD1)" holds. As there is nothing to prove in case D satisfies the 4-point condition, we may safely assume that four distinct elements x, y, u, v with, say, $D(x, y) + D(u, v) < D(x, u) + D(y, v)$ and, hence, in particular, $\alpha^D(xy:uv) > 0$, exist in X. Also, we may note, using (7.9) and induction on $|supp(\nu^D)|$, that (TSD2) holds for the split-prime part D_{sp} of D whenever it holds for D, and we may also assume that D is split prime implying that $n \geq 6$ must hold as $\alpha^D(xy:uv) > 0$ together with (TSD2) implies that $\alpha^D(axy:uv) > 0$ or $\alpha^D(xy:auv) > 0$ must hold for every $a \in A$. In addition, by induction on n, we may also assume that the restriction $D^{(x)}$ of D to $(X - x) \times (X - x)$ is totally split decomposable for every $x \in X$. Thus, given any $x \in X$, there must exist a split $A|B$ of $X - x$ with $\alpha^D(A|B) > 0$ while, for any $a \in A$, either $\alpha^D(x + A - a|B) > 0$ or $\alpha^D(A - a|x + B) > 0$ and, for any $b \in B$, either $\alpha^D(x + A|B - b) > 0$ or $\alpha^D(A|x + B - b) > 0$ must hold.

In particular, choosing x, A, and B so that B has maximal cardinality, $\alpha^D(x + A - a|B) > 0$ must hold for all $a \in A$.

However, if $\alpha^D(x + A - a|B), \alpha^D(x + A - a'|B) > 0$ were to hold for two distinct elements $a, a' \in A$, (S2) — with A replaced by $x + A$ and a_1, a_2, a_3 by x, a, a' — would imply

$$\alpha^D(x + A : B)$$

$$= \min\{\alpha^D(A : B), \alpha^D(x + A - a : B), \alpha^D(x + A - a' : B)\} > 0$$

in contradiction to our assumption that D is split prime. So, $|A| = 1$ and $|B| = n - 2 \geq 4$ must hold in this case. Yet, there can also be only one $b \in B$ with $\alpha^D(A|x + B - b) > 0$, implying that $\alpha^D(x + A|B - b) > 0$ must hold for at least three distinct elements $b = b_1, b_2, b_3$ in B. So, (S2) — with A replaced by B, a_1, a_2, a_3 by b_1, b_2, b_3 and B by $x + A$ — would imply

$$\alpha^D(x + A : B) = \min \begin{Bmatrix} \alpha^D(x + A : B - b_1) \\ \alpha^D(x + A : B - b_2) \\ \alpha^D(x + A : B - b_3) \end{Bmatrix} > 0,$$

again in contradiction to our assumption that D is split prime, a final contradiction. □

The following result implies that ν^D is also consistent for treelike metrics:

Theorem 7.12 *Given any metric D and any split system ν with $\nu(S) \leq \nu_D(S)$ for every split $S \in \Sigma(X)$, the bivariate map $D' := D - D_\nu$ is a metric and one has $\nu_{D'}(S) = \nu_D(S) - \nu(S)$ for every split S of X. In particular, one has $\nu_D = \nu^D$ for every treelike metric $D \in \mathbb{T}(X)$.*

Proof Clearly, as $\nu(S) \leq \nu_D(S) \leq \nu^D(S)$ holds for every split S of X, the map D' must be a metric and one has $\nu^{D'}(S) = \nu^D(S) - \nu(S)$ for every split

S of X. To show that also $\nu_{D'}(S) = \nu_D(S) - \nu(S)$ holds for every split $S \in \Sigma(X)$, we may, as in the proof of Theorem 7.8, assume that $supp(\nu)$ consists of a single split $S = A|B$ only and denote the number $\nu(S)$ by ρ in which case we have to show that $\nu_{D'}(S) = \nu_D(S) - \rho$ holds for that split S and that $\nu_{D'}(S') = \nu_D(S')$ holds for any other split S' of X.

However, we can argue as above that

$$D'_+(aa' : bb') = \begin{cases} D_+(aa' : bb') - \rho & \text{if } S(a) = S(a') \neq S(b) = S(b'), \\ D_+(aa' : bb') & \text{otherwise,} \end{cases} \tag{7.12}$$

holds for all $a, a', b, b' \in X$. Indeed, if $S(a) = S(a') \neq S(b) = S(b')$ holds, we have $D'(a, a') + D'(b, b') = D(a, a') + D(b, b') \leq D(a, b) + D(a', b') - 2\rho = D'(a, b) + D'(a', b')$ and, therefore,

$$\begin{aligned} &2D'(aa' : bb') \\ &\quad = D'(a, b) + D'(a', b') - D'(a, a') - D'(b, b') \\ &\quad = D(a, b) + D(a', b') - 2\rho - D(a, a') - D(b, b') \\ &\quad = 2D(aa' : bb') - 2\rho \geq 0 \end{aligned}$$

and, hence, $D'_+(aa' : bb') = D_+(aa' : bb') - \rho$, as claimed. Otherwise, we may assume, just as above, that either $S(a) = S(a') = S(b) = S(b')$, or $S(a) = S(a') = S(b) \neq S(b')$ or $S(a) = S(b) \neq S(a') = S(b')$ or $S(a) = S(b') \neq S(a') = S(b)$ holds. Again, our claim holds for obvious reasons in the first and the second case.
In the third case, we have $D'(a, a') + D'(b, b') = D(a, a') + D(b, b') - 2\rho$ and $D'(a, b) + D'(a', b') = D(a, b) + D(a', b')$, as well as

$$D(a, b) + D(a', b') \leq D(a, a') + D(b, b') - 2\rho$$

and, therefore, $D'(a, b) + D'(a', b') \leq D'(a, a') + D'(b, b')$ as well as $D'_+(aa' : bb') = D_+(aa' : bb') = 0$.

And in the last case, we have $D'(a, a') + D'(b, b') = D(a, a') + D(b, b') - 2\rho$ and $D'(a, b) + D'(a', b') = D(a, b) + D(a', b') - 2\rho$. So, our claim holds again for obvious reasons.

Together, this establishes (7.12) implying that $\nu_{D'}(S') \leq \nu_D(S')$ must hold for every split $S' \in \Sigma(X)$, and that equality must hold in case $S' = A'|B'$ is distinct from S: This is obvious in case $S' \notin supp(\nu_D)$. Otherwise, we may choose $a_1, a_2 \in A'$ and $b_1, b_2 \in B'$ so that $\Sigma(a_1a_2|b_1b_2) \cap supp(\nu_D) = \{S'\}$ and, therefore, also

$$\nu_D(S') = D(a_1a_2 : b_1b_2) = D_+(a_1a_2 : b_1b_2)$$

holds. By our choice of a_1, a_2, b_1, and b_2, we cannot have $S(a_1) = S(a_2) \neq S(b_1) = S(b_2)$. So, we must have

$$D'_+(a_1a_2 : b_1b_2) = D_+(a_1a_2 : b_1b_2) \leq D_+(aa' : bb') = D'_+(aa' : bb')$$

for all $a, a' \in A'$ and $b, b' \in B'$ for which $S(a) = S(a') \neq S(b) = S(b')$ does not hold. And even if $S(a) = S(a') \neq S(b) = S(b')$ does hold, $D'_+(a_1a_2 : b_1b_2) = D_+(a_1a_2 : b_1b_2) = \nu_D(S')$ cannot exceed $D'_+(aa' : bb') = D_+(aa' : bb') - \rho$: Indeed, we may assume, without loss of generality, that some element $x \in X$ with, say, $x \in A \cap B'$ exists implying, in view of (7.1), that

$$\begin{aligned} D'(aa' : bb') &= D(aa' : bb') - \rho \\ &= D(ax : bb') - \rho + D(aa' : xb') \\ &\geq \nu_D(S) - \rho + \nu_D(S') \\ &\geq \nu_D(S') = D(a_1a_2 : b_1b_2) = D'_+(a_1a_2 : b_1b_2) > 0 \end{aligned}$$

holds. Together, this implies that

$$\begin{aligned} \nu_D(S') = D(a_1a_2 : b_1b_2) = \min_{\substack{a,a' \in A' \\ b,b' \in B'}} D'(aa' : bb') = D'_+(a_1a_2 : b_1b_2) \\ = \nu_{D'}(S') \end{aligned}$$

indeed holds, as claimed.

Finally, suppose D is a treelike metric. Then, by definition, there exists an edge-weighted X-tree $\mathcal{T} = (V, E, \omega, \varphi)$ such that $D = D_{\mathcal{T}} = D_\nu$, where $\nu := \nu_{\mathcal{T}}$ is the weighted split system induced by $\mathcal{T}$. By Theorem 3.4 the split system $supp(\nu)$ is compatible and, therefore, $\nu = \nu_D$ must hold by Theorem 7.1. At the same time, $supp(\nu)$ is weakly compatible, which, by Theorem 7.9, implies that $\nu = \nu^D$, and, therefore, $\nu_D = \nu^D$ must hold, as required. □

In this context, it is also worth recalling from [10] that, given a metric D, a split S of X, and a non-negative real number ρ, one has $\rho \leq \nu^D(S)$ if and only if (i) the map $D' := D - \rho\, D_S$ is a metric and (ii) every map $f \in T(D)$ is a sum of a map f_1 in $T(D')$ and a map f_2 in $T(\rho\, D_S)$ and that, in consequence, this implies:

Theorem 7.13 *The decomposition $D = D_{\mathrm{sp}} + \sum_{S \in supp(\nu^D)} \nu^D(S) D_S$ is the unique decomposition of D into a split-prime metric, say D_0, and a sum, say, $\sum_{S \in \Sigma} \rho_S D_S$ of non-vanishing split metrics $\rho_S D_S$, S in some split system Σ, for which there exists, for every map $f \in T(D)$, a map $f_0 \in T(D_0)$*

and a Σ-indexed family $(f_S)_{S\in\Sigma}$ of maps $f_S \in T(\rho_S D_S)$ such that $f = f_0 + \sum_{S\in\Sigma} f_S$ holds.

Interestingly, using this theorem, it can be shown that Condition (ii) of Theorem 5.17 is also equivalent to asserting that the metric D is totally split-decomposable. The full details of the proof of this equivalence may be found in [58], but the key steps are as follows: If D is totally split-decomposable and f is a map in $T(D)$, we may write $f = \sum_{S\in supp(\nu^D)} f_S$ as in Theorem 7.13. It is obvious that in case $\dim[f]_D = 1$, there must be some split $S_0 = A_0|B_0$ in $supp(\nu^D)$ with $\dim[f_{S_0}]_{\nu^D(S_0)D_{S_0}} = 1$ and, therefore,

$$\begin{aligned} K(f|D) &= \bigcap_{S\in supp(\nu^D)} K(f_S|\nu^D(S)D_S) \\ &\subseteq K(f_{S_0}|\nu^D(S_0)D_{S_0}) \\ &= \big\{\{a,b\} : a \in A_0 \text{ and } b \in B_0\big\}, \end{aligned}$$

so that, in particular, $K(f|D)$ is bipartite. Since $\dim[f]_D = 1$, it follows by Lemma 5.11 that f is connected, i.e. Condition (ii) in Theorem 5.17 holds.

Conversely, suppose that D is not totally split-decomposable. Then it can be shown that there must be some subset $Z := \{x, y, u, v, t\}$ of X of cardinality 5 and some map $f \in T(D|_Z)$ with

$$K(f|D|_Z) = \{\{x, y\}, \{v, u\}, \{u, t\}, \{t, v\}\},$$

and that this map can be recursively modified to give a map $f' \in T(D)$ with $\dim[f']_D = 1$ and $K(f|D|_Z)$ a subgraph of $K(f'|D)$. But, using Lemma 5.11 again, this implies that $K(f'|D)$ cannot be connected (as, containing the edges $\{v, u\}$, $\{u, t\}$, and $\{t, v\}$, it cannot be bipartite). Thus, Condition (ii) in Theorem 5.17 does not hold, as required.

Actually, it was the observation presented in Theorem 7.13, derived during a car ride to the first conference of the newly formed **International Federation of Classification Societies** in Aachen in 1987, that kicked off the development of "split-decomposition theory" to which a considerable part of this book is devoted.

We now show that, even though the split system Σ^D can contain considerably more splits than the split system Σ_D (which, as it is compatible, has cardinality at most $2n - 3$), the number of splits in Σ^D is bounded by $\binom{n}{2}$, a fact that was used in [10] to show that a simple recursive, yet potentially exponential algorithm for computing ν^D for an arbitrary metric D is, in fact, polynomial (of order n^6).

Corollary 7.14 *For every metric D, the split system Σ^D contains at most $\binom{n}{2}$ splits.*

Proof Clearly, the dimension of $\mathcal{M}(X) = \mathcal{M}(X, \mathbb{R})$, the vector space of all symmetric bivariate maps from $X \times X$ into $\mathbb{R}$ that vanish on the diagonal, is $\binom{n}{2}$. Therefore it suffices to show that, for any weakly compatible split system $\Sigma \subseteq \Sigma(X)$, the metrics in $\{D_S : S \in \Sigma\}$ form a linearly independent subset of $\mathcal{M}(X)$, a result that is also of considerable independent interest (see e.g., [33] and the references quoted there).

To see that this is the case, let $(\lambda_S)_{S\in\Sigma}$ be a Σ-indexed family of real numbers such that $\sum_{S\in\Sigma} \lambda_S D_S = \mathbf{0}$ holds. Define the weighted split systems $\nu^+, \nu^- : \Sigma(X) \to \mathbb{R}_{\geq 0}$ by putting $\nu^+(S) := \max\{\lambda_S, 0\}$ and $\nu^-(S) := \nu^+(S) - \lambda_S$ for every $S \in \Sigma$ and $\nu^+(S) = \nu^-(S) := 0$ for all other splits $S \in \Sigma(X)$. Then, we have $D_{\nu^+} = D_{\nu^-}$, which, in view of the fact that both, $supp(\nu^+)$ and $supp(\nu^-)$, are weakly compatible, implies, by Theorem 7.9, that $\nu^+ = \nu^-$ holds. But then we must have $supp(\nu^+) = supp(\nu^-) = \emptyset$, since, by construction, $supp(\nu^+) \cap supp(\nu^-) = \emptyset$ holds, which implies that $\lambda_S = 0$ holds for all $S \in \Sigma$, as required. □

We conclude this section by noting that every circular split system is weakly compatible. In addition, we note (see e.g., [41]) that the class of *circular metrics*, i.e., totally split decomposable metrics D for which Σ^D is circular, coincides with the class of the so-called *Kalmanson metrics* [106], i.e., metrics defined on a finite set X for which there exists a linear ordering "$\prec$" of X such that $\max\{D(i, j) + D(i', j'), D(i, j') + D(i', j)\} \leq D(i, i') + D(j, j')$ holds for all $i, i, j, j' \in X$ with $i \prec j \prec i' \prec j'$. These observations have been exploited within the algorithms for constructing split networks called NeighborNet [34] and QNet [85]. NeighborNet is essentially a heuristic approach to iteratively construct a circular metric that approximates a given input metric while QNet takes as input a set of quartets rather than a metric. More details concerning QNet may be found in Chapter 10.

7.3 From weighted split systems to bivariate maps

In the first two sections of this chapter, we have described some constructions that associate metrics to weighted split systems. Now, we investigate in some more detail the linear map

$$\lambda = \lambda_X : \mathbb{R}^{\Sigma(X)} \to \mathcal{M}(X) : \nu \mapsto \sum_{S\in\Sigma(X)} \nu(S)\, D_S$$

from the real vector space $\mathbb{R}^{\Sigma(X)}$ of all $\mathbb{R}$-weighted split systems ν (that is, maps from $\Sigma(X)$ into $\mathbb{R}$) into the real vector space $\mathcal{M}(X)$ introduced in the proof of Corollary 7.14 that consists of all symmetric bivariate maps from $X \times X$ into $\mathbb{R}$ that vanish on the diagonal.

Let $\ker(\lambda)$ denote the kernel of the map λ, that is, the set of those $\nu \in \mathbb{R}^{\Sigma(X)}$ with $\lambda(\nu) = \mathbf{0}$. In the following lemma, we present some surprising properties of that kernel. Define, for every $x \in X$, every abelian group $\mathcal{A}$, and every map $\nu \in \mathcal{A}^{\Sigma(X)}$, the map $\nu_x \in \mathcal{A}^{\Sigma(X-x)}$ by $\nu_x(A'|B') = \nu(A'+x|B') + \nu(A'|B'+x)$ for every split $A'|B'$ of $X - x$. Then, we have:

Lemma 7.15

(i) *If, for some $\nu \in \mathcal{A}^{\Sigma(X)}$, ν_x vanishes for every $x \in X$, then there exists some $\alpha \in \mathcal{A}$ such that $\nu(S) = \pm\,\alpha$ for every split $S \in \Sigma(X)$. More specifically, if n is odd, then $2\alpha = 0$ holds and ν is constant. And if n is even, then ν is constant on all splits $S = A|B$ with $|A| \equiv |B| \equiv 0 \mod 2$, and it is constant, but of the opposite sign on all splits $S = A|B$ with $|A| \equiv |B| \equiv 1 \mod 2$, i.e., there is some $\alpha \in \mathcal{A}$ with*

$$\nu(A|B) = \begin{cases} \alpha & \text{if } |A| \text{ is even,} \\ -\alpha & \text{if } |A| \text{ is odd,} \end{cases} \tag{7.13}$$

for all $A|B \in \Sigma(X)$.

(ii) *If ν is a non-vanishing map in* $\ker(\lambda)$, *then* $supp(\nu)$ *contains at least seven distinct splits. And if, furthermore, the values $\nu(S)$ of ν are integers for all $S \in \Sigma(X)$, but not all of them are even integers, then at least seven of them are odd.*

Proof (i) Let x be an arbitrary element of X. As, by assumption, ν_x vanishes, we must have

$$\nu(A|B) = -\nu(A - x|B + x) \tag{7.14}$$

for every split $A|B$ of X with $x \in A$ and $|A| \geq 2$. Now, consider two arbitrary splits $S_1 = A_1|B_1, S_2 = A_2|B_2 \in \Sigma(X)$. Without loss of generality, we may assume $A_1 \cap A_2 \neq \emptyset$. We transform the split $A_1|B_1$ first into the split $A_1 \cap A_2|B_1 \cup B_2$ by removing the elements in $A_1 - A_2$, one at a time, from A_1 and adding them to B_1, and then, using similar moves of single elements, into the split $A_2|B_2$. But this implies, by Equation (7.14), that $\nu(S_2) = \pm\,\nu(S_1)$ must hold, as required.

In addition, if n is even, $|A|$ and $|B|$ are either both even or both odd for every split $A|B$ of X and, clearly, moving a single element from A to B changes parity — so, (7.13) must hold. And, if n is odd, consider two distinct trivial splits $S_x = \{x\}|(X - x)$ and $S_y = \{y\}|(X - y)$. Then, we have $\nu(S_x) = \nu(S_y)$, since we can obtain S_y in two moves from S_x, but at the same time we have $\nu(S_x) = -\nu(S_y)$, since we can obtain S_y in $n-2$ moves from S_x, which implies that $\nu(S) = 0$ must hold for all $S \in \Sigma(X)$.

(ii) We prove (ii) by induction on the cardinality of X. Clearly, there is nothing to prove if $n \leq 3$ holds as $\Sigma(X)$ is compatible in this case which implies that $\ker(\lambda)$ consists of the "all-zero map" $\mathbf{0}$, only.

Now, assume $n \geq 4$ and note that $\nu_x \in \ker(\lambda_{(X-x)})$ as well as $|supp(\nu_x)| \leq |supp(\nu)|$ holds for every $\nu \in \ker(\lambda_X)$ and every $x \in X$.

So, by induction, either ν_x vanishes for every $x \in X$ or we have $7 \leq |supp(\nu_x)|$ for at least one $x \in X$ and, therefore, also $7 \leq |supp(\nu)|$, as claimed. However, the former implies that, in view of (i), there must exist some $\alpha \in \mathbb{R}$ such that $\nu(S) = \pm\alpha$ holds for every $S \in \Sigma(X)$ and, hence, $supp(\nu) = \Sigma(X)$ in view of $\nu \neq \mathbf{0}$. But then, $|supp(\nu)| = |\Sigma(X)| = 2^{n-1} - 1 \geq 7$ must hold also in this case in view of $n \geq 4$.

Finally, assume that all values $\nu(S)$ of ν are integers, but that not all of them are even integers. Then, a similar induction argument shows that also the number $|\nu|_2 := |\{S \in \Sigma(X) : \nu(S) \text{ is odd}\}|$ must be at least 7 in case it is non-zero: Indeed, $|\nu_x|_2 \leq |\nu|_2$ must hold for every $x \in X$. And applying (i) for $\mathcal{A} := \mathcal{F}_2$, we see that $|\nu_x|_2$ vanishes for all $x \in X$ if and only if the map $\overline{\nu}$ from $\Sigma(X)$ into $\mathcal{F}_2$ obtained by concatenating $\nu : \Sigma(X) \to \mathbb{Z}$ with the canonical map $\mathbb{Z} \to \mathcal{F}_2$, is constant. □

Note that the bound in Lemma 7.15 is tight. To see this, define, for any partition of X into four non-empty subsets A_1, A_2, A_3, A_4, the map $\nu = \nu_{A_1,A_2,A_3,A_4} \in \mathbb{Z}^{\Sigma(X)}$ from $\Sigma(X)$ into $\mathbb{Z}$ by $\nu(A_i|X - A_i) := 1$ for $i = 1, \ldots, 4$, $\nu(A_i \cup A_j|X - (A_i \cup A_j)) := -1$ for all i, j with $1 \leq i < j \leq 3$, and $\nu(S) := 0$ for every other split S in $\Sigma(X)$, and note that $\nu \in \ker(\lambda)$ as well as $supp(\nu) = 7$ holds.

Without too much trouble, it can also be shown that $\ker(\lambda)$ is generated by the split systems ν_{A_1,A_2,A_3,A_4} where A_1, A_2, A_3, A_4 runs through all partitions of X into exactly four distinct sets.

We conclude our discussion of the kernel of λ by pointing out an intriguing connection to coding theory (for more on coding theory, see e.g., [128]): To this end, recall that a (*binary*) *code* of length m, $m \in \mathbb{N}$, is a subset C of $\{0, 1\}^{\mathcal{X}}$ for some set $\mathcal{X}$ of cardinality m. The code C is called *linear* if, considered as a subset of $\mathcal{F}_2^{\mathcal{X}}$, it is an $\mathcal{F}_2$-linear subspace of $\mathcal{F}_2^{\mathcal{X}}$.

(a)

D	a	b	c	d
a	0	3	4	3
b	3	0	3	4
c	4	3	0	3
d	3	4	3	0

(b) (c)

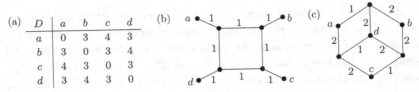

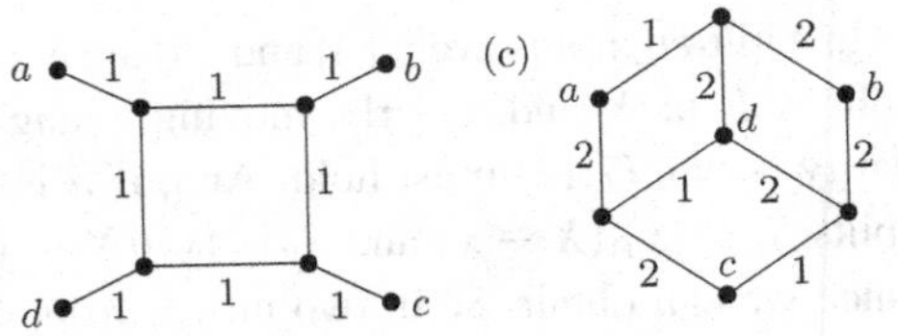

Figure 7.1 A metric D on $X = \{a, b, c, d\}$ and two split networks representing distinct weighted split systems whose image under the map λ is D.

For $c, c' \in \{0, 1\}^{\mathcal{X}}$ define $D_{\mathcal{X}}(c, c') := |\{i \in \mathcal{X} : c(i) \neq c'(i)\}|$, and for every $c \in \{0, 1\}^{\mathcal{X}}$ and every $t \in \mathbb{N}$, put $B_C(c|t) := \{c^* \in C : D_{\mathcal{X}}(c, c^*) \leq t\}$. A code $C \subseteq \{0, 1\}^{\mathcal{X}}$ is *t-error correcting* for some $t \in \mathbb{N}$ if $B_C(c|t) \cap B_C(c'|t) = \emptyset$ holds for any two distinct elements $c, c' \in C$.

Clearly, if C is a linear binary code, it is t-error correcting for every t with $|supp(c)| \geq 2t + 1$ for every non-zero element $c \in C$.

Now associating, to every $\nu \in \mathbb{Z}^{\Sigma(X)}$, the element $c^\nu : \Sigma(X) \to \{0, 1\}$ defined by $c^\nu(S) := 0$ if $\nu(S)$ is even and $c^\nu(S) := 1$ if $\nu(S)$ is odd, consider the set $C_X := \{c^\nu : \nu \in \ker(\lambda|_{\mathbb{Z}^{\Sigma(X)}})\}$. Then, we have:

Theorem 7.16 *The set C_X is a linear* 3*-error correcting binary code.*

Proof Clearly, C_X is a linear binary code. Thus, it is 3-error correcting in view of the last assertion in Lemma 7.15. □

We remark that C_X is a member of the well-known family of *Reed–Muller codes* [111, ch. 13]. Note also that, since the kernel of λ is non-empty for $n > 3$, it is not surprising that there exist metrics $D \in \mathbb{M}(X)$ for which there are at least two distinct weighted split systems ν_1 and ν_2 in $\mathbb{R}^{\Sigma(X)}$, and even in $\mathbb{R}^{\Sigma(X)}_{\geq 0}$, such that $\lambda(\nu_1) = D = \lambda(\nu_2)$ holds (an example of such a metric along with two distinct weighted split systems giving rise to D is depicted in Figure 7.1, where the support of both weighted split systems is represented by an edge-weighted split network). In other words, the restriction of λ to $\mathbb{R}^{\Sigma(X)}_{\geq 0}$ is not an injective map into the set $\mathbb{M}(X)$. Moreover, for $n > 4$, there exist metrics D that are not contained in $\lambda\left(\mathbb{R}^{\Sigma(X)}_{\geq 0}\right)$, that is, λ is not surjective. The problem to decide whether a given metric D is contained in $\lambda\left(\mathbb{R}^{\Sigma(X)}_{\geq 0}\right)$ is known as the *ℓ_1-embeddability problem*, which is known to be NP-complete [5]. More details on this and other aspects of this problem can be found, for example, in [48].

In the remainder of this section, we will shed some more light on split systems $\Sigma \subseteq \Sigma(X)$ for which, at least, the restriction of λ to those $\nu \in \mathbb{R}^{\Sigma(X)}$ with

$supp(\nu) \subseteq \Sigma$ is an injective map into the set $\mathbb{M}(X)$. Note that this is equivalent to requiring that the set $\{D_S : S \in \Sigma\}$ of split metrics forms a linearly independent subset of $\mathcal{M}(X)$. We call such a split system Σ an *$\mathcal{M}$-independent split system.* We have already seen in the proof of Corollary 7.14 that every weakly compatible split system is $\mathcal{M}$-independent. There are, however, also non-weakly compatible split systems that are $\mathcal{M}$-independent — for example, the support of the weighted split system depicted in Figure 7.1(c).

Before we give some more interesting examples of such split systems, we first present a characterization of $\mathcal{M}$-independent split systems. To this end, we define, for any split system $\Sigma \subseteq \Sigma(X)$, the matrix $\mathbf{A}(\Sigma) = (a_{S,S'})$, whose rows and columns are both indexed by the splits in Σ, with $a_{S,S'} := |A \cap A'||B \cap B'| + |A \cap B'||B \cap A'|$ for all splits $S = A|B$ and $S' = A'|B'$ in Σ. The following fact was noted in [33]:

Theorem 7.17 *A split system $\Sigma \subseteq \Sigma(X)$ is $\mathcal{M}$-independent if and only if* $\det(\mathbf{A}(\Sigma)) \neq 0$ *holds.*

Proof Clearly, it suffices to show that Σ is $\mathcal{M}$-independent if and only if the rank of $\mathbf{A}(\Sigma)$ is $|\Sigma|$. To this end, let $\mathbf{B} := (b_{\{x,y\},S})$ denote the $\{0, 1\}$-matrix whose rows are indexed by the 2-subsets of X and whose columns are indexed by the splits in Σ, with $b_{\{x,y\},S} := D_S(x, y)$ for all $\{x, y\} \in \binom{X}{2}$ and all $S \in \Sigma$. Note that Σ is $\mathcal{M}$-independent if and only if the rank of $\mathbf{B}$ is $|\Sigma|$, which holds if and only if the rank of the matrix $\mathbf{B}^T\mathbf{B}$ is $|\Sigma|$. Hence, it suffices to show that $\mathbf{B}^T\mathbf{B} = \mathbf{A}(\Sigma)$ holds. So, let $S = A|B$ and $S' = A'|B'$ be arbitrary splits in Σ. Then, we have

$$\begin{aligned}
a_{S,S'} &= |A \cap A'||B \cap B'| + |A \cap B'||B \cap A'| \\
&= \left|\left\{\{x, y\} \in \binom{X}{2} : S(x) \neq S(y) \text{ and } S'(x) \neq S'(y)\right\}\right| \\
&= \sum_{\{x,y\}\in\binom{X}{2}} D_S(x, y) \cdot D_{S'}(x, y) \\
&= \sum_{\{x,y\}\in\binom{X}{2}} b_{\{x,y\},S} \cdot b_{\{x,y\},S'},
\end{aligned}$$

as required. □

More details about $\mathcal{M}$-independent split systems can also be found in [33]. In that paper, further interesting examples of such split systems are described, namely the *affine* split systems. These are split systems $\Sigma \subseteq \Sigma(X)$ for which there exists (i) a map $\xi : X \to \mathbb{R}^2$ that associates, to every element $x \in X$, a

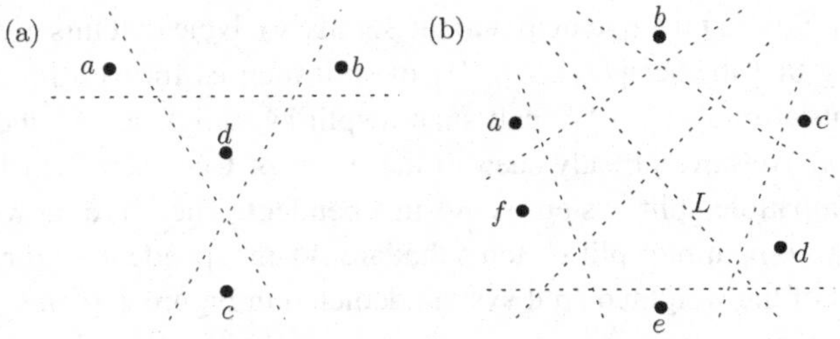

Figure 7.2 (a) A representation of the split system in Figure 7.1(c) by points and straight lines in the plane. (b) A circular split system, the straight line L representing, for example, the split $\{a, e, f\}|\{b, c, d\}$.

pair $\xi(x) = (\xi_1(x), \xi_2(x)) \in \mathbb{R}^2$, and (ii) a map $\gamma : \Sigma \to \mathbb{R}^3$ that associates to every split $S \in \Sigma$ a 3-tuple $\gamma(S) = (\gamma_1, \gamma_2, \gamma_3) \in \mathbb{R}^3$ so that

$$S = \{x \in X : \xi_1(x)\gamma_1 + \xi_2(x)\gamma_2 + \gamma_3 < 0\}| \\ \{x \in X : \xi_1(x)\gamma_1 + \xi_2(x)\gamma_2 + \gamma_3 > 0\}$$

holds. Note that the maps ξ and γ associate a point in $\mathbb{R}^2$ to every element in X and a straight line in $\mathbb{R}^2$ to every split $S \in \Sigma$ such that the split of X induced by the straight line in $\mathbb{R}^2$ is the same as the split S. For example, in Figure 7.2(a) such a representation of the split system pictured in Figure 7.1(c) is depicted, implying that, in particular, this split system is affine. Note also that every circular split system is affine: The map ξ can be chosen so that it arranges the points on a circle in $\mathbb{R}^2$ respecting the linear order of X that must exist according to the definition of a circular split system (see, for example, Figure 7.2(b)).

Using simple geometric "general position" arguments, it can be shown that, for every affine split system $\Sigma \subseteq \Sigma(X)$, there exists an affine split system $\Sigma' \subseteq \Sigma(X)$ such that $\Sigma \subseteq \Sigma'$ and $|\Sigma'| = \binom{n}{2}$ hold. We call an affine split system $\Sigma \subseteq \Sigma(X)$ with $|\Sigma| = \binom{n}{2}$ *full*. Remarkably, as is shown in [33], every full affine split system (and, therefore, every affine split system) is $\mathcal{M}$-independent. The idea for proving this is to first establish that, if a split system $\Sigma \subseteq \Sigma(X)$ satisfies the *pairwise separation property*, that is, if there exist, for any two distinct elements $x, y \in X$, disjoint subsets $A, B \subseteq X - \{x, y\}$ such that the splits $A+x|B+y$ and $A+y|B+x$ as well as the splits $A \cup \{x, y\}|B$ if $B \neq \emptyset$ and $A|B \cup \{x, y\}$ if $A \neq \emptyset$ are contained in Σ, the associated split metrics D_S, $S \in \Sigma$, generate — and, thus, must be a basis of — $\mathcal{M}(X)$ if $|\Sigma| \leq \binom{n}{2}$ and, thus, $|\Sigma| = \binom{n}{2}$ holds. It then suffices to observe that every full affine split system satisfies the pairwise separation property.

It is also worth noting that affine split systems are a proper subclass of split systems that arise in the theory of *oriented matroids* [25]: Indeed, oriented matroids of rank 3 can be represented by arrangements of so-called *pseudolines*, and it was observed in [33] that "full" split systems arising from pseudoline arrangements satisfy, just as full affine split systems, the pairwise separation property and, thus, are $\mathcal{M}$-independent.

7.4 The Buneman complex and the tight span

Looking at the diagrams in Figure 1, the reader might wonder how the maps that we have presented in this chapter may be related to the constructions that we have described earlier. For example, the Buneman graph discussed in Chapter 4 provides a way to generate a network from a split system, and the tight span discussed in Chapter 5 gives a way to obtain a network from a metric. So, if that metric is of the form $D = D_\Sigma := \sum_{S \in \Sigma} D_S$ for some split system $\Sigma \subseteq \Sigma(X)$, one might ask how $\mathcal{B}(\Sigma)$ and $T(D_\Sigma)$ may be related.

In this section, we will present some intriguing results concerning the relationship between $\mathcal{B}(\Sigma)$ and $T(D_\Sigma)$. Note that all of the definitions and results below can be generalized naturally so as to work for arbitrary weighted split systems $\nu \in \mathbb{R}_{\geq 0}^{\Sigma(X)}$ (cf. [60, 62]), rather than just subsets Σ of $\Sigma(X)$. We restrict our discussion here to just this situation only to simplify the presentation.

First, let us consider how $\mathcal{B}(\Sigma)$ and $T(D_\Sigma)$ are related in case Σ is compatible. Recall that, by Theorem 4.8(iv), the (labeled) Buneman graph $\mathcal{T} := (\mathcal{B}(\Sigma), \varphi_\Sigma)$ is, up to canonical isomorphism, the unique X-tree $\mathcal{T}$ with $\Sigma_\mathcal{T} = \Sigma$. And, by Theorem 5.14, the labeled network $\mathcal{N}_{D_\Sigma}$ that is derived from the tight span $T(D_\Sigma)$ (ignoring the edge weights) is also isomorphic to $\mathcal{T}$ — so, these two X-labeled networks must be "canonically" isomorphic in this case.

Inspired by this observation, it was shown in [62] that it is also possible to establish a relationship between $\mathcal{B}(\Sigma)$ and the tight span $T(D_\Sigma)$ for more general split systems Σ. To describe this relationship, it is useful to first define a generalization of the Buneman graph.

To this end, let $\Sigma \subseteq \Sigma(X)$ be a split system, and associate, to every vertex ϕ of $\mathcal{B}(\Sigma)$, the map

$$\mu = \mu_\phi : \mathcal{C}(\Sigma) \to \{0, 1\}$$

by putting, for every $A \in \mathcal{C}(\Sigma) = \bigcup_{S \in \Sigma} S$, $\mu(A) := 0$ if $\phi\big(A|(X - A)\big) = A$ holds, and $\mu(A) := 1$ else. Then, it is not hard to see that the conditions (BG1) and (BG2) defining the vertex set of $\mathcal{B}(\Sigma)$ translate into the following conditions for the associated maps $\mu \in \{0,1\}^{\mathcal{C}(\Sigma)}$:

(i) For all $A \in \mathcal{C}(\Sigma)$, $\mu(A) + \mu(X - A) = 1$ must hold.
(ii) For all $A, B \in \mathcal{C}(\Sigma)$, $\mu(A) > 0$, $\mu(B) > 0$, and $A \cup B = X$ implies $B = X - A$.

Therefore, it is natural to consider the set $\overline{\mathcal{B}}(\Sigma)$, consisting of all those maps $\mu \in \mathbb{R}_{\geq 0}^{\mathcal{C}(\Sigma)}$ that satisfy Conditions (i) and (ii). This set or, more precisely, this polytopal complex is known as the *Buneman complex* [62]. It is straightforward to check that the 1-skeleton of $\overline{\mathcal{B}}(\Sigma)$ coincides with the Buneman graph $\mathcal{B}(\Sigma)$ [55].

We now present the first result that hints how to relate $\overline{\mathcal{B}}(\Sigma)$ to $T(D_\Sigma)$. This relies on the key map $\psi = \psi_\Sigma : \overline{\mathcal{B}}(\Sigma) \to \mathbb{R}_{\geq 0}^X$ that assigns, to every $\mu \in \overline{\mathcal{B}}(\Sigma)$, the function $f_\mu \in \mathbb{R}_{\geq 0}^X$ defined by putting $f_\mu(x) := \sum_{S \in \Sigma} \mu(S(x))$ for every $x \in X$.

Lemma 7.18 *For every split system $\Sigma \subseteq \Sigma(X)$, the image $\psi(\overline{\mathcal{B}}(\Sigma))$ is contained in $P(D_\Sigma)$.*

Proof According to the definition of $P(D_\Sigma)$ (cf. Chapter 5), we have to show that $f_\mu(x) + f_\mu(y) \geq D_\Sigma(x, y)$ holds for all $\mu \in \overline{\mathcal{B}}(\Sigma)$ and all $x, y \in X$. So, let μ be an arbitrary map in $\overline{\mathcal{B}}(\Sigma)$ and let x and y be two arbitrary elements in X. Then, we have

$$\begin{aligned} f_\mu(x) + f_\mu(y) &= \sum_{S \in \Sigma} \mu(S(x)) + \sum_{S \in \Sigma} \mu(S(y)) \\ &\geq \sum_{\substack{S \in \Sigma, \\ S(x) \neq S(y)}} (\mu(S(x)) + \mu(S(y))) \\ &= |\{S \in \Sigma : S(x) \neq S(y)\}| \\ &= D_\Sigma(x, y), \end{aligned}$$

as required. □

Now, it is not hard to see that if Σ is compatible, then ψ maps $\overline{\mathcal{B}}(\Sigma)$ bijectively onto $T(D_\Sigma)$. However, there are (necessarily non-compatible) split systems for which the image of ψ is not contained in $T(D_\Sigma)$. For example, let $X = \{a, b, c, d\}$ and $\Sigma \subseteq \Sigma(X)$ consisting of the three splits $\{a, b\}|\{c, d\}$,

$\{a, c\}|\{b, d\}$, and $\{a, d\}|\{b, c\}$ already considered in Figure 7.1(c). It can be easily checked that the map μ that, for every $A \in \mathcal{C}(\Sigma)$, assigns 1 to A if $a \in A$ and 0 otherwise, is contained in $\overline{\mathcal{B}}(\Sigma)$. However, we have $f_\mu(a) = 3$ and $f_\mu(x) = 1$ for all $x \in X - a$, and so $f_\mu(a) + f_\mu(x) = 4 > D_\Sigma(a, x)$ holds for all $x \in X - a$, implying that f_μ is not contained in $T(D_\Sigma)$.

Thus, it is quite remarkable that Theorem 7.13 can be used to characterize precisely those split systems Σ for which ψ maps $\overline{\mathcal{B}}(\Sigma)$ surjectively onto $T(D_\Sigma)$ (for a proof of this result, see [60]):

Theorem 7.19 *Let $\Sigma \subseteq \Sigma(X)$ be a split system. The map $\psi : \overline{\mathcal{B}}(\Sigma) \to T(D_\Sigma)$ is surjective (i.e., $\psi(\overline{\mathcal{B}}(\Sigma)) = T(D_\Sigma)$ holds) if and only if Σ is weakly compatible.*

In view of this result, it is natural to ask under what conditions ψ is also injective. To see that, in general, this is not the case, consider the set $X = \{a_1, a_2, a_3, a_4, a_5, a_6\}$ and the split system $\Sigma \subseteq \Sigma(X)$ consisting of the splits $\{a_1, a_2, a_3\}|\{a_4, a_5, a_6\}$, $\{a_1, a_3, a_4\}|\{a_2, a_5, a_6\}$, $\{a_1, a_4, a_5\}|\{a_2, a_3, a_6\}$, and $\{a_1, a_2, a_5\}|\{a_3, a_4, a_6\}$. It is easy to check that Σ is weakly compatible and that no two distinct splits in Σ are compatible. Moreover, the map $\mu_1 : \mathcal{C}(\Sigma) \to \{0, 1\}$ that assigns 1 to the subsets $\{a_1, a_2, a_3\}$, $\{a_1, a_4, a_5\}$, $\{a_2, a_5, a_6\}$, and $\{a_3, a_4, a_6\}$ and 0 to all other subsets in $\mathcal{C}(\Sigma)$ is contained in $\overline{\mathcal{B}}(\Sigma)$. Similarly, the map $\mu_2 : \mathcal{C}(\Sigma) \to \{0, 1\}$, defined by putting $\mu_2(A) := 1 - \mu_1(A)$ for all $A \in \mathcal{C}(\Sigma)$, is also contained in $\overline{\mathcal{B}}(\Sigma)$, and, clearly, $\mu_1 \neq \mu_2$ holds. But we have $f_{\mu_1}(x) = 2 = f_{\mu_2}(x)$ for all $x \in X$.

Interestingly, the split system used in this last example can be viewed as the prototype of a family of subconfigurations that must be excluded from a weakly compatible split system in order to ensure that ψ is injective. More precisely, define a split system $\Sigma = \{S_1, S_2, S_3, S_4\} \subseteq \Sigma(X)$ consisting of four distinct splits to be *octahedral* if there exists a partition of X into six non-empty subsets $A_1, A_2, \ldots, A_6$ such that

$$S_1 = (A_1 \cup A_2 \cup A_3)|(A_4 \cup A_5 \cup A_6),$$

$$S_2 = (A_1 \cup A_3 \cup A_4)|(A_2 \cup A_5 \cup A_6),$$

$$S_3 = (A_1 \cup A_4 \cup A_5)|(A_2 \cup A_3 \cup A_6),$$

$$S_4 = (A_1 \cup A_2 \cup A_5)|(A_3 \cup A_4 \cup A_6)$$

holds. Note that it is called octahedral as we can obtain such split systems by labeling the vertices of the octahedron by X and considering the splits obtained

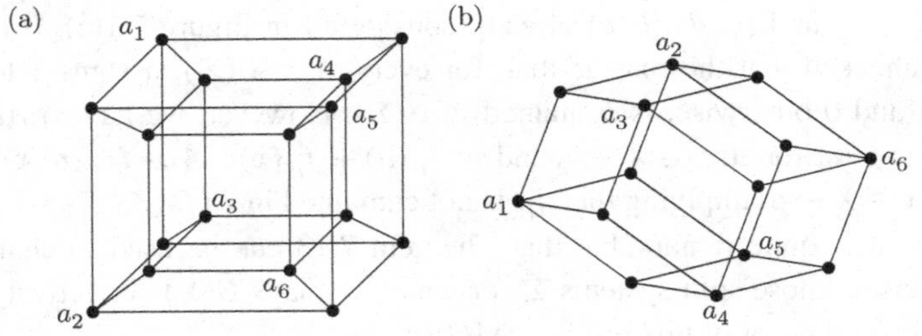

Figure 7.3 (a) The 1-skeleton of the Buneman complex of the octahedral split system Σ on $X = \{a_1, a_2, \ldots, a_6\}$ considered in the text. (b) The 1-skeleton of the tight span $T(D_\Sigma)$ for this octahedral split system.

by taking the labels of the faces of the octahedron and their complements. Then, we have:

Theorem 7.20 *Let $\Sigma \subseteq \Sigma(X)$ be a split system. The map ψ is a bijection between $\overline{\mathcal{B}}(\Sigma)$ and $T(D_\Sigma)$ if and only if Σ is weakly compatible and does not contain an octahedral split system.*

Note that for the octahedral split system Σ on $X = \{a_1, a_2, \ldots, a_6\}$ considered above, the map ψ clearly does not respect the cell complex structure of $\overline{\mathcal{B}}(\Sigma)$ and $T(D_\Sigma)$ (i.e., it does not map cells to cells), as the number of 0-dimensional cells of $\overline{\mathcal{B}}(\Sigma)$ and $T(D_\Sigma)$ are not the same (cf. Figure 7.3). However, it can be shown that the map ψ *does* respect the cell complex structures of $\overline{\mathcal{B}}(\Sigma)$ and $T(D_\Sigma)$ in case it is a bijection. For more details on this, and a proof of Theorem 7.20 for arbitrary weighted split systems, see [62].

8

Maps to and from quartet systems

In the previous chapter, we have considered relationships between metrics and split systems, corresponding to the double arrow labeled ⑦ in the outer triangle of Figure 1. We now explore the relationships indicated by the additional double arrows in that triangle.

First, we shall focus on constructions that directly associate a quartet system to a split system and, conversely, a split system to a quartet system. In particular, we study how these constructions transform properties of a given split system into properties of the corresponding quartet system, and those of a quartet system into properties of the corresponding split system. After this, we present a construction that allows one to directly go from quartet systems to metrics. And, finally, we focus on some results concerning transitive quartet systems that arise naturally in this context.

We keep denoting by X a finite set which we — to exclude trivial cases — assume to have cardinality $n \geq 4$ and to which all further concepts and constructions refer, by $\Sigma(X)$, $\Sigma_{triv}(X)$, and $\Sigma^*(X)$ the set of all, of all trivial, and of all non-trivial splits of X, respectively, and by $\Sigma_{\text{part}}(X)$ the collection of all partial splits $A|B$ of X. And we denote by $\Sigma^*_{\text{part}}(X)$ the collection of all "non-trivial" partial splits $A|B$ of X, i.e., all partial splits $A|B$ of X with $|A|$, $|B| \geq 2$ so that $\Sigma^*(X) = \Sigma(X) \cap \Sigma^*_{\text{part}}(X)$ always holds.

8.1 A Galois connection between split and quartet systems

We begin by presenting a simple way to relate split systems and quartet systems that was described in [12]. Given any partial split $A|B \in \Sigma_{\text{part}}(X)$, we denote by $\mathcal{Q}(A|B) := \{q \in \mathcal{Q}(X) : q \preceq A|B\}$ the quartet system consisting of all quartets that are extended by $A|B$. Further, we put $\mathcal{Q}(\Sigma) := \bigcup_{A|B \in \Sigma} \mathcal{Q}(A|B)$ for every collection Σ of partial splits of X, the system of all quartets in $\mathcal{Q}(X)$

that are extended by at least one split in Σ. Clearly, we have $\mathcal{Q}(\Sigma) = \{A|B \in \mathcal{Q}(X) : \Sigma(A|B) \cap \Sigma \neq \emptyset\}$ for every split system $\Sigma \subseteq \Sigma(X)$.

Conversely, we define, for every quartet system $\mathcal{Q} \subseteq \mathcal{Q}(X)$, a (partial) split $A|B$ of X to be a (partial) *$\mathcal{Q}$-split* if $A|B \in \Sigma^*(X)$ ($A|B \in \Sigma^*_{\text{part}}(X)$) and $\mathcal{Q}(A|B) \subseteq \mathcal{Q}$ holds, we denote the collection of all partial $\mathcal{Q}$-splits by $\widehat{\mathcal{Q}}$ and its subset consisting of all $\mathcal{Q}$-splits by $\Sigma(\mathcal{Q})$, we will say that two elements a, $b \in X$ form a *$\mathcal{Q}$-cherry* if

$$S = S_{ab} := \{a, b\}|(X - \{a, b\}) \in \Sigma(\mathcal{Q})$$

holds, and we denote, for every subset Y of X, by $\mathcal{Q}|_Y$ the set of all quartets in a quartet system $\mathcal{Q} \subseteq \mathcal{Q}(X)$ contained in $\mathcal{Q}(Y)$. We also recall that $\mathcal{Q}$ is called thin if $|\mathcal{Q}|_Y| \leq 1$ holds for every 4-subset Y of X, we define $\mathcal{Q}$ to be a *cover* (for X) if, conversely, $|\mathcal{Q}|_Y| \geq 1$ holds for every 4-subset Y of X, and we define $\mathcal{Q}$ to be a *simple* or, respectively, a *double* cover if $|\mathcal{Q}|_Y| = 1$ or, respectively, $|\mathcal{Q}|_Y| = 2$ holds for every 4-subset Y of X.

Note that in case $\mathcal{Q} := \mathcal{Q}_\mathcal{T}$ and $\Sigma := \Sigma_\mathcal{T} \cap \Sigma^*(X)$ holds for some phylogenetic X-tree $\mathcal{T}$, we have $\mathcal{Q}(\Sigma) = \mathcal{Q}$ because any quartet in $\mathcal{Q}$ extends to a split in $\Sigma_\mathcal{T}$. And we have $\Sigma(\mathcal{Q}) = \Sigma$ because $\Sigma(\mathcal{Q})$ is compatible implying that, in view of Lemma 7.2, there exists a quartet q in $\mathcal{Q}(S)$ for every split $S \in \Sigma(\mathcal{Q})$ such that S is the only split in $\Sigma(\mathcal{Q})$ with $q \in \mathcal{Q}(S)$. Together, this implies that also $\Sigma(\mathcal{Q}(\Sigma)) = \Sigma$ and $\mathcal{Q}(\Sigma(\mathcal{Q})) = \mathcal{Q}$ must hold in this case. In particular, two elements $a, b \in X$ form a $\mathcal{Q}_\mathcal{T}$-cherry if and only if they form a cherry in $\mathcal{T}$ and the vertex $v_\mathcal{T}(a) = v_\mathcal{T}(b)$ has degree 3.

More generally, Lemma 7.2 implies that every partial $\mathcal{Q}$-split $A|B \in \widehat{\mathcal{Q}}$ extends to a split in Σ (and, hence, is a split in Σ if it is a $\mathcal{Q}$-split): Indeed, choosing any quartet q in $\mathcal{Q}(A|B)$ such that $A|B$ is the only $A \cup B$-split in $\Sigma(\mathcal{Q}|_{A\cup B})$ with $q \in \mathcal{Q}(A|B)$, one must have $A|B \preceq S$ for every split $S \in \Sigma$ with $q \in \mathcal{Q}(S)$.

These observations suggest studying, in analogy to the well-known correspondence between subsets of groups and fields in Galois theory, the following *Galois connection* between arbitrary subsets of $\mathcal{Q}(X)$ and $\Sigma^*(X)$:

Theorem 8.1 *Given any split system $\Sigma \subseteq \Sigma^*(X)$ and any quartet system $\mathcal{Q} \subseteq \mathcal{Q}(X)$, the following hold:*

(i) $\mathcal{Q}(\Sigma) \subseteq \mathcal{Q} \iff \Sigma \subseteq \Sigma(\mathcal{Q})$,
(ii) $\Sigma \subseteq \Sigma(\mathcal{Q}(\Sigma))$ *and* $\mathcal{Q}(\Sigma(\mathcal{Q})) \subseteq \mathcal{Q}$,
(iii) $\Sigma = \Sigma(\mathcal{Q}(\Sigma))$ *holds if and only if* Σ *is of the form* $\Sigma = \Sigma(\mathcal{Q}_0)$ *for some quartet system* $\mathcal{Q}_0 \subseteq \mathcal{Q}(X)$,
(iv) $\mathcal{Q} = \mathcal{Q}(\Sigma(\mathcal{Q}))$ *holds if and only if* $\mathcal{Q}$ *is of the form* $\mathcal{Q} = \mathcal{Q}(\Sigma_0)$ *for some split system* $\Sigma_0 \subseteq \Sigma^*(X)$.

Proof (i) is obvious and (ii) follows from (i) by putting $\mathcal{Q} := \mathcal{Q}(\Sigma)$ and $\Sigma := \Sigma(\mathcal{Q})$. (iii) and (iv) are well-known and easily established consequences of (i) and (ii) (see e.g., [70] for a short discussion of the general abstract context within the framework of partially ordered sets): If $\Sigma = \Sigma(\mathcal{Q}(\Sigma))$ holds, we have, of course, $\Sigma = \Sigma(\mathcal{Q}_0)$ for $\mathcal{Q}_0 := \mathcal{Q}(\Sigma)$. Conversely, if $\Sigma = \Sigma(\mathcal{Q}_0)$ holds for some quartet system $\mathcal{Q}_0 \subseteq \mathcal{Q}(X)$, we have $\mathcal{Q}(\Sigma) = \mathcal{Q}(\Sigma(\mathcal{Q}_0)) \subseteq \mathcal{Q}_0$ and, hence, $\Sigma \subseteq \Sigma(\mathcal{Q}(\Sigma)) \subseteq \Sigma(\mathcal{Q}_0) = \Sigma$, as required. Similarly, if $\mathcal{Q} = \mathcal{Q}(\Sigma(\mathcal{Q}))$ holds for some quartet system $\mathcal{Q} \subseteq \mathcal{Q}(X)$, then $\mathcal{Q} = \mathcal{Q}(\Sigma_0)$ holds for the split system $\Sigma_0 := \Sigma(\mathcal{Q})$. And if there exists some split system $\Sigma_0 \subseteq \Sigma^*(X)$ with $\mathcal{Q} = \mathcal{Q}(\Sigma_0)$, then we have $\Sigma_0 \subseteq \Sigma(\mathcal{Q}(\Sigma_0)) = \Sigma(\mathcal{Q})$ and, hence, $\mathcal{Q} = \mathcal{Q}(\Sigma_0) \subseteq \mathcal{Q}(\Sigma(\mathcal{Q})) \subseteq \mathcal{Q}$, again as required. □

Particular instances of this correspondence are given in the next theorem:

Theorem 8.2 *Given, as above, a split system $\Sigma \subseteq \Sigma^*(X)$ and a quartet system $\mathcal{Q} \subseteq \mathcal{Q}(X)$, the following hold:*

(i) *We have $\Sigma = \Sigma(\mathcal{Q}(\Sigma))$ whenever the following three-*split condition *holds:*

(3S) *There exists, for any three splits $A_1|B_1,\ A_2|B_2,\ A_3|B_3 \in \Sigma$ with $|A_1 \cap A_2 \cap A_3|,\ |(B_1 \cap B_2) \cup (B_1 \cap B_3) \cup (B_2 \cap B_3)| \geq 2$, some split $A|B \in \Sigma$ with $A_1 \cap A_2 \cap A_3 \subseteq A$ and $(B_1 \cap B_2) \cup (B_1 \cap B_3) \cup (B_2 \cap B_3) \subseteq B$.*

More precisely, this assertion holds if and only if every partial $\mathcal{Q}(\Sigma)$-split $A|B$ extends to a split $S \in \Sigma$ (and, hence, belongs to Σ in case the partial $\mathcal{Q}(\Sigma)$-split $A|B$ is actually a split).

(ii) *We have $\mathcal{Q} = \mathcal{Q}(\Sigma(\mathcal{Q}))$ whenever the following* local split-extendability condition *holds:*

(LSE) *For every partial $\mathcal{Q}$-split $A|B \in \widehat{\mathcal{Q}}$ with $|A|,\ |B| \leq 3$ and for all $x \in X$, either $\mathcal{Q}(x + A|B)$ or $\mathcal{Q}(A|x + B)$ is a partial $\mathcal{Q}$-split.*

More precisely, this assertion holds if and only if every partial $\mathcal{Q}$-split extends to a $\mathcal{Q}$-split.

(iii) *Moreover, if $\mathcal{Q} = \mathcal{Q}(\Sigma)$ and $\Sigma = \Sigma(\mathcal{Q})$ holds, then Σ satisfies the three-split condition if and only if $\mathcal{Q}$ satisfies the local split-extendability condition in which case Σ is the only split system contained in $\Sigma^*(X)$ with $\mathcal{Q} = \mathcal{Q}(\Sigma)$ satisfying the three-split condition, and $\mathcal{Q}$ is the only quartet system with $\Sigma = \Sigma(\mathcal{Q})$ satisfying the local split-extendability condition.*

Proof (i) Clearly, if every partial $\mathcal{Q}(\Sigma)$-split extends to a $\mathcal{Q}(\Sigma)$-split, then, given any three splits $A_1|B_1,\ A_2|B_2,\ A_3|B_3 \in \Sigma$, this holds in particular for the partial split $A_0|B_0$ defined by $A_0 := A_1 \cap A_2 \cap A_3$ and $B_0 := (B_1 \cap B_2) \cup (B_1 \cap B_3) \cup (B_2 \cap B_3)$ in case $|A_0|, |B_0| \geq 2$ as $aa'|bb' \in \mathcal{Q}(A_0|B_0)$ implies $aa'|bb' \in \mathcal{Q}(A_1|B_1)$ or $aa'|bb' \in \mathcal{Q}(A_2|B_2)$ or $aa'|bb' \in \mathcal{Q}(A_3|B_3)$. Conversely, if Assertion (3S) holds and if $\mathcal{Q}(A_0|B_0) \subseteq \mathcal{Q}(\Sigma)$ holds for some partial split $A_0|B_0 \in \Sigma^*_{\text{part}}(X)$, then $\Sigma(A_0|B_0) \cap \Sigma \neq \emptyset$ must hold: Indeed, otherwise

(a) choose an inclusion-minimal pair A', B' of disjoint subsets of X with $A' \subseteq A_0$, $B' \subseteq B_0$, $|A'|, |B'| \geq 2$, and $\Sigma(A'|B') \cap \Sigma = \emptyset$,
(b) note that, in view of $\mathcal{Q}(A_0|B_0) \subseteq \mathcal{Q}(\Sigma)$, we may assume that, say, $|B'| \geq 3$ holds,
(c) choose three distinct elements $b_1, b_2, b_3 \in B'$,
(d) note that, for each $i = 1, 2, 3$, there must exist a split $S_i = A_i|B_i$ with $A' \subseteq A_i$ and $B' - b_i \subseteq B_i$ and, hence, $b_i + A' \subseteq A_i$ and $B' - b_i \subseteq B_i$,
(e) and that, in consequence, there must exist some split $A|B \in \Sigma$ with $A_1 \cap A_2 \cap A_3 \subseteq A$ and $(B_1 \cap B_2) \cup (B_1 \cap B_3) \cup (B_2 \cap B_3) \subseteq B$ and, hence, with $A' \subseteq A_1 \cap A_2 \cap A_3 \subseteq A$ and $B' \subseteq (B_1 \cap B_2) \cup (B_1 \cap B_3) \cup (B_2 \cap B_3) \subseteq B$ — the latter in view of $B' - \{b_1, b_2, b_3\} \subseteq B_1 \cap B_2 \cap B_3$ and $b_3 \in B_1 \cap B_2$, $b_1 \in B_2 \cap B_3$, and $b_2 \in B_1 \cap B_3$ — clearly in contradiction to $\Sigma(A'|B') \cap \Sigma = \emptyset$.

(ii) It is obvious that a quartet system $\mathcal{Q} \subseteq \mathcal{Q}(X)$ satisfies Assertion (LSE) whenever every partial $\mathcal{Q}$-split extends to a $\mathcal{Q}$-split. Conversely, if $\mathcal{Q}$ satisfies Assertion (LSE) and $A_0|B_0$ is a partial $\mathcal{Q}$-split, we have to show that $A_0|B_0$ extends to a $\mathcal{Q}$-split. So, assume that $A|B$ is a maximal partial $\mathcal{Q}$-split with $A_0 \subseteq A$, $B_0 \subseteq B$. If there were some $x \in X - (A \cup B)$, there would exist some $a_1 \in A$ and some $b_1, b_2 \in B$ with $a_1x|b_1b_2 \notin \mathcal{Q}$ implying, in view of Assertion (LSE), that $\mathcal{Q}(a_1aa'|xb_1b_2b) \subseteq \mathcal{Q}$ must hold for all $a, a' \in A$ and $b \in B$ and that, therefore, $\mathcal{Q}(A|x + B) \subseteq \mathcal{Q}$ would hold, a contradiction.

(iii) Finally, if $\mathcal{Q} = \mathcal{Q}(\Sigma)$ and $\Sigma = \Sigma(\mathcal{Q})$ holds for some quartet system $\mathcal{Q} \subseteq \mathcal{Q}(X)$ and some split system $\Sigma \subseteq \Sigma^*(X)$, then Σ satisfies Assertion (3S) if and only if every partial $\mathcal{Q}$-split extends to a $\mathcal{Q}$-split if and only if $\mathcal{Q}$ satisfies Assertion (LSE). In this case, Σ is clearly the only split system contained in $\Sigma^*(X)$ with $\mathcal{Q} = \mathcal{Q}(\Sigma)$ satisfying Assertion (3S) as $\Sigma' = \Sigma(\mathcal{Q}) = \Sigma$ must hold for any such split system Σ', and $\mathcal{Q}$ is the only quartet system with $\Sigma = \Sigma(\mathcal{Q})$ satisfying Assertion (LSE) as $\mathcal{Q}' = \mathcal{Q}(\Sigma) = \mathcal{Q}$ must hold for any such quartet system $\mathcal{Q}'$. □

It follows that there exists a polynomial time algorithm to decide whether, for a given quartet system $\mathcal{Q} \subseteq \mathcal{Q}(X)$, there is a split system $\Sigma \subseteq \Sigma^*(X)$ with $\mathcal{Q} = \mathcal{Q}(\Sigma)$ satisfying Assertion (3S) as such quartet systems are characterized by a 7-point condition, that is, we just need to check whether a certain condition holds for every 7-subset of X. This is not immediately clear from just Assertion (3S) since Σ can contain an exponential number of splits. For example, if $\mathcal{Q} := \mathcal{Q}(X)$, then $\mathcal{Q} = \mathcal{Q}(\Sigma)$ holds for the split system $\Sigma := \Sigma^*(X)$ which clearly satisfies this condition, yet contains $2^{n-1} - n - 1$ splits. This example also demonstrates that, in case $n > 4$, such quartet systems can be induced by more than one split system: Indeed, given any element $x \in X$, we also have $\mathcal{Q}(X) = \mathcal{Q}(\Sigma^x)$ for the split system $\Sigma^x := \{A|B \in \Sigma(X) : |A| = 2, x \in B\}$.

We now show that, for quartet systems induced by weakly compatible split systems, a 6-point condition in fact suffices:

Theorem 8.3 *One has* $\Sigma = \Sigma(\mathcal{Q}(\Sigma))$ *for every weakly compatible split system* $\Sigma \subseteq \Sigma^*(X)$. *Moreover, a quartet system* $\mathcal{Q}$ *is of the form* $\mathcal{Q} = \mathcal{Q}(\Sigma)$ *for some weakly compatible split system* $\Sigma \subseteq \Sigma^*(X)$ *if and only if it satisfies Assertion* (LSE) *and* $|\mathcal{Q}|_Y| \leq 2$ *holds for every* 4*-subset* Y *of* X *if and only if the latter holds and* $\mathcal{Q}$ *satisfies the following* very local split-extendability condition:

(VLSE) *For every partial* $\mathcal{Q}$*-split* $A|B \in \widehat{\mathcal{Q}}$ *with* $|A|+|B| \leq 5$ *and all* $x \in X$, *either* $\mathcal{Q}(x + A|B)$ *or* $\mathcal{Q}(A|x + B)$ *is a partial* $\mathcal{Q}$*-split.*

in which case also $\Sigma = \Sigma(\mathcal{Q})$ *must, of course, hold for that weakly compatible split system* $\Sigma \subseteq \Sigma^*(X)$ *with* $\mathcal{Q} = \mathcal{Q}(\Sigma)$.

Proof For the first fact, we present two distinct proofs:

(1) By definition, a split system $\Sigma \subseteq \Sigma^*(X)$ is weakly compatible if and only if the set $\mathcal{Q}(\Sigma)|_Y$ contains, for every 4-subset Y of X, at most two distinct quartets.

Thus, if $\mathcal{Q}$ is any quartet system $\mathcal{Q} \subseteq \mathcal{Q}(X)$ with $|\mathcal{Q}|_Y| \leq 2$ for every 4-subset Y of X, the associated split system $\Sigma(\mathcal{Q})$ must be weakly compatible in view of Theorem 8.1(ii). Thus, given any weakly compatible split system $\Sigma \subseteq \Sigma^*(X)$, the union Σ' of $\Sigma(\mathcal{Q}(\Sigma))$ and all trivial splits of X must also be weakly compatible. So, we may invoke Lemma 7.10, to conclude that, given any split $S = A|B \in \Sigma(\mathcal{Q}(\Sigma))$, there must exist some $a, a' \in A$ and $b, b' \in B$ with $\Sigma(aa'|bb') \cap \Sigma' = \{S\}$ and, therefore, also $a \neq a'$, $b \neq b'$, and $aa'|bb' \in \mathcal{Q}(S) \subseteq \mathcal{Q}(\Sigma) = \bigcup_{S' \in \Sigma} \mathcal{Q}(S')$. So, $S \in \Sigma$ must hold as S is the only split in Σ' with $aa'|bb' \in \mathcal{Q}(S)$.

(2) Alternatively, we may note that every weakly compatible split system $\Sigma \subseteq \Sigma^*(X)$ satisfies Assertion (3S): Indeed, if $A_1|B_1$, $A_2|B_2$, $A_3|B_3$ are any three distinct splits in such a split system Σ with $A_1 \cap A_2 \cap A_3 \neq \emptyset$, then $(B_1 \cap B_2) \cup (B_1 \cap B_3) \cup (B_2 \cap B_3)$ must be contained in either B_1, B_2, or B_3 as

$A_1 \cap A_2 \cap A_3 \neq \emptyset$ implies that either $A_1 \cap B_2 \cap B_3 = \emptyset$ or $B_1 \cap A_2 \cap B_3 = \emptyset$ or $B_1 \cap B_2 \cap A_3 = \emptyset$ holds.

To establish the remaining claims, assume that $\mathcal{Q} = \mathcal{Q}(\Sigma)$ holds for some weakly compatible split system $\Sigma \subseteq \Sigma^*(X)$. Then, as noted in (1) above, $|\mathcal{Q}|_Y| \leq 2$ must hold for every 4-subset Y and $\mathcal{Q}$ must satisfy Assertion (LSE) since, as noted in (2), Σ satisfies Assertion (3S).

So, it suffices to note that a quartet system $\mathcal{Q}$ that satisfies Assertion (VLSE) and for which $|\mathcal{Q}|_Y| \leq 2$ holds for every 4-subset Y satisfies also Assertion (LSE): Otherwise, there would exist some partial $\mathcal{Q}$-split $A|B$ with $|A| = |B| = 3$, some $x \in X - (A \cup B)$, and elements $a, a_1, a_2 \in A$ and $b, b_1, b_2 \in B$ with $a_1 \neq a_2$, $b_1 \neq b_2$, $xa|b_1b_2 \notin \mathcal{Q}$, and $a_1a_2|xb \notin \mathcal{Q}$ in which case, in view of our assumption that Assertion (VLSE) holds, also $A = \{a, a_1, a_2\}$ and $B = \{b, b_1, b_2\}$ must hold. This implies further that also the two sets $\mathcal{Q}(xaa_1a_2|bb_1)$ and $\mathcal{Q}(xaa_1a_2|bb_2)$ must be contained in $\mathcal{Q}$ (as $a_1a_2|xb \notin \mathcal{Q}$ holds) and that $\mathcal{Q}(aa_1a_2|xb_1b_2)$ must also be contained in $\mathcal{Q}$ (as $ax|b_1b_2 \notin \mathcal{Q}$ holds) which, in turn, implies also that $\mathcal{Q}(xaa_1b_2|bb_1) \subseteq \mathcal{Q}$ and $\mathcal{Q}(xaa_1b_1|bb_2) \subseteq \mathcal{Q}$ must hold (as $\mathcal{Q}(xaa_1|bb_1b_2) \subseteq \mathcal{Q}$ does not hold) and that $\mathcal{Q}(ba_1a_2|xb_1b_2)$ must hold (as $\mathcal{Q}(a_1a_2|bxb_1b_2) \subseteq \mathcal{Q}$ does not hold).

So, we would have $a_1b_2|bb_1, a_1b_1|bb_2, a_1b|b_1b_2 \in \mathcal{Q}$, in contradiction to our assumption that $|\mathcal{Q}|_Y| \leq 2$ holds for any 4-subset Y and, hence, in particular for $Y := \{a_1, b, b_1, b_2\}$. □

Note that some 6-point condition is necessary for characterizing quartet systems $\mathcal{Q}$ that are of the form $\mathcal{Q}(\Sigma)$ for some weakly compatible split system Σ. For example, consider the quartet system $\mathcal{Q} \subseteq \mathcal{Q}(X)$ for $X := \{a, a', b, b', c, c'\}$ that is the union of the following quartet systems:

$$\mathcal{Q}(ab|a'b'cc'),\quad \mathcal{Q}(ab'|a'bcc'),\quad \mathcal{Q}(a'b|ab'cc'),\quad \mathcal{Q}(a'b'|abcc'),$$
$$\mathcal{Q}(cc'|aa'b),\quad \mathcal{Q}(cc'|aa'b').$$

Then, it is not hard to check that, for each $x \in X$, the collection $\mathcal{Q}|_{X-x}$ is induced by a weakly compatible split system $\Sigma^{(x)}$ of $(X - x)$-splits while the collection of all $\mathcal{Q}$-splits is not induced by any weakly compatible split system Σ of X-splits. Indeed, by symmetry, it suffices to check the first claim for $x := a$, b, or c. In the first case, we have $\mathcal{Q}|_{X-a} = \mathcal{Q}(\Sigma^{(a)})$ for $\Sigma^{(a)} := \{a'b|b'cc', a'b'|bcc'\}$. In the second case, we have $\mathcal{Q}|_{X-b} = \mathcal{Q}(\Sigma^{(b)})$ for $\Sigma^{(b)} := \{ab'|a'cc', a'b'|acc', aa'b'|cc'\}$. And in the third case, we have $\mathcal{Q}|_{X-c} = \mathcal{Q}(\Sigma^{(c)})$ for $\Sigma^{(c)} := \{ab|a'b'c', ab'|a'bc', a'b|ab'c', a'b'|abc'\}$. And there can be no weakly compatible split system Σ of X-splits with $\mathcal{Q} = \mathcal{Q}(\Sigma)$ as $\mathcal{Q}$ does not even satisfy Assertion (VLSE): The split $cc'|aa'b$ can neither be extended to $b'cc'|aa'b$ (because of $b'c|aa' \notin \mathcal{Q}$) nor to $cc'|aa'bb'$ (because of $cc'|bb' \notin \mathcal{Q}$).

Note also that Assertion (LSE) need not hold for the partial $\mathcal{Q}$-splits associated with a quartet system $\mathcal{Q}$ of the form $\mathcal{Q} = \mathcal{Q}(\Sigma)$ for some split system Σ that satisfies $\Sigma = \Sigma(\mathcal{Q}(\Sigma))$: For example, consider $X := \{a, b, c, c', d, d'\}$ and the split system $\Sigma \subseteq \Sigma(X)$ consisting of the three splits $ac|bc'dd'$, $ac'|bcdd'$, and $add'|bcc'$, and put $\mathcal{Q} := \mathcal{Q}(\Sigma)$. Then, $\Sigma = \Sigma(\mathcal{Q})$ must hold: Indeed, every $\mathcal{Q}$-split S must be of the form $S = A|B$ with $a \in A$ and $b \in B$ because $\mathcal{Q}$ does not contain any quartet of the form $ab|xy$, $B \cap \{c, c'\}$ cannot be empty as $\mathcal{Q}$ does not contain any quartet of the form $xb|cc'$ and $d \in B \iff d' \in B$ must hold as $\mathcal{Q}$ does not contain any quartet of the form $bd|d'x$. However, the partial $\mathcal{Q}$-split $acc'|dd'$ can clearly not be extended to a $\mathcal{Q}$-split of X.

Finally note that a split system $\Sigma \subseteq \Sigma^*(X)$ that satisfies Assertion (3S) does not need to be weakly compatible: For example, the split system $\Sigma^*(X)$ consisting of all non-trivial splits of X is not weakly compatible, but clearly satisfies Assertion (3S).

8.2 A map from quartets to metrics

In the previous chapter, we implicitly discussed some maps that associate a quartet system to a metric (for example, for D a metric, the quartet system consisting of those quartets $q = ab|cd$ with $D_+(\{a, b\} : \{c, d\}) > 0$). We will now describe a map in the reverse direction that allows one to go directly from quartet systems to symmetric bivariate maps.

To this end, we define for any quartet system $\mathcal{Q} \subseteq \mathcal{Q}(X)$ and any 2-subset $Y = \{a, b\}$ of X, the graph (cf. Figure 8.1)

$$G_{(\mathcal{Q},Y)} := (X, E_{(\mathcal{Q},Y)})$$

with vertex set X and edge set

$$E_{(\mathcal{Q},Y)} := \left\{\{c, d\} \in \binom{X}{2} : \{c, d\} = Y \text{ or } \{c, d\} \cap Y = \emptyset \text{ and } ac|bd,\ ad|bc \notin \mathcal{Q}\right\},$$

and consider the map

$$D = D_{\mathcal{Q}} : X \times X \to \mathbb{R}_{\geq 0} : (x, y) \mapsto \begin{cases} |\pi_0(G_{(\mathcal{Q},\{x,y\})})| & \text{if } x \neq y, \\ 0 & \text{otherwise.} \end{cases}$$

Clearly, the map $D_{\mathcal{Q}}$ is always symmetric and $D_{\mathcal{Q}}(x, x) = 0$ holds, by definition, for every $x \in X$. However, $D_{\mathcal{Q}}$ need not be a metric since the

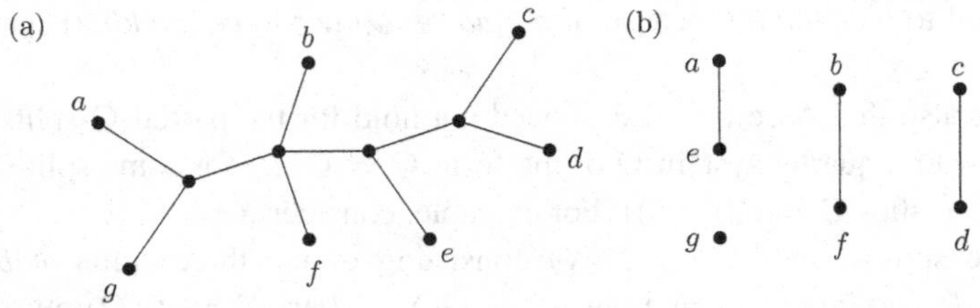

Figure 8.1 (a) A phylogenetic $\{a, b, \ldots, g\}$-tree $\mathcal{T}$ with a vertex of degree 4. (b) The graph $G_{(\mathcal{Q},\{a,e\})}$ for the quartet system $\mathcal{Q} := \mathcal{Q}_{\mathcal{T}}$.

triangle inequality need not hold. For example, given any graph $G = (X, E)$ with vertex set X and edge set E, define $\mathcal{Q} = \mathcal{Q}_G$ to be the set of those quartets $ab|cd \in \mathcal{Q}(X)$ for which $\{a, b\}$ and $\{c, d\}$ are disjoint edges in E. Then, given any two disjoint 2-subsets $\{a, b\}, \{c, d\} \in \binom{X}{2}$, one has

$$\{c, d\} \in E_{(\mathcal{Q}_G,\{a,b\})} \iff ac|bd, ad|bc \notin \mathcal{Q}_G$$

and, therefore,

$$\{c, d\} \notin E_{(\mathcal{Q}_G,\{a,b\})} \iff \{a, c\}, \{b, d\} \in E \quad \text{or} \quad \{a, d\}, \{b, c\} \in E$$

or, equivalently,

$$\{c, d\} \notin E_{(\mathcal{Q}_G,\{a,b\})} \iff c \in N_G(a) \,\&\, d \in N_G(b) \quad \text{or} \quad d \in N_G(a) \,\&\, c \in N_G(b).$$

In consequence, we have $\{c, d\} \in E_{(\mathcal{Q}_G,\{a,b\})}$ in that case if and only if either $c, d \notin N_G(a)$ or $c, d \notin N_G(b)$ or $c \notin N_G(a) \cup N_G(b)$ or $d \notin N_G(a) \cup N_G(b)$ holds. So, given any element $c \in X - \{a, b\}$ that is neither contained in $N_G(a)$ nor in $N_G(b)$, one has $\{c, d\} \in E_{(\mathcal{Q}_G,\{a,b\})}$ for every $d \in X - \{a, b, c\}$ implying that $D_{\mathcal{Q}_G}(a, b) = 2$ holds for any two distinct elements $a, b \in X$ with $X \neq \{a, b\} \cup N_G(a) \cup N_G(b)$.

Furthermore, if $X = \{a, b\} \cup N_G(a) \cup N_G(b)$ holds, the above analysis implies that $\{c, d\} \in E_{(\mathcal{Q}_G,\{a,b\})}$ holds for some 2-subset $\{c, d\}$ in $\binom{X-\{a,b\}}{2}$ if and only if $c, d \notin N_G(a)$ or $c, d \notin N_G(b)$ holds. Thus, putting

$$N_G(a \backslash b) := N_G(a) - (N_G(b) + b) \text{ and } N_G(b \backslash a) := N_G(b) - (N_G(a) + a),$$

we have

$$E_{(\mathcal{Q}_G,\{a,b\})} = \{\{a, b\}\} \cup \binom{N_G(a \backslash b)}{2} \cup \binom{N_G(b \backslash a)}{2}$$

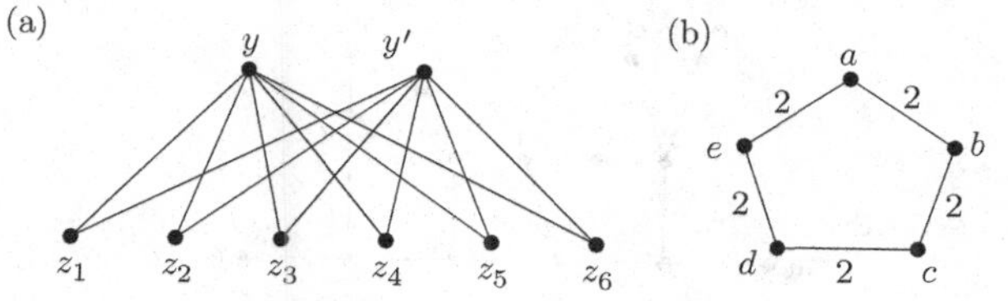

Figure 8.2 (a) Based on this graph, we can define a quartet system $\mathcal{Q}$ such that $D_{\mathcal{Q}}$ is not a metric. (b) The metric D induced by this edge-weighted graph is not treelike, but $D = D_{\mathcal{Q}_D}$ holds.

and, therefore

$$D_{\mathcal{Q}_G}(a, b) = 1 + |N_G(a) \cap N_G(b)| + \begin{cases} 0 & \text{if } N_G(a \backslash b) = N_G(b \backslash a) = \emptyset, \\ 2 & \text{if } N_G(a \backslash b), N_G(b \backslash a) \neq \emptyset, \\ 1 & \text{otherwise}, \end{cases}$$

in this case.

In particular, if — for example — X is the disjoint union of a 2-set Y and a k-set Z with $k \geq 3$, and $G = (X, E)$ is the complete bipartite graph with edge set $E := \{\{y, z\} : y \in Y, z \in Z\}$ (see e.g., Figure 8.2(a)), then $D_{\mathcal{Q}_G}(y, y') = 1 + k$ holds for the two distinct elements $y, y' \in Y$, and $D_{\mathcal{Q}_G}(y, z) = D_{\mathcal{Q}_G}(y', z) = 3$ holds for every $z \in Z$. So, $D_{\mathcal{Q}_G}$ does not satisfy the triangle inequality for every $k \geq 6$.

Even though $D_{\mathcal{Q}}$ is not always a metric, we now show that, for quartet systems arising from (unweighted) phylogenetic X-trees, we do get back the corresponding treelike metric. Indeed, defining the quartet system $\mathcal{Q}_D$ for a metric D by

$$\mathcal{Q}_D := \left\{ ab|cd \in \mathcal{Q}(X) : D(a, b) + D(c, d) < \min \left\{ \begin{array}{l} D(a, c) + D(b, d) \\ D(a, d) + D(b, c) \end{array} \right\} \right\},$$

we in fact obtain a bijective correspondence between quartet systems and metrics that arise from phylogenetic X-trees:

Theorem 8.4 *Suppose that $\mathcal{T} = (V, E)$ is a simple phylogenetic X-tree, and put $\mathcal{Q} := \mathcal{Q}_{\mathcal{T}}$ and $D := D_{\mathcal{T}}$. Then the following four assertions hold:*

(i) $D_{\mathcal{Q}} = D$, (ii) $\mathcal{Q}_D = \mathcal{Q}$, (iii) $\mathcal{Q}_{D_{\mathcal{Q}}} = \mathcal{Q}$, and (iv) $D_{\mathcal{Q}_D} = D$.

Proof (i) Let a and b be two arbitrary distinct elements in X. For every interior vertex v of $\mathcal{T}$ on the path $\mathbf{p}_{\mathcal{T}}(a, b)$ let X_v denote the set of those elements $x \in X$ for which the path $\mathbf{p}_{\mathcal{T}}(v, x)$ has no edge in common with $\mathbf{p}_{\mathcal{T}}(a, b)$. In Figure 8.3, for example, we have $X_v = \{c_6, c_7, c_8\}$. Note that, for any two

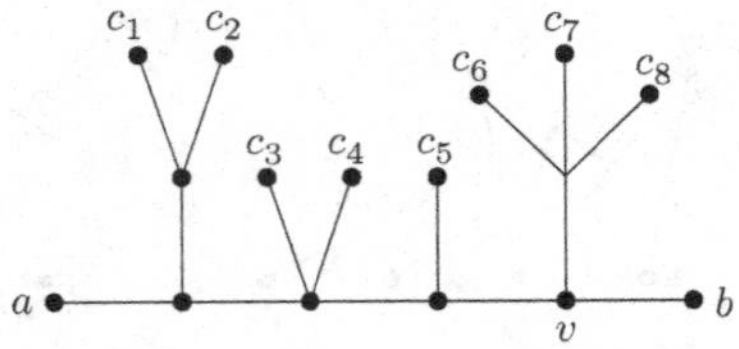

Figure 8.3 A phylogenetic X-tree $\mathcal{T}$ on $X = \{a, b, c_1, \ldots, c_8\}$. The subtrees branching off the path $\mathbf{p}_{\mathcal{T}}(a, b)$ are reminiscent of the trees in an apple orchard (see also [131, p. 12]).

distinct elements $c, c' \in X - \{a, b\}$, $\{c, c'\}$ is an edge of the graph $G_{(\mathcal{Q},\{a,b\})}$ if and only if $\{c, c'\} \subseteq X_v$ holds for some interior vertex v of $\mathcal{T}$ on the path $\mathbf{p}_{\mathcal{T}}(a, b)$. Hence, $G_{(\mathcal{Q},\{a,b\})}$ is the disjoint union of the cliques with vertex sets $\{a, b\}$ and X_v, v an interior vertex on $\mathbf{p}_{\mathcal{T}}(a, b)$. So, in particular, the number of connected components of $G_{(\mathcal{Q},\{a,b\})}$ equals the number of edges on $\mathbf{p}_{\mathcal{T}}(a, b)$, implying that $D_{\mathcal{Q}}(a, b) = D_{\mathcal{T}}(a, b)$ holds, as required.

(ii) Given any quartet $ab|cd \in \mathcal{Q}(X)$, we have $ab|cd \in \mathcal{Q}_{\mathcal{T}}$ if and only if the paths $\mathbf{p}_{\mathcal{T}}(a, b)$ and $\mathbf{p}_{\mathcal{T}}(c, d)$ have no vertex in common, which is the case if and only if $D(a, b) + D(c, d) < D(a, c) + D(b, d) = D(a, d) + D(b, c)$ and, hence (as $D = D_{\mathcal{T}}$ satisfies the 4-point condition), if and only if $ab|cd \in \mathcal{Q}_D$ holds.

The assertions (iii) and (iv) now follow immediately from (i) and (ii). □

It is obvious that $D = D_{\mathcal{Q}_D}$ does not hold for all metrics D. However, there are non-treelike metrics for which this last equality does hold. For example, consider the metric D induced by the edge-weighted graph in Figure 8.2(b) on its vertex set $X = \{a, b, c, d, e\}$. It is straightforward to check that $D_{\mathcal{Q}_D} = D$ holds. It is thus of interest to ask for which metrics D the equality $D = D_{\mathcal{Q}_D}$ holds. It seems to be difficult to answer this question in general.

8.3 Transitive quartet systems

Next, recall that a quartet system $\mathcal{Q} \subseteq \mathcal{Q}(X)$ has been dubbed "transitive" if $ab|ce \in \mathcal{Q}$ holds for any five distinct elements $a, b, c, d, e \in X$ with $ab|cd$, $ab|de \in \mathcal{Q}$ or, equivalently, if the connected components of $G_{(\mathcal{Q},\{a,b\})}$ are all cliques, for any two distinct elements $a, b \in X$.

In this section, we will present some intriguing results regarding transitive simple and double covers. Clearly, if a, b, c, d, e are five distinct elements in X and if $\mathcal{Q} \subseteq \mathcal{Q}(X)$ is any transitive cover, then

$$abc|de \in \widehat{\mathcal{Q}} \Rightarrow ab|cde \in \widehat{\mathcal{Q}} \text{ or } ac|bde \in \widehat{\mathcal{Q}} \text{ or } cb|ade \in \widehat{\mathcal{Q}} \quad (8.1)$$

holds as $ab|cd \in \mathcal{Q}$ or $ac|bd \in \mathcal{Q}$ or $bc|ad \in \mathcal{Q}$ must hold. We will write $ab|c|de \in \widehat{\mathcal{Q}}$ if $abc|de, ab|cde \in \widehat{\mathcal{Q}}$ holds.

Next, we present a result from [10] (see also [88]):

Theorem 8.5 *Given any quartet system $\mathcal{Q} \subseteq \mathcal{Q}(X)$, the following assertions are equivalent:*

(i) *There exists some circular split system $\Sigma \subseteq \Sigma^*(X)$ of maximal cardinality $\binom{n}{2} - n = \frac{n(n-3)}{2}$ with $\mathcal{Q} = \mathcal{Q}(\Sigma)$.*
(ii) *$\mathcal{Q}$ is a transitive double cover for X.*
(iii) *$\mathcal{Q} \subseteq \mathcal{Q}(X)$ is a double cover for X and there exists a linear order "$<$" on X such that $x_1x_2|x_3x_4, x_2x_3|x_4x_1 \in \mathcal{Q}$ holds for all $x_1, x_2, x_3, x_4 \in X$ with $x_1 < x_2 < x_3 < x_4$.*

Proof (i) $\Rightarrow$ (ii): That the quartet system $\mathcal{Q}(\Sigma)$ associated to a circular split system $\Sigma \subseteq \Sigma^*(X)$ of maximal cardinality is a transitive double cover can be checked easily by just restricting one's attention to the case $n = 5$.

(ii) $\Rightarrow$ (iii): Now, assume that $\mathcal{Q}$ is a transitive double cover. To find a linear order "$<$" on X such that $x_1x_2|x_3x_4, x_2x_3|x_4x_1 \in \mathcal{Q}$ holds for all $x_1, x_2, x_3, x_4 \in X$ with $x_1 < x_2 < x_3 < x_4$, we define, for any two distinct elements a, b in X, the binary relations "$<_{a\to b}$" and "$\sim_{a,b}$" on X by

$$x <_{a\to b} y \iff b \neq x \neq y \neq a \text{ and } x = a \text{ or } y = b \text{ or } ay|bx \notin \mathcal{Q} \quad (8.2)$$

and

$$x \sim_{a,b} y \iff x = y \text{ or } |\{x, y, a, b\}| = 4 \text{ and } ab|xy \in \mathcal{Q} \quad (8.3)$$

for all $x, y \in X$. Note first that the binary relation "$<_{a\to b}$" is a partial order, i.e., we have

$$x <_{a\to b} y \quad \text{and} \quad y <_{a\to b} z \Rightarrow x <_{a\to b} z \quad (8.4)$$

for all $x, y, z \in X$: This is obvious in case $x = a$ or $z = b$. And if x, y, z are any three distinct elements in $X - \{a, b\}$ and, say, $ax|yz \in \mathcal{Q}$ holds, our claim $az|bx \notin \mathcal{Q}$ follows from $ax|by, ax|yz, ab|xy, ab|yz \in \mathcal{Q}$ as this implies by transitivity that $ax|bz, ab|xz \in \mathcal{Q}$ holds. Otherwise, we must have $az|xy \in \mathcal{Q}$ in which case our claim follows, again by transitivity, from the fact that $y <_{a\to b} z$ implies $az|by \notin \mathcal{Q}$ and, hence, $az|bx \notin \mathcal{Q}$.

Furthermore, we have $x_1x_2|x_3x_4 \in \mathcal{Q}$ and $x_2x_3|x_4x_1 \in \mathcal{Q}$ for all $x_1, x_2, x_3, x_4 \in X$ with $x_1 <_{a\to b} x_2 <_{a\to b} x_3 <_{a\to b} x_4$: Indeed, if $x_1 = a$ and $x_4 = b$ hold, we have $ax_3|bx_2 \notin \mathcal{Q}$ and, therefore, $x_1x_2|x_3x_4 = ax_2|x_3b, x_2x_3|x_4x_1 = x_2x_3|ab \in \mathcal{Q}$. If, say, $x_1 = a$ and $x_4 \neq b$ hold, we have $x_1x_2|x_3b \in \mathcal{Q}$

and $x_1x_2|x_4b \in \mathcal{Q}$ and, therefore, $x_1x_2|x_3x_4 \in \mathcal{Q}$, and we have $ax_2|x_4b$, $ax_3|x_4b \in \mathcal{Q}$ and, therefore, $x_2x_3|x_4b \in \mathcal{Q}$ which — together with $x_2x_3|ab = x_2x_3|bx_1 \in \mathcal{Q}$ — implies that also $x_2x_3|x_4x_1 \in \mathcal{Q}$ holds. And if $x_1 \neq a$ and $x_4 \neq b$ hold, we have $ax_1|x_3x_4$, $ax_2|x_3x_4 \in \mathcal{Q}$ and, therefore, $x_1x_2|x_3x_4 \in \mathcal{Q}$, and we have $x_2x_3|x_4a$ and $ax_1|x_2x_3 \in \mathcal{Q}$ and, therefore, $x_2x_3|x_4x_1 \in \mathcal{Q}$ also in this case.

And restricting the binary relation "$<_{a\to b}$" to an equivalence class C of the equivalence relation $\sim_{a,b}$ contained in $X - \{a, b\}$ yields a linear order on $C \cup \{a, b\}$ as, given any two distinct $\sim_{a,b}$-equivalent elements x, y in $X - \{a, b\}$, we have $ax|by \in \mathcal{Q} \iff ay|bx \notin \mathcal{Q}$ and, therefore $x <_{a\to b} y \iff y \not<_{a\to b} x$.

Thus, it remains to observe that there exist always two distinct elements a', $b' \in X$ for which the elements in $X - \{a', b'\}$ form a single $\sim_{a',b'}$-equivalence class. However, if a, b, c, d are any four distinct elements in X, one clearly has

$$a \not\sim_{b,c} d \iff ab|dc, \quad ac|bd \in \mathcal{Q} \iff c \sim_{a,b} d \quad \text{and} \quad c <_{a\to b} d$$

implying in particular that for $a' := c$ and $b' := b$, all elements in $X - \{a', b'\}$ are $\sim_{a',b'}$-equivalent to a whenever c is a maximal element in $X - \{a, b\}$ relative to "$<_{a\to b}$". Altogether, these observations establish that (ii) implies (iii).

(iii) $\Rightarrow$ (i): This is obvious. □

Clearly, the result above implies that a split system $\Sigma \subseteq \Sigma^*(X)$ is a circular split system of maximal cardinality $\binom{n}{2} - n$ if and only if the restriction of Σ to every subset Y of cardinality at most 5 is a circular split system of maximal cardinality $|\binom{Y}{2}| - |Y| = 5$. And it is obvious that a split system $\Sigma \subseteq \Sigma^*(X)$ is weakly compatible if and only if the restriction of Σ to every subset Y of cardinality 4 is circular. So, one may ask whether there is some natural number $k \geq 5$ such that a split system $\Sigma \subseteq \Sigma^*(X)$ is circular if and only if the restriction of Σ to every subset Y of cardinality at most k is circular?

Thus, it is worth noting that no such k can exist and that, in consequence, there is also no finite list of "k-point conditions" that characterize those quartet systems $\mathcal{Q}$ with $\mathcal{Q} = \mathcal{Q}(\Sigma)$ for some arbitrary circular split system $\Sigma \subseteq \Sigma^*(X)$ (as $\Sigma = \Sigma(\mathcal{Q}(\Sigma))$ holds for every circular — and, thus, weakly compatible — split system).

Consider for example, for every $n \geq 4$, the split system $\Sigma^{(n)} \subseteq \Sigma(\langle n\rangle)$ consisting of the $n-2$ splits $\{j, j+1\}|\langle n\rangle - \{j, j+1\}$, $j \in \langle n-2\rangle$, and the split $\{1, n-1\}|\langle n\rangle - \{1, n-1\}$. An example of such a split system for $n = 7$ is given in Figure 8.4. It is not hard to see that none of these split systems $\Sigma^{(n)}$ is circular while, for every $x \in \langle n\rangle$, the restriction of $\Sigma^{(n)}$ to the subset $\langle n\rangle - x$ is circular (see [130], and also [86] for further details).

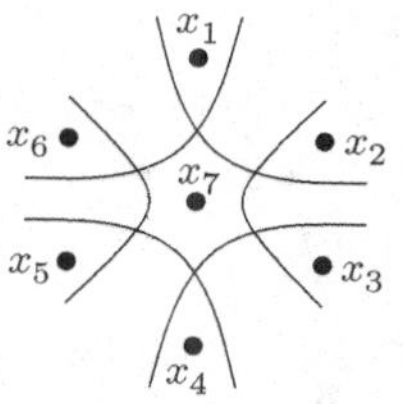

Figure 8.4 The split system Σ_7 described in the text. The splits in Σ_7 are indicated by curves.

We now turn our attention to simple transitive covers. First, given any two sets X and X', we define two quartet systems $\mathcal{Q} \subseteq \mathcal{Q}(X)$ and $\mathcal{Q}' \subseteq \mathcal{Q}(X')$ to be *isomorphic* and write $\mathcal{Q} \simeq \mathcal{Q}'$ if there exists a bijection $\psi : \bigcup supp(\mathcal{Q}) \to \bigcup supp(\mathcal{Q}')$ such that $\mathcal{Q}' = \psi(\mathcal{Q}) := \{\psi(a)\psi(b)|\psi(c)\psi(d) : ab|cd \in \mathcal{Q}\}$ holds.

Note next that, given any two distinct elements $a, b \in X$ and any transitive quartet system $\mathcal{Q} \subseteq \mathcal{Q}(X - a)$, there exists a unique minimal transitive extension $\mathcal{Q}^{b \leftarrow a}$ of $\mathcal{Q}$ for which the two elements a, b form a $\mathcal{Q}$-cherry, *viz.* the union of $\mathcal{Q}$, the set $\left\{ab|xy : \{x, y\} \in \binom{X-\{a,b\}}{2}\right\}$ (to make a, b form a $\mathcal{Q}^{b \leftarrow a}$-cherry), and the set $\{az|xy : x, y, z \in X - \{a, b\}, bz|xy \in \mathcal{Q}\}$ (to retain transitivity). Note also that $\mathcal{Q}^{b \leftarrow a}$ is thin or a simple transitive cover for X if and only if $\mathcal{Q}$ is thin or a simple transitive cover for $X - a$, respectively. In particular, whenever $\mathcal{Q} \subseteq \mathcal{Q}(X)$ is a thin and transitive quartet system, and some two elements $a, b \in X$ form a $\mathcal{Q}$-cherry, then $\mathcal{Q}$ must coincide with $(\mathcal{Q}|_{X-a})^{b \leftarrow a}$.

And recall that the split system $\Sigma(\mathcal{Q}) \subseteq \Sigma^*(X)$ associated with any thin quartet system $\mathcal{Q} \subseteq \mathcal{Q}(X)$ must be compatible. In consequence, there must exist, for every such quartet system $\mathcal{Q}$, some phylogenetic X-tree $\mathcal{T} = \mathcal{T}_{\mathcal{Q}}$ — uniquely determined by $\mathcal{Q}$ up to canonical isomorphism — such that $\Sigma^*_{\mathcal{T}} := \Sigma^*(X) \cap \Sigma_{\mathcal{T}}$ coincides with $\Sigma(\mathcal{Q})$ and, therefore, also $\mathcal{Q}(\Sigma(\mathcal{Q}))$ with $\mathcal{Q}(\Sigma^*_{\mathcal{T}}) = \mathcal{Q}_{\mathcal{T}}$. In particular, one has $\mathcal{Q} = \mathcal{Q}(\Sigma(\mathcal{Q}))$ for a thin quartet system $\mathcal{Q} \subseteq \mathcal{Q}(X)$ if and only if $\mathcal{Q} = \mathcal{Q}_{\mathcal{T}}$ holds for the phylogenetic X-tree $\mathcal{T}_{\mathcal{Q}}$ if and only if this holds for some appropriate phylogenetic X-tree $\mathcal{T}$ and hence, in view of Theorem 3.7, if and only if one has $\mathcal{Q}|_Y = \mathcal{Q}(\Sigma(\mathcal{Q}|_Y))$ for every 5-subset Y of X. And $\mathcal{Q} = \mathcal{Q}(\Sigma(\mathcal{Q}))$ holds for a simple cover $\mathcal{Q}$ for X if and only if $\mathcal{T}_{\mathcal{Q}}$ is a binary X-tree.

So, if $\mathcal{Q} \subseteq \mathcal{Q}(X)$ is thin and transitive, but not saturated, we can — in contrast to what we just learned about transitive double covers — neither expect that every quartet in $\mathcal{Q}$ can be extended to a split in $\Sigma(\mathcal{Q})$ nor that $\mathcal{T}_{\mathcal{Q}}$ is a binary X-tree — even if $\mathcal{Q}$ is not only thin, but a simple cover for X.

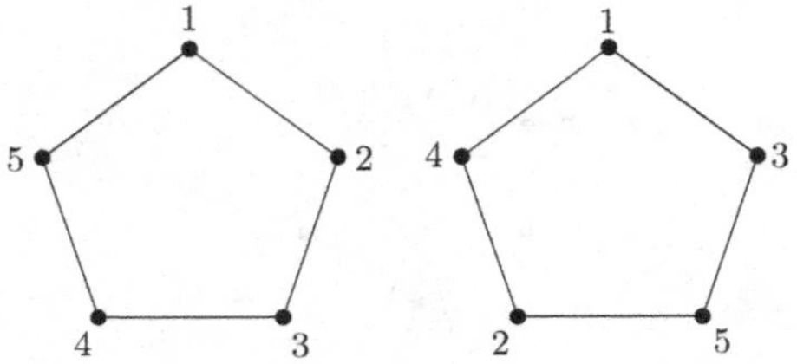

Figure 8.5 This figure illustrates geometric and group-theoretic features of the quartet system $\mathcal{Q}_5$: As $\mathcal{Q}_5$ can be obtained from either one of the two pentagons with vertex set $\langle 5 \rangle$ by choosing all quartets $ab|cd \in \mathcal{Q}(\langle 5 \rangle)$ for which the two straight lines connecting a with b and c with d are parallel to each other, the automorphism group $Aut(\mathcal{Q}_5)$ of $\mathcal{Q}_5$ is the subgroup of order 20 of the full symmetric group $\mathfrak{S}_5$ on $\langle 5 \rangle$ generated by the permutations (12345), (25)(34), and (2354).

And indeed, the quartet system $\mathcal{Q}_5 := \{12|35, 13|45, 14|23, 15|24, 25|34\} \subseteq \mathcal{Q}(\langle 5 \rangle)$ is easily seen to be a simple transitive cover for the 5-set $\langle 5 \rangle$ for which not even a single $\mathcal{Q}_5$-split exists — and the same holds for the quartet system $\mathcal{Q}_6 \subseteq \mathcal{Q}(\langle 6 \rangle)$ that can be obtained from $\mathcal{Q}_5$ by extending this quartet system by the five pairs of quartets of the form $ab|e6, cd|e6 \in \mathcal{Q}(\langle 6 \rangle)$ for which $\langle 5 \rangle = \{a, b, c, d, e\}$ and $ab|cd \in \mathcal{Q}_5$ holds.

Remark 8.6 *The two quartet systems $\mathcal{Q}_5$ and $\mathcal{Q}_6$ have some simple, yet somewhat mathematically intriguing geometric and group-theoretic features:*

Given a 5-set X, $\mathcal{Q}_5 \simeq \mathcal{Q}$ holds for a quartet system $\mathcal{Q} \subseteq \mathcal{Q}(X)$ if and only if there exists a regular pentagon P with vertex set X such that $\mathcal{Q}$ coincides with the quartet system $\mathcal{Q}_P$ consisting of all quartets $ab|cd \in \mathcal{Q}(X)$ for which the two straight lines connecting a with b and c with d are parallel to each other (cf. Figure 8.5).

It follows that the automorphism group $Aut(\mathcal{Q})$ of any such quartet system $\mathcal{Q}$ consisting of all permutations ψ of X with $\mathcal{Q} = \psi(\mathcal{Q})$ contains exactly 20 elements and acts sharply transitively on the set of "flagged 2-subsets" of X, i.e., the pairs a, A with $a \in A \subseteq X$ and $|A| = 2$.

And given a 6-set X, $\mathcal{Q}_6 \simeq \mathcal{Q}$ holds for a quartet system $\mathcal{Q} \subseteq \mathcal{Q}(X)$ if and only if one can label the six axes of fivefold symmetry of a regular icosahedron by the elements of X in such a way that $\mathcal{Q}$ consists exactly of all quartets $ab|cd$ for which the plane containing the axes labeled a and b is orthogonal to the plane containing the axes labeled c and d (cf. Figure 8.6).

In particular, the group of proper symmetries of the icosahedron acting by permutations on those six axes can be viewed as a subgroup (of index 2) of the

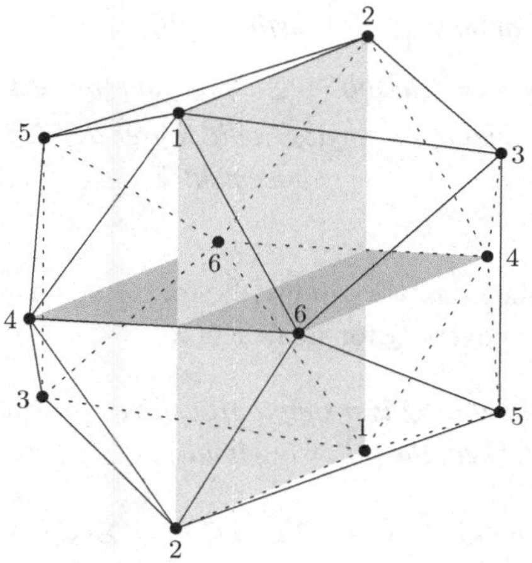

Figure 8.6 This figure illustrates geometric and group-theoretic features of the quartet system $\mathcal{Q}_6$: Labeling the pairs of vertices that define the six axes of fivefold symmetry of a regular icosahedron by the elements of $\langle 6 \rangle$ appropriately, it can easily be verified that $\mathcal{Q}_6$ consists exactly of all quartets $ab|cd$ for which the plane containing the axes labeled a and b is orthogonal to the plane containing the axes labeled c and d. For example, the two planes indicated in the figure correspond to the quartet $12|46$. It follows that the automorphism group $Aut(\mathcal{Q}_6)$ of $\mathcal{Q}_6$ acts transitively on $\langle 6 \rangle$ and, hence, has order 120 as its stabilizer subgroup $Fix_6(Aut(\mathcal{Q}_6))$ relative to the element 6 must be isomorphic, by restriction to $\langle 5 \rangle$, to $Aut(\mathcal{Q}_5)$.

full automorphism group $Aut(\mathcal{Q})$ *of any such quartet system* $\mathcal{Q}$ *which is one of the various subgroups of order* 120 *of the full symmetric group* $\mathfrak{S}_6$ *on* $\langle 6 \rangle$.

In view of the facts observed above, one may expect that things could only get worse if X gets even larger. However, the thorough (though sometimes quite technical) analysis of such quartet systems presented by Jan Weyer-Menkhoff in his thesis [131] shows that, essentially, the opposite is the case: For every simple transitive cover $\mathcal{Q} \subseteq \mathcal{Q}(X)$ for some n-set X with $n \geq 4$ and $\mathcal{Q}_5, \mathcal{Q}_6 \not\preceq \mathcal{Q}$, there exists, at least, one $\mathcal{Q}$-split. And the deviations of an arbitrary simple transitive cover from a simple transitive cover $\mathcal{Q}$ with $\mathcal{Q} = \mathcal{Q}(\Sigma(\mathcal{Q}))$ can all be "described" in terms of these two exceptional simple transitive covers without any single $\mathcal{Q}$-split (see Theorem 8.10 for details). Here, we'll present some of the principal results from [131], yet relying on somewhat simplified proofs. We first deal with the following basic theorem:

Theorem 8.7 *Given any simple transitive cover* $\mathcal{Q} \subseteq \mathcal{Q}(X)$ *for an n-set X with* $n \geq 4$, *the following hold:*

(i) *$\Sigma(\mathcal{Q})$ contains at least* $\left\lceil \frac{n-6}{4} \right\rceil$ *distinct splits.*

(ii) *One has* $\mathcal{Q}(\Sigma(\mathcal{Q})) = \{aa'|bb' \in \mathcal{Q} : \forall_{x \in X} xaa'|bb' \in \widehat{\mathcal{Q}} \text{ or } aa'|xbb' \in \widehat{\mathcal{Q}}\}$.

(iii) *The interior vertices of the phylogenetic X-tree* $\mathcal{T}_{\mathcal{Q}}$ *defined — up to canonical isomorphism — by the requirement* $\Sigma^*_{\mathcal{T}} = \Sigma(\mathcal{Q})$, *all have degree either* 3, 5, *or* 6.

To establish this theorem, we will use the following characterization of certain simple transitive covers $\mathcal{Q}$ for small sets X:

Lemma 8.8 *Assume that $\mathcal{Q}$ is a simple transitive cover for a set X of cardinality $n = 5$, 6, or 7. Then, the following hold:*

(i) *In case $n = 5$, either $|\Sigma(\mathcal{Q})| = 2$ and $\mathcal{Q} = \mathcal{Q}(\Sigma(\mathcal{Q}))$ holds or there exists not even a single $\mathcal{Q}$-split, that is, one has $\mathcal{Q} = \widehat{\mathcal{Q}}$ in which case $\mathcal{Q} \simeq \mathcal{Q}_5$ holds.*

(ii) *In case $n = 6$, either $\mathcal{Q} = \mathcal{Q}(\Sigma(\mathcal{Q}))$, $\mathcal{Q} \simeq \mathcal{Q}_6$, or $\mathcal{Q} \simeq \mathcal{Q}_{5\leftarrow 6} := (\mathcal{Q}_5)^{5\leftarrow 6}$ holds.*

Furthermore, we have $\mathcal{Q} \simeq \mathcal{Q}_6$ if and only if $\mathcal{Q} = \widehat{\mathcal{Q}}$ holds or, equivalently, if and only if $\mathcal{Q}|_{X-x} \simeq \mathcal{Q}_5$ holds for every $x \in X$ in which case $\mathcal{Q}$ can be obtained, for every $x \in X$, from $\mathcal{Q}|_{X-x}$ by extending this quartet system by the five pairs of quartets of the form $ab|ex$, $cd|ex \in \mathcal{Q}(X)$ for which $X - x = \{a, b, c, d, e\}$ and $ab|cd \in \mathcal{Q}|_{X-x}$ holds. In particular, $\mathcal{Q}$ must coincide with $\mathcal{Q}_6$ in case $X = \langle 6 \rangle$, $\mathcal{Q} = \widehat{\mathcal{Q}}$, and $\mathcal{Q}|_{\langle 5 \rangle} = \mathcal{Q}_5$ holds.

And $\mathcal{Q} \simeq \mathcal{Q}_{5\leftarrow 6}$ holds if and only if $\mathcal{Q}(\Sigma(\mathcal{Q})) \neq \mathcal{Q} \neq \widehat{\mathcal{Q}}$ holds in which case there exists a unique $\mathcal{Q}$-cherry a, b, the corresponding split S_{ab} is the only split in $\Sigma(\mathcal{Q})$, and $\mathcal{Q}|_{X-x} \simeq \mathcal{Q}_5$ holds for some $x \in X$ if and only if x is part of this $\mathcal{Q}$-cherry. Furthermore, two distinct elements a, $b \in X$ form that $\mathcal{Q}$-cherry in this case if and only if there exist three distinct elements y, y', $y'' \in X - \{a, b\}$ with $ab|yy'y'' \in \widehat{\mathcal{Q}}$.

In particular, if $\mathcal{Q}$ is a simple transitive cover for a 6-set X, there exists some $x \in X$ with $\mathcal{Q}|_{X-x} \simeq \mathcal{Q}_5$, and there exist four distinct elements y, y_1, y_2, $y_3 \in X - x$ with $xy|y_1y_2y_3 \in \widehat{\mathcal{Q}}$, then we have $\mathcal{Q} = (\mathcal{Q}|_{X-x})^{y\leftarrow x} \simeq \mathcal{Q}_{5\leftarrow 6}$ and the two elements x, y form the unique cherry for $\mathcal{Q}$.

(iii) *And if, in case $n = 7$, there exists some $x \in X$ with $\mathcal{Q}|_{X-x} \simeq \mathcal{Q}_6$, then $\mathcal{Q} \simeq \mathcal{Q}_{6\leftarrow 7} := (\mathcal{Q}_6)^{6\leftarrow 7}$ holds in which case there exists a unique $\mathcal{Q}$-cherry a, b, the corresponding split S_{ab} is the only split in $\Sigma(\mathcal{Q})$, we have, of course, $\mathcal{Q} = (\mathcal{Q}|_{X-a})^{b\leftarrow a}$, and $\mathcal{Q}|_{X-x} \simeq \mathcal{Q}_6$ holds for some $x \in X$ if and only if $x = a$ or $x = b$ holds.*

Proof Note first that, given any transitive cover $\mathcal{Q}$ for X and a non-trivial partial split $A|B \in \Sigma^*_{part}(X)$ of X with $|B| = 3$, we have

$$A|B \in \widehat{\mathcal{Q}} \Rightarrow \exists_{b \in B}(A+b)|(B-b) \in \widehat{\mathcal{Q}}. \tag{8.5}$$

Indeed, choosing any $a \in A$, there must exist some $b \in B$ with $ab|b'b'' \in \mathcal{Q}$ for the two elements $b', b'' \in B$ with $\{b', b''\} = B - b$ which, together with $aa'|b'b'' \in \mathcal{Q}$ for all $a' \in A - a$ implies our claim.

In particular, if $n=5$ holds and $\mathcal{Q}$ is a transitive cover for X with $|\Sigma(\mathcal{Q})| \le 1$, we must have $|\Sigma(\mathcal{Q})| = 0$ implying also that, in case $ab|cd \in \mathcal{Q}$ for four distinct elements $a, b, c, d \in X$, we cannot have $ab|ce \in \mathcal{Q}$ for the fifth element $e \in X$ and that, in consequence,

$$ab|cd \in \mathcal{Q} \Rightarrow ae|bc \in \mathcal{Q} \quad \text{or} \quad ac|be \in \mathcal{Q}$$

holds for any five distinct elements $a, b, c, d, e \in X$ in this case. So, to establish (i), we may assume now that $\mathcal{Q}$ is a simple transitive cover for the set $X := \langle 5 \rangle$ and that, without loss of generality, $12|35 \in \mathcal{Q}$ and $14|23 \in \mathcal{Q}$ holds. Then, as neither $24|35 \in \mathcal{Q}$ nor $23|45 \in \mathcal{Q}$ can hold, we must have $34|52 \in \mathcal{Q}$. And, for similar reasons, we must also have $45|13, 51|24 \in \mathcal{Q}$, implying that, in this case, $\mathcal{Q}$ actually coincides with $\mathcal{Q}_5$.

It follows in particular that, given any simple transitive cover $\mathcal{Q}$ for an arbitrary finite set X of cardinality at least 5, one has either $\mathcal{Q} = \mathcal{Q}(\Sigma(\mathcal{Q}))$, or there exists a 5-point subset Y of X with $\mathcal{Q}|_Y \simeq \mathcal{Q}_5$.

Thus, to establish (ii), we may assume that $X = \langle 6 \rangle$ and $\mathcal{Q}|_{\langle 5 \rangle} = \mathcal{Q}_5$ holds. If $\mathcal{Q} = \widehat{\mathcal{Q}}$ holds, that is, there is not even a single partial $\mathcal{Q}$-split $A|B$ with $|A|=2$ and $|B|=3$, $\mathcal{Q}=\mathcal{Q}_6$ must hold as $\mathcal{Q}$ must contain the five pairs of quartets of the form $ab|e6, cd|e6 \in \mathcal{Q}(\langle 6 \rangle)$ for which $\langle 5 \rangle = \{a, b, c, d, e\}$ and $ab|cd \in \mathcal{Q}_5$ holds: Otherwise, if $ab|cd \in \mathcal{Q}_5$ and, say, $ab|e6 \notin \mathcal{Q}$ were to hold, we may assume that $ae|b6 \in \mathcal{Q}$ holds. However, we have also either $ae|bc \in \mathcal{Q}_5$ or $ac|be \in \mathcal{Q}_5$ and $ae|bd \in \mathcal{Q}_5$ or $ad|be \in \mathcal{Q}_5$. Yet, our assumption that $ae|b6 \in \mathcal{Q} = \widehat{\mathcal{Q}}$ holds implies that neither $ae|bc \in \mathcal{Q}_5$ nor $ae|bd \in \mathcal{Q}_5$ can hold. So, we would have $ac|be \in \mathcal{Q}_5$ and $ad|be \in \mathcal{Q}_5$ which is also impossible.

Now assume that there exists a partial $\mathcal{Q}$-split $A|B$ with $|A| = 2$ and $|B| = 3$ and therefore, in view of (8.5), also some $b \in B$ with $(A+b)|(B-b) \in \widehat{\mathcal{Q}}$. We claim that, in this case, one such partial $\mathcal{Q}$-split $A|B$ with $6 \in A$ must exist. Otherwise, we may — by symmetry relative to $Aut(\mathcal{Q})$ — assume that, say, $13|456 \in \widehat{\mathcal{Q}}$ holds and, therefore, also $136|45 \in \widehat{\mathcal{Q}}$ in view of (8.5). Further, either $12|36$ or $23|16$ or $13|26$ is in $\mathcal{Q}$. However, in the first case, we would get $12|56 \in \mathcal{Q}$ in view of $12|35 \in \mathcal{Q}_5$ and, then, also $123|56 \in \widehat{\mathcal{Q}}$ in view of $13|56 \in \mathcal{Q}(13|456) \subseteq \mathcal{Q}$ in contradiction to our assumption that no partial $\mathcal{Q}$-split $A|B$ with $|A| = 2$, $|B| = 3$, and $6 \in A$ exists. In the second case, we would get $23|56 \in \mathcal{Q}$ in view of $15|23 \in \mathcal{Q}_5$ and, then, again $123|56 \in \widehat{\mathcal{Q}}$ in view of $13|56 \in \mathcal{Q}$. And in the last case, we would get $13|2456 \in \Sigma(\mathcal{Q})$ in view of $13|456 \in \widehat{\mathcal{Q}}$ in contradiction to $14|23 \in \mathcal{Q}_5$.

So, again by symmetry relative to $Aut(\mathcal{Q})$, we may assume that, say, $123|56 \in \widehat{\mathcal{Q}}$ holds. We claim that $1234|56$ must be a $\mathcal{Q}$-split in this case. Indeed, $12|35$, $13|45 \in \mathcal{Q}_5$ implies $12|356$, $13|456 \in \widehat{\mathcal{Q}}$. Further, $23|46$, $24|36$, or $34|26$ is contained in $\mathcal{Q}$. In the first case, we would get $23|456 \in \widehat{\mathcal{Q}}$ in contradiction to $25|34 \in \mathcal{Q}_5$. In the second case, we would get $14|36 \in \widehat{\mathcal{Q}}$ in view of $12|36 \in \mathcal{Q}(12|356) \subseteq \mathcal{Q}$ in contradiction to $13|46 \in \mathcal{Q}(13|456) \subseteq \mathcal{Q}$. So, $34|26 \in \mathcal{Q}$ and therefore, in view of $34|25 \in \mathcal{Q}_5$, also $34|56 \in \mathcal{Q}$ and, hence, in view of $123|56 \in \widehat{\mathcal{Q}}$ also $1234|56 \in \widehat{\mathcal{Q}}$ and, thus, $\mathcal{Q} = \mathcal{Q}_{5\leftarrow 6}$ must hold in this last case, as claimed.

It follows that, in this case, $\mathcal{Q}|_{X-x} \simeq \mathcal{Q}_5$ holds if and only if $x = 5$ or $x = 6$ holds because $(\langle 4\rangle - x)|56 \in \Sigma(\mathcal{Q}|_{\langle 6\rangle - x}) \neq \emptyset$ holds for each $x \in \langle 4\rangle$, and that no other $\mathcal{Q}$-split except S_{56} can exist: Indeed, we must have $|A \cap (\langle 6\rangle - 6)| = 1$ as well as $|A \cap (\langle 6\rangle - 5)| = 1$ for every $\mathcal{Q}$-split $S = A|B$ with $|A| \leq |B|$ in view of $\Sigma(\mathcal{Q}|_{\langle 6\rangle - 6}) = \Sigma(\mathcal{Q}|_{\langle 6\rangle - 5}) = \emptyset$ which can hold only in case $A = \{5, 6\}$, as claimed.

Thus, if $\mathcal{Q}$ is a simple transitive cover for a 6-set X and there exists some $x \in X$ with $\mathcal{Q}|_{X-x} \simeq \mathcal{Q}_5$ and four distinct elements $y, y_1, y_2, y_3 \in X - x$ with $xy|y_1y_2y_3 \in \widehat{\mathcal{Q}}$, we must have $\mathcal{Q} \simeq \mathcal{Q}_{5\leftarrow 6}$ and the two elements x, y must form the unique $\mathcal{Q}$-cherry: Indeed, our assumptions imply $\mathcal{Q}(\Sigma(\mathcal{Q})) \neq \mathcal{Q} \neq \widehat{\mathcal{Q}}$ and, hence, $\mathcal{Q} \simeq \mathcal{Q}_{5\leftarrow 6}$. Further, $\mathcal{Q}|_{X-x} \simeq \mathcal{Q}_5$ implies that x must be one of the two elements a, b that form the unique $\mathcal{Q}$-cherry for $\mathcal{Q}$. And if the other element y' in that $\mathcal{Q}$-cherry were distinct from y, it must also be distinct from y_1, y_2, and y_3 as $xy'|yy'' \in \mathcal{Q}$ would hold for all $y'' \in X - \{x, y, y'\}$. Thus, $xy'|y_1y_2y_3 \in \widehat{\mathcal{Q}}$ would, together with $xy|y_1y_2y_3 \in \widehat{\mathcal{Q}}$, imply that also $xyy'|y_1y_2y_3 \in \Sigma(\mathcal{Q})$ would hold in contradiction to $\Sigma(\mathcal{Q}) = \{S_{xy'}\}$.

Finally, to establish (iii), we may assume, without loss of generality, that $X = \langle 7\rangle$ and $\mathcal{Q}|_{\langle 6\rangle} = \mathcal{Q}_6$ holds. Then, given any $x \in \langle 6\rangle$, we have

$$(\mathcal{Q}|_{X-x})|_{(X-x)-7} = (\mathcal{Q}|_{X-7})|_{(X-7)-x} = \mathcal{Q}_6|_{\langle 6\rangle - x} \simeq \mathcal{Q}_5 \tag{8.6}$$

and, therefore, $\mathcal{Q}|_{X-x} \simeq \mathcal{Q}_6$ or $\mathcal{Q}|_{X-x} \simeq \mathcal{Q}_{5\leftarrow 6}$. Furthermore, (8.6) implies in view of (ii) that, for every $x \in \langle 6\rangle$ with $\mathcal{Q}|_{X-x} \simeq \mathcal{Q}_{5\leftarrow 6}$, there must exist some element $y \in \langle 6\rangle - x$ with $y7|(X - x) - \{y, 7\} \in \Sigma(\mathcal{Q}|_{X-x})$ and, hence, $y7|\langle 6\rangle - \{x, y\} \in \widehat{\mathcal{Q}}$. Thus, if also $\mathcal{Q}|_{X-y} \simeq \mathcal{Q}_{5\leftarrow 6}$ were to hold for one such element $y \in \langle 6\rangle - x$, there would also exist some $z \in \langle 6\rangle - y$ (not necessarily distinct from x) with $z7|\langle 6\rangle - \{y, z\} \in \widehat{\mathcal{Q}}$ and, therefore, with $yz|B \in \widehat{\mathcal{Q}}$ for $B := (\langle 6\rangle - \{x, y\}) \cap \langle 6\rangle - \{y, z\} = \langle 6\rangle - \{x, y, z\}$, a set of cardinality at least 3. This, however, is impossible in view of our assumption $\mathcal{Q}|_{\langle 6\rangle} = \mathcal{Q}_6$.

So, there must exist some $x \in \langle 6\rangle$ with $\mathcal{Q}|_{X-x} \simeq \mathcal{Q}_6$ and, without loss of generality, we may assume that this holds for $x := 6$. It follows from (ii) that $\mathcal{Q}|_{X-6} = \mathcal{Q}|_{\langle 5\rangle+7}$ can be obtained from the quartet system $\mathcal{Q}_5$ the same way that $\mathcal{Q}_6$ is obtained form $\mathcal{Q}_5$, that is, by extending $\mathcal{Q}_5$ by the five pairs of quartets of the form $ab|e7, cd|e7 \in \mathcal{Q}|_{\langle 5\rangle+7}$ for which $\langle 5\rangle = \{a, b, c, d, e\}$ and $ab|cd \in \mathcal{Q}|_5$ holds. So, $ab|c7 \in \mathcal{Q}$ holds for any three distinct elements a, b, $c \in \langle 5\rangle$ with $ab|c6 \in \mathcal{Q}_6$ which readily implies that $\mathcal{Q}$ coincides with $\mathcal{Q}_{6\leftarrow 7}$ in this case. Direct inspection now yields the remaining claims. Altogether, this establishes Lemma 8.8. □

To further exploit the last arguments, assume that $\mathcal{Q} \subseteq \mathcal{Q}(X)$ is a simple transitive cover for some set X of cardinality at least 5, and that Y is a 5-subset of X with $\mathcal{Q}|_Y \simeq \mathcal{Q}_5$. For every $y \in Y$, put

$$A(y, Y) := \{x \in X : x = y \quad \text{or} \quad x \notin Y \quad \text{and} \quad xy|Y - y \in \widehat{\mathcal{Q}}\},$$

and put $B(y, Y) := X - A(y, Y)$ and $B(Y) := \bigcap_{y\in Y} B(y, Y)$. Clearly, we have

(i) $x \in A(y, Y) - y \iff \mathcal{Q}|_{Y+x} = (\mathcal{Q}|_Y)^{y\leftarrow x} \simeq \mathcal{Q}_{5\leftarrow 6}$ and, therefore, also $\mathcal{Q}|_{(Y-y)+x} \simeq \mathcal{Q}_5$ for every $y \in Y$ and $x \in A(y, Y) - y$,
(ii) $A(y, Y) \cap A(y', Y) = \emptyset$ for any two distinct elements y, $y' \in Y$ as $xy|y'y'' \in \mathcal{Q}$ and $xy'|yy'' \in \mathcal{Q}$ cannot hold simultaneously for any three distinct elements $y, y', y'' \in Y$,
(iii) $B(Y) = \{x \in X : \mathcal{Q}|_{Y+x} \simeq \mathcal{Q}_6\}$, and
(iv) $x \in A(y, Y)$ for every $x \in X - Y$ for which three distinct elements y_1, y_2, $y_3 \in Y - y$ with $xy|y_1y_2y_3 \in \widehat{\mathcal{Q}}$ exist in view of the last assertion in Lemma 8.8(ii).

In particular, given any two distinct elements y, $y' \in Y$, we have $A(y', Y) = A(y', (Y - y) + x)$ for all $x \in A(y, Y)$ as, denoting the three distinct elements in $Y - \{y, y'\}$ by y_1, y_2, y_3, we have $A(y', Y) - y' = \{x' \in X : x'y'|y_1y_2y_3 \in \widehat{\mathcal{Q}}\} = A(y', (Y - y) + x) - y'$. And we have $xy|x'y' \in \mathcal{Q}$ for all $x \in A(y, Y) - y$ and $x' \in A(y', Y) - y'$ as we have just seen that these assumptions imply $x' \in A(y', (Y - y) + x)$ and, hence, $x'y'|xy_1y_2y_3 \in \widehat{\mathcal{Q}}$ in addition to $x'y'|yy_1y_2y_3 \in \widehat{\mathcal{Q}}$ and, therefore, also $x'y'|xyy_1y_2y_3 \in \widehat{\mathcal{Q}}$. By symmetry, we have also $xy|x'y'y_1y_2y_3 \in \widehat{\mathcal{Q}}$ and therefore, denoting the four distinct elements in $Y - \{y, y'\}$ by y_1, y_2, y_3, y_4, also $xy|x'y_1y_2y_3y_4 \in \widehat{\mathcal{Q}}$ for all $x \in A(y, Y) - y$ and $x' \in \bigcup_{i=1,2,3,4}(A(y_i, Y) - y_i)$.

Furthermore, this holds also for all $x \in A(y, Y) - y$ and $x' \in B(Y)$, that is, we have $A(y, Y)|B(y, Y) \in \Sigma(\mathcal{Q})$ for every $y \in Y$ with $|A(y, Y)| > 1$: Indeed, $x' \in B(Y)$ implies $\mathcal{Q}|_{Y+x'} \simeq \mathcal{Q}_6$. Hence, putting $Y' := Y \cup \{x, x'\}$ and

$\mathcal{Q}' := \mathcal{Q}|_{Y'}$, and noting that $\mathcal{Q}'|_{Y'-x} = \mathcal{Q}|_{Y+x'} \simeq \mathcal{Q}_6$, $\mathcal{Q}'|_{Y'-x'} = \mathcal{Q}|_{Y+x} \simeq \mathcal{Q}_{5\leftarrow 6}$, and $xy|y_1y_2y_3y_4 \in \widehat{\mathcal{Q}}$ and, therefore, $\mathcal{Q}'|_{Y'-y_i} \not\simeq \mathcal{Q}_6$ holds for all $i = 1, 2, 3, 4$, we may apply Assertion (iii) of Lemma 8.8 to derive that the other element z in Y' distinct from x for which $\mathcal{Q}'|_{Y'-z} \simeq \mathcal{Q}_6$ holds (that must exist in view of Lemma 8.8(iii)) must be the element y. So, the two elements x, y must form the only $\mathcal{Q}'$-cherry, that is, $xy|x'y_1y_2y_3y_4 \in \widehat{\mathcal{Q}}$ must indeed hold, as claimed.

And we have $aa'|bb' \in \mathcal{Q}$ for any four distinct elements $a, a', b, b' \in X$ with $a, a' \in A(Y) := \bigcup_{y\in Y} A(y, Y)$ and $b, b' \in B(Y)$: Indeed, $yx|bb' \in \mathcal{Q}$ holds for every $y \in Y$ and $x \in A(y, Y) - y$ in view of $A(y, Y)|B(y, Y) \in \Sigma(\mathcal{Q})$. And putting $Y' := Y \cup \{b, b'\}$ and $\mathcal{Q}' := \mathcal{Q}|_{Y'}$, we have $\mathcal{Q}'|_{Y'-b} = \mathcal{Q}|_{Y+b'} \simeq \mathcal{Q}_6$ and $\mathcal{Q}'|_{Y'-b'} = \mathcal{Q}|_{Y+b} \simeq \mathcal{Q}_6$ and, therefore, $y'y''|bb' \in \mathcal{Q}$ for any two distinct elements $y', y'' \in Y = Y' - \{b, b'\}$, again by Lemma 8.8(iii). Together, transitivity now implies that $aa'|bb' \in \mathcal{Q}$ must hold for any two distinct elements $a, a' \in A(Y)$, as claimed.

In other words, any 5-subset Y of X with $\mathcal{Q}|_Y \simeq \mathcal{Q}_5$ gives rise to a partition Π_Y of X into either the five non-empty disjoint subsets $\{A(y, Y) : y \in Y\}$ of X (in case $B(Y) = \emptyset$) or into the five non-empty disjoint subsets $\{A(y, Y) : y \in Y\}$ of X and the set $B(Y)$ (in case $B(Y) \neq \emptyset$) such that $C|X - C \in \Sigma(\mathcal{Q})$ holds for every subset $C \in \Pi_Y$ of cardinality at least 2.

These observations clearly imply

Lemma 8.9 *There exists a $\mathcal{Q}$-split for every simple transitive cover $\mathcal{Q} \subseteq \mathcal{Q}(X)$ unless $\mathcal{Q} \simeq \mathcal{Q}_5$ or $\mathcal{Q} \simeq \mathcal{Q}_6$ holds.*

Finally, the following concept is crucial for establishing Theorem 8.7: Given a quartet system $\mathcal{Q} \subseteq \mathcal{Q}(X)$, we define a *$\mathcal{Q}$-tree* to be a pair $(\mathcal{T}, \mathcal{Q}_\star)$ that consists of a simple phylogenetic X-tree $\mathcal{T} = (V, E)$ and a family $\mathcal{Q}_\star = (\mathcal{Q}_\star(v))_{v\in\mathcal{T}_\star}$ of "local" quartet systems $\mathcal{Q}_\star(v) \subseteq \mathcal{Q}(E(v))$ indexed by the vertices in the set $\mathcal{T}_\star$ of interior vertices of $\mathcal{T}$ of degree at least 4 and defined, for each $v \in \mathcal{T}_\star$, on the set $E(v)$ of edges incident to v such that $\mathcal{Q}$ coincides with the subset $\mathcal{Q}(\mathcal{T}, \mathcal{Q}_\star)$ of $\mathcal{Q}(X)$ that consists of exactly all those quartets $aa'|bb' \in \mathcal{Q}(X)$ that are either contained in $\mathcal{Q}_\mathcal{T}$ or for which the 4-set $A := \{a, a', b, b'\}$ is contained in the set $\overline{\mathcal{Q}}_\mathcal{T}$ introduced in Chapter 2 (that is, none of the three quartets $aa'|bb'$, $ab|a'b'$, and $ab'|a'b$ is contained in $\mathcal{Q}_\mathcal{T}$) and $e^{v\to a}e^{v\to a'}|e^{v\to b}e^{v\to b'} \in \mathcal{Q}_\star(v)$ holds, for $v := v_\mathcal{T}(A)$, for the four (necessarily distinct) edges $e^{v\to a}, e^{v\to a'}, e^{v\to b}, e^{v\to b'} \in E(v)$ that separate the vertex v from a, a', b, and b', respectively. For every vertex $v \in \mathcal{T}_\star$, we denote by $\mathcal{Q}_\star^\uparrow(v)$ the collection of all such quartets in $\mathcal{Q}(X)$. Figure 8.7(a) presents an example of a $\mathcal{Q}$-tree where, with $\mathcal{Q}_\star(v) := \{e_1e_4|e_2e_3\}$ for the only vertex

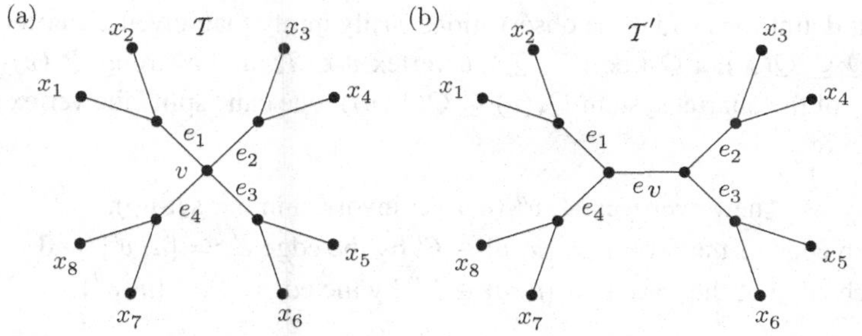

Figure 8.7 (a) A phylogenetic X-tree $\mathcal{T}$ with $X = \{x_1, \ldots, x_8\}$. Together with the one-member family $\mathcal{Q}_\star$ given by $\mathcal{Q}_\star(v) = \{e_1e_4|e_2e_3\}$ for the only vertex v of $\mathcal{T}$ of degree at least 4, it forms a $\mathcal{Q}$-tree for the quartet system $\mathcal{Q} := \mathcal{Q}_{\mathcal{T}'}$ associated to the phylogenetic X-tree $\mathcal{T}'$ depicted in (b). Note that for $\mathcal{T}'$ we have $\mathcal{Q}_\star^\uparrow(v) = \{x_ix_j|x_kx_l : i \in \{1,2\}, j \in \{7,8\}, k \in \{3,4\}, l \in \{5,6\}\}$. (b) The X-tree $\mathcal{T}'$ obtained from $\mathcal{T}$ by splitting the vertex v.

$v \in \mathcal{T}_\star$, we have $\mathcal{Q}_\star^\uparrow(v) = \{x_ix_j|x_kx_l : i \in \{1,2\}, j \in \{7,8\}, k \in \{3,4\}, l \in \{5,6\}\}$ and, hence, $\mathcal{Q}(\mathcal{T}, \mathcal{Q}_\star) = \mathcal{Q}_{\mathcal{T}'}$ for the phylogenetic X-tree $\mathcal{T}'$ depicted in Figure 8.7(b).

Clearly, there exists, for every quartet system $\mathcal{Q} \subseteq \mathcal{Q}(X)$, an essentially unique "coarsest" $\mathcal{Q}$-tree, *viz.* the $\mathcal{Q}$-tree $(\mathcal{T}, \mathcal{Q}_\star)$ whose underlying X-tree is the simple star tree $\mathcal{T} := (X + *, \{\{x, *\} : x \in X\})$ for X with vertex set $V := X + *$ and edge set $E := \{\{x, *\} : x \in X\}$, together with the one-member family of local quartet systems $\mathcal{Q}_\star = (\mathcal{Q}_\star(v))_{v \in \mathcal{T}_\star}$ defined by $\mathcal{Q}_\star(*) := \{\{a, *\}\{b, *\}|\{c, *\}\{d, *\} : ab|cd \in \mathcal{Q}\}$ for the only vertex $v = *$ in $\mathcal{T}_\star$.

Further, $\mathcal{Q}$ is transitive, a cover, or a simple cover if and only if there exists a $\mathcal{Q}$-tree such that all the local quartet systems $\mathcal{Q}_\star(v)$ listed in the family $(\mathcal{Q}_\star(v))_{v \in \mathcal{T}_\star}$ of local quartet systems are transitive, covers, or simple covers, respectively, if and only if this holds for every $\mathcal{Q}$-tree.

Next, we define a $\mathcal{Q}$-split $S = A|B \in \Sigma(\mathcal{Q})$ to be a *strong $\mathcal{Q}$-split* if $ab|a'b' \notin \mathcal{Q}$ holds for any four distinct elements $a, a' \in A$ and $b, b' \in B$, and $a'a''|bb' \in \mathcal{Q}$ implies $aa''|bb' \in \mathcal{Q}$ for all five distinct elements $a, a' \in A$ and $a'', b, b' \in B$.

Clearly, every $\mathcal{Q}$-split is a strong $\mathcal{Q}$-split whenever $\mathcal{Q}$ is a simple transitive cover. It is also obvious that, given a $\mathcal{Q}$-tree $\mathcal{T}$, every split in $\Sigma_{\mathcal{T}}^*$ is not only a $\mathcal{Q}$-split, but a strong $\mathcal{Q}$-split, and that any two strong $\mathcal{Q}$-splits are compatible.

And it is easy to see, too, that, given a quartet system $\mathcal{Q} \subseteq \mathcal{Q}(X)$, a $\mathcal{Q}$-tree $(\mathcal{T}, \mathcal{Q}_\star)$, and a vertex $v \in \mathcal{T}_\star$, an $E(v)$-split $E'|E''$ is a (strong) $\mathcal{Q}_\star(v)$-split whenever the associated X-split $\{x' \in X : e^{v \to x'} \in E'\}|\{x'' \in X : e^{v \to x''} \in E''\}$ is a (strong) $\mathcal{Q}$-split.

Our definitions and these observations easily imply that, given a quartet system $\mathcal{Q} \subseteq \mathcal{Q}(X)$, a $\mathcal{Q}$-tree $(\mathcal{T}, \mathcal{Q}_\star)$, a vertex $v \in \mathcal{T}_\star$, and a strong $\mathcal{Q}_\star(v)$-split $E'|E''$ of the quartet system $\mathcal{Q}_\star(v) \subseteq \mathcal{Q}(E(v))$, one can "split" the vertex v by replacing

- v by two "new" vertices v', v'' (not yet involved in our set-up),
- each edge of the form $e = \{u, v\} \in E'$ by the edge $e' := \{u, v'\}$, and
- each edge of the form $e = \{u, v\} \in E''$ by the edge $e'' := \{u, v''\}$,

and adding the edge $e_v := \{v', v''\}$ to also connect the vertices v' and v'' with each other. This way, we obtain a new X-tree $\mathcal{T}'$ which, in addition, is also a $\mathcal{Q}$-tree relative to the family $\mathcal{Q}'_\star = (\mathcal{Q}'(u'))_{u' \in \mathcal{T}'_\star}$ defined by putting $\mathcal{Q}'(u') := \mathcal{Q}_\star(v)$ for all $u' \in \mathcal{T}'_\star - \{v', v''\} = \mathcal{T}_\star - v$ while one defines the local quartet system $\mathcal{Q}'(v')$ in case $\deg_{\mathcal{T}'}(v') = |E'| + 1 > 3$ by

$$\mathcal{Q}'(v') := \left\{e'_1 e'_2 \,\middle|\, f'_1 e_v : e_1, e_2, f_1 \in E' \text{ and } \exists_{f_2 \in E''}\, e_1 e_2 \middle|\, f_1 f_2 \in \mathcal{Q}_\star(v)\right\}$$
$$\cup \left\{e'_1 e'_2 \,\middle|\, f'_1 f'_2 : e_1 e_2 \middle|\, f_1 f_2 \in \mathcal{Q}_\star(v)|_{E'}\right\},$$

and the local quartet system $\mathcal{Q}'(v'')$ (if required, i.e., if $\deg_{\mathcal{T}'}(v'')$ exceeds 3) analogously. This construction is illustrated in Figure 8.7. It easily yields:

Theorem 8.10 *There exists, for any quartet system $\mathcal{Q} \subseteq \mathcal{Q}(X)$, a "finest" $\mathcal{Q}$-tree $(\mathcal{T}, \mathcal{Q}_\star)$ uniquely determined by $\mathcal{Q}$ up to canonical isomorphism, i.e., a $\mathcal{Q}$-tree with a maximal number of edges and vertices that is characterized, up to canonical isomorphism, by the fact that $\Sigma_\mathcal{T}$ coincides with the collection $\Sigma_{strong}(\mathcal{Q})$ of all strong $\mathcal{Q}$-splits or, equivalently, by the fact that none of the local quartet systems $\mathcal{Q}_\star(v)$ $(v \in \mathcal{T}_\star)$ admits a strong $\mathcal{Q}_\star(v)$-split.*

Proof Given a phylogenetic X-tree $\mathcal{T} = (V, E, \varphi)$ with $\Sigma_\mathcal{T} = \Sigma_{strong}(\mathcal{Q})$ (which exists as this set is compatible), it suffices — in view of the fact that, given any $\mathcal{Q}$-tree $(\mathcal{T}', \mathcal{Q}'_\star)$, every $\mathcal{T}'$-split is necessarily a strong $\mathcal{Q}$-split and, hence, a $\mathcal{T}$-split — to construct a family $\mathcal{Q}_\star$ of local quartet systems $\mathcal{Q}_\star(v)$ $(v \in \mathcal{T}_\star)$ such that $(\mathcal{T}, \mathcal{Q}_\star)$ is a $\mathcal{Q}$-tree. However, this holds indeed for the family $\mathcal{Q}_\star := (\mathcal{Q}_\star(v))_{v \in \mathcal{T}_\star}$ with

$$\mathcal{Q}_\star(v) := \{e^{v \to a} e^{v \to a'} | e^{v \to b} e^{v \to b'} : aa'|bb' \in \mathcal{Q} \text{ and } v = v_\mathcal{T}(\{a, a', b, b'\})\}$$

for each $v \in \mathcal{T}_\star$: By definition of $\mathcal{T}$ and $\mathcal{Q}_\star$, we have $\mathcal{Q} \subseteq \mathcal{Q}(\mathcal{T}, \mathcal{Q}_\star)$. However, we even have $\mathcal{Q} = \mathcal{Q}(\mathcal{T}, \mathcal{Q}_\star)$ — as required — because $aa'|bb' \in \mathcal{Q}(\mathcal{T}, \mathcal{Q}_\star)$ implies that either $aa'|bb' \in \mathcal{Q}_\mathcal{T} \subseteq \mathcal{Q}$ or $\{a, a', b, b'\} \in \overline{\mathcal{Q}}_\mathcal{T}$ and $v := v_\mathcal{T}(\{a, a', b, b'\}) \in \mathcal{T}_\star$ holds. Furthermore, if the latter holds, there exist $a_1, a'_1, b_1, b'_1 \in X$ with $e^{v \to x} = e^{v \to x_1}$ for all $x = a, a', b, b'$, $a_1 a'_1 | b_1 b'_1 \in \mathcal{Q}$, $\{a_1, a'_1, b_1, b'_1\} \in \overline{\mathcal{Q}}_\mathcal{T}$, and $v = v_\mathcal{T}(\{a_1, a'_1, b_1, b'_1\})$. Thus, treating the elements

a, a', b, b' one after the other, it suffices to note that this implies $aa'_1|b_1b'_1 \in \mathcal{Q}$ which follows from $a_1a'_1|b_1b'_1 \in \mathcal{Q}$ and the fact that the $\mathcal{T}$-split induced by the edge $e^{v\to a} = e^{v\to a_1}$ is, by assumption, a strong $\mathcal{Q}$-split and that it separates a, a_1 from a'_1, b_1, b'_1. □

Consider, for example, again the phylogenetic X-tree $\mathcal{T}$ in Figure 8.7(a). Note that $\mathcal{T}$, while not being the finest $\mathcal{Q}$-tree for the quartet system $\mathcal{Q}$ considered there, it is, up to canonical isomorphism, the unique finest $\mathcal{Q}'$-tree for the quartet system $\mathcal{Q}' := \mathcal{Q}_{\mathcal{T}} \cup \mathcal{Q}(x_1x_2x_7x_8|x_3x_4x_5x_6) \cup \mathcal{Q}(x_1x_2x_3x_4|x_5x_6x_7x_8)$ and we have $\mathcal{Q}'_v = \{e_1e_4|e_2e_3, e_1e_2|e_3e_4\}$, that is, $E(v)$ does not admit a strong $\mathcal{Q}'_v$-split. The reader is encouraged to check the details.

It is now easy to also establish Theorem 8.7: Indeed, assume that $\mathcal{Q} \subseteq \mathcal{Q}(X)$ is a simple transitive cover and that $\mathcal{T} = (V, E, \varphi)$ is a phylogenetic X-tree with $\Sigma^*_{\mathcal{T}} = \Sigma(\mathcal{Q})$. Then, $\deg_{\mathcal{T}}(v) \in \{5, 6\}$ must hold for every interior vertex $v \in \mathcal{T}_\star$ as $\mathcal{T}$ is — in view of $\Sigma(\mathcal{Q}) = \Sigma_{strong}(\mathcal{Q})$ — the underlying X-tree of a "finest" $\mathcal{Q}$-tree $(\mathcal{T}, \mathcal{Q}_\star)$ which implies that the corresponding local quartet systems $\mathcal{Q}_\star(v)$, $v \in \mathcal{T}_\star$, are necessarily simple transitive covers that do not admit a $\mathcal{Q}_\star(v)$-split. So, $\deg_{\mathcal{T}}(v) = |E(v)| \in \{5, 6\}$ must hold for every interior vertex $v \in \mathcal{T}_\star$ as claimed, and $\mathcal{Q}_\star(v) \simeq \mathcal{Q}_5$ or $\mathcal{Q}_\star(v) \simeq \mathcal{Q}_6$ must hold in case $\deg_{\mathcal{T}}(v) = 5$ or 6, respectively.

In particular, we must have $2|E| = n + 3|V^{(3)}| + 5|V^{(5)}| + 6|V^{(6)}|$ as well as $|E| + 1 = |V| = n + |V^{(3)}| + |V^{(5)}| + |V^{(6)}|$ for the sets $V^{(i)}$ of vertices of degree i in the tree $\mathcal{T}$ by Lemma 1.2(iii) and, therefore, also $n + 3|V^{(3)}| + 5|V^{(5)}| + 6|V^{(6)}| = 2n + 2|V^{(3)}| + 2|V^{(5)}| + 2|V^{(6)}| - 2$ or, equivalently, $n - 2 = |V^{(3)}| + 3|V^{(5)}| + 4|V^{(6)}|$. In consequence, the cardinality $|E| - n = |V^{(3)}| + |V^{(5)}| + |V^{(6)}| - 1$ of interior edges of $\mathcal{T}$ — and, hence, the number of splits in $\Sigma(\mathcal{Q})$ — must be bounded from below by $\frac{n-2}{4} - 1 = \frac{n-6}{4}$, as claimed, while this bound is sharp if and only if $n - 2$ is divisible by 4 and *all* interior vertices of $\mathcal{T}$ have degree exactly 6.

Finally, it is obvious that, given any quartet system $\mathcal{Q} \subseteq \mathcal{Q}(X)$, the quartet system $\mathcal{Q}(\Sigma(\mathcal{Q}))$ is, of course, contained in the quartet system

$$\mathcal{Q}_{ext} := \{aa'|bb' \in \mathcal{Q} : \forall_{x\in X} xaa'|bb' \in \widehat{\mathcal{Q}} \text{ or } aa'|xbb' \in \widehat{\mathcal{Q}}\}$$

consisting of all *locally $\mathcal{Q}$-extendable quartets* in $\mathcal{Q}$. To see that equality holds in case $\mathcal{Q}$ is a simple transitive cover, it suffices to note that no quartet in $\mathcal{Q} = \mathcal{Q}_5$ or $\mathcal{Q} = \mathcal{Q}_6$ is locally $\mathcal{Q}$-extendable. Hence, given any "finest" $\mathcal{Q}$-tree $(\mathcal{T}, \mathcal{Q}_\star)$ with $\mathcal{T} = (V, E, \varphi)$ and $\Sigma^*_{\mathcal{T}} = \Sigma(\mathcal{Q})$, a quartet in $\mathcal{Q}$ is not locally $\mathcal{Q}$-extendable if it is contained in $\mathcal{Q}^{\uparrow}_\star(v)$ for some interior vertex v of degree 5 or 6 with $\mathcal{Q}_\star(v) \simeq \mathcal{Q}_5$ or $\mathcal{Q}_\star(v) \simeq \mathcal{Q}_6$. In turn, this implies that all locally

$\mathcal{Q}$-extendable quartets must be contained in the complement $\mathcal{Q}-\bigcup_{v\in V,\deg(v)\geq 5}$ $\mathcal{Q}_\star^\uparrow(v)$ of $\bigcup_{v\in V,\deg(v)\geq 5}\mathcal{Q}_\star^\uparrow(v)$ and, therefore, in $\mathcal{Q}(\Sigma(\mathcal{Q}))$ as claimed. □

We conclude this chapter by noting that, based on the results regarding simple transitive covers presented above, one strategy for building trees from an arbitrary simple cover $\mathcal{Q}\subseteq\mathcal{Q}(X)$ could be to first switch some quartets $ab|cd$ in $\mathcal{Q}$ to either $ac|bd$ or $ad|bc$ to obtain, in a step-by-step fashion, a simple transitive cover $\mathcal{Q}'$ in a "most parsimonious way" — that is, by introducing as few changes as possible.

This can be done by applying appropriate variants of algorithms developed for the closely related problem of detecting community structures in networks (see e.g., [40, 81, 116]).

More specifically, one may proceed as follows: Given any finite graph $G=(V,E)$ with $E\subseteq\binom{V}{2}$, let for instance $Cl(G)$ denote the cardinality of a smallest subset F of $\binom{V}{2}$ for which the graph $G\Delta F:=(V,E\Delta F)$ whose edge set is the the symmetric difference $E\Delta F$ of E and F is a disjoint union of cliques. Then, given any simple cover $\mathcal{Q}\subseteq\mathcal{Q}(X)$ as above, any 4-subset $A=\{a,b,c,d\}$ of X with, say, $q:=ab|cd\in\mathcal{Q}$, and any 2-subset Y of A, one may consider the number $Cl(G_{(\mathcal{Q},Y)})$, check whether the sum $\sum_{Y\in\binom{A}{2}}Cl(G_{(\mathcal{Q},Y)})$ decreases or increases when $\mathcal{Q}$ is replaced by $\mathcal{Q}_{q\to ac|bd}:=\mathcal{Q}-q+\{ac|bd\}$ or by $\mathcal{Q}_{q\to ad|bc}:=\mathcal{Q}-q+\{ad|bc\}$, and then replace $\mathcal{Q}$ by that quartet system $\mathcal{Q}_{q'\to a'b'|c'd'}$ (where q' runs over all quartets in $\mathcal{Q}$) for which this sum decreases as much as possible — hopefully by 4, the upper limit for that decrease. Alternatively, one may check, for any 4-subset $A=\{a,b,c,d\}$ with $q=ab|cd\in\mathcal{Q}$, whether, by some other measure of "modularity", the six graphs $G_{(\mathcal{Q}_{q\to ac|bd},Y)}$ $(Y\in\binom{A}{2})$ or $G_{(\mathcal{Q}_{q\to ad|bc},Y)}$ $(Y\in\binom{A}{2})$ resemble a disjoint union of cliques more closely than the original six graphs $G_{(\mathcal{Q},Y)}$ and, this way, try to identify "wrong" quartets in $\mathcal{Q}$ and to replace them by the "correct" quartets.

This approach worked quite well already for a number of "real" biological examples. Remarkably, in those examples, most simple transitive covers that were obtained in this way, were actually quite, or even perfectly, "treelike".

More generally, the number of changes required to obtain a simple transitive cover from an arbitrary quartet system $\mathcal{Q}$ can give some indication about how far $\mathcal{Q}$ is from a quartet system induced by an X-tree, that is, about the number of "inconsistencies" in $\mathcal{Q}$. In Chapter 10, we shall explore various ways of measuring inconsistencies in split systems, distances, and quartet systems in more detail, and provide applications of such measures for constructing trees and networks.

9

Rooted trees and the Farris transform

In previous chapters, we have considered methods for representing the evolutionary relationships between a finite set X of n species in terms of X-trees. However, biologists are often also interested in representing the *phylogenetic history* of the species in X. This is commonly represented in form of a *dated rooted X-tree* $\mathcal{T} = (V, A, \varphi, \tau)$, that is, a quadruple consisting of

(R1) the vertex set V and the arc set A of a finite rooted tree $\underline{\mathcal{T}} := (V, A)$, also called the *underlying rooted tree* of $\mathcal{T}$, together with
(R2) a *labeling map* $\varphi : X \to V$ such that its image $\varphi(X)$ contains (at least) all vertices $v \in V$ of out-degree 0 as well as all those vertices v of out-degree 1 that are distinct from the root $\mathfrak{r}_{\mathcal{T}} := \mathfrak{r}_{\underline{\mathcal{T}}}$, and
(R3) a *dating map* τ, i.e., a map $\tau : V \to \mathbb{R}$ that is strictly monotonically decreasing, that is, with $\tau(v) > \tau(v')$ for every arc $(v, v') \in A$.

In addition, we define a *rooted X-tree* $\mathcal{T}$ (without any reference to dating) to be a triple $\mathcal{T} = (V, A, \varphi)$ consisting of the vertex set V and the arc set A of a finite rooted tree (V, A), also denoted by $\underline{\mathcal{T}}$ and called the *underlying rooted tree* of $\mathcal{T}$, together with a labeling map φ for which (R2) holds. As before, we will freely use the terms introduced in previous chapters (for directed graphs as well as for unrooted X-trees) also for rooted X-trees. An example of a dated rooted X-tree is pictured in Figure 9.1.

In any such representation, each vertex v of $\mathcal{T}$ except perhaps its root is supposed to represent the putative *last common ancestor* of exactly all those species $x \in X$ for which $v \preceq_{\mathcal{T}} \varphi(x)$ holds (while this holds for the root if and only if its degree exceeds 1). In addition, supposing that $\tau = 0$ represents the present and that, accordingly (as we do not deal with the future), $\tau(v) \geq 0$ holds for all $v \in V$,

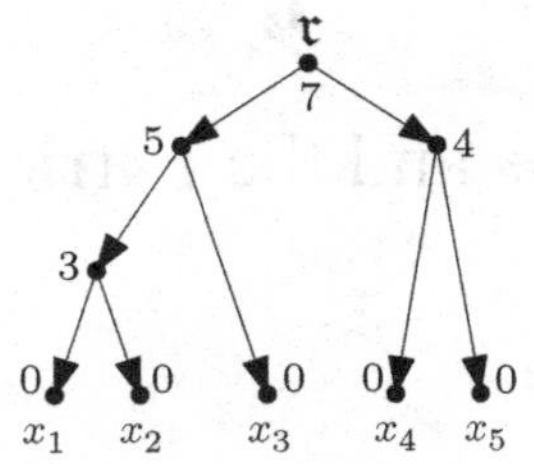

Figure 9.1 An example of a dated rooted $\{x_1, \ldots, x_5\}$-tree.

(i) one has $\deg_{\mathcal{T}}(\varphi(x)) = 1$ and $\tau(\varphi(x)) = 0$ for some $x \in X$ if and only if x represents an extant species in X,

(ii) $\tau(v)$ is supposed to represent, usually quite roughly, the date at which — going backwards in time and counting in whatever time units seem appropriate — that common ancestor represented by v lived on this planet or, equivalently, the amount of time that passed since then,

(iii) while, unless the root $\mathfrak{r}_{\mathcal{T}}$ is the only vertex of a dated rooted X-tree $\mathcal{T}$, $\tau(\mathfrak{r}_{\mathcal{T}}) > 0$ must hold in this case.

However, reflecting the fact that there is no obvious natural way to declare some point in time to be *the* point "0", and also to simplify our conceptual set-up and our arguments, we will in general not insist that $\tau(v) \geq 0$ holds for every $v \in V$, or $\tau(\varphi(x)) = 0$ for every $x \in X$.

In analogy to associating a split system, a quartet system, and a metric to an (edge-weighted) X-tree, we can associate a *cluster system*, a *triplet system*, and a symmetric bivariate map to a (dated) rooted X-tree:

Indeed, if $\mathcal{T} = (V, A, \varphi)$ is a rooted X-tree and v is any vertex of $\mathcal{T}$, define

$$C(v) = C_{\mathcal{T}}(v) := \{x \in X : v \preceq_{\mathcal{T}} \varphi(x)\}$$

and put

$$\mathcal{C}_{\mathcal{T}} := \{C(v) : v \in V\}.$$

Note, for later use, that $X \in \mathcal{C}_{\mathcal{T}}$ holds in view of $X = C_{\mathcal{T}}(\mathfrak{r}_{\mathcal{T}})$ and that

$$\emptyset \neq C(v) \tag{9.1}$$

and

$$v \preceq_{\mathcal{T}} v' \iff C(v') \subseteq C(v) \tag{9.2}$$

holds for any two vertices $v, v' \in V$ in view of our degree assumptions in (R2).

Further, using the symbol $ab|c$, for any three elements $a, b, c \in X$, as a shorthand for the ordered pair $(\{a, b\}, \{c\})$ of subsets of X consisting of the two subsets $\{a, b\}$ and $\{c\}$ of X, we will refer to such a pair $ab|c = (\{a, b\}, \{c\})$ in $\mathbb{P}(X) \times \mathbb{P}(X)$ also as a *triplet* or, more specifically, an X-triplet in case $a, b \neq c$ holds. We denote the collection of all X-triplets by $\mathcal{R}(X)$, and we define $\mathcal{R}_\mathcal{T}$ to denote the subset of $\mathcal{R}(X)$ consisting of exactly all such ordered pairs $ab|c$ for which the (undirected) paths connecting $\varphi(a)$ with $\varphi(b)$ and $\varphi(c)$ with $\mathfrak{r}_\mathcal{T}$ in the underlying undirected tree $\underline{\mathcal{T}}^\circ = (V, A^\circ)$ do not even have a single vertex in common. Note that $aa|b \in \mathcal{R}_\mathcal{T}$ holds for any two distinct elements a, b in X if and only if $\mathcal{T} = (V, A, \varphi)$ is a *phylogenetic rooted X-tree*, that is, if and only if φ is a bijection between X and the set of *leaves of the rooted X-tree* (V, A), that is, the set $V_{out}^{(0)}(\mathcal{T})$ of vertices $v \in V$ of outdegree 0.

And if $\mathcal{T} = (V, A, \varphi, \tau)$ is a dated rooted X-tree, put

$$D_\mathcal{T}(x, y) := 2\tau\big(\mathbf{lca}_\mathcal{T}(x, y)\big)$$

for all $x, y \in X$ where $\mathbf{lca}_\mathcal{T}(x, y) \in V$ denotes the "last common ancestor of x and y represented by $\mathcal{T}$", that is, the median $med_\mathcal{T}(x, y, \mathfrak{r}_\mathcal{T})$ of $\varphi(x), \varphi(y)$, and $\mathfrak{r}_\mathcal{T}$ relative to $\underline{\mathcal{T}}^\circ$ or, as well, the vertex where the two directed paths from $\mathfrak{r}_\mathcal{T}$ to $\varphi(x)$ and $\varphi(y)$ start separating. So, $D_\mathcal{T}(x, y)$ records twice the time that has passed since this last common ancestor of x and y lived on this planet while $D_\mathcal{T}(x, x) = 0$ holds for some $x \in X$ if and only if x is an extant species, i.e., $\tau\big(\varphi(x)\big) = 0$ holds. The factor 2 is used for calibration purposes as will become clear in Section 9.4: In case $\tau\big(\varphi(x)\big) = \tau\big(\varphi(y)\big) = 0$, the factor 2 ensures that $D_\mathcal{T}(x, y)$ is the total amount of time that "separates" the two species x and y.

In analogy with Theorems 2.4, 2.5, and 2.7 in which we showed that splits, metrics, and quartets, respectively, can be used to encode X-trees, we shall show first in this chapter that $\mathcal{C}_\mathcal{T}$, $\mathcal{R}_\mathcal{T}$, and $D_\mathcal{T}$ can all be used to encode a (dated) rooted X-tree, and that the resulting cluster and triplet systems and symmetric bivariate maps can be characterized in a fashion similar to the characterizations presented in Theorems 3.1, 3.3, and 3.7.

Furthermore, we will also explore the various interrelationships between these encodings as summarized in Figure 9.2 and, using the so-called *Farris transform*, their relations to the encodings of (unrooted) X-trees considered before.

We keep denoting by X a finite set of cardinality n, generally assumed as above to be a set of species, to which all further concepts and constructions shall refer.

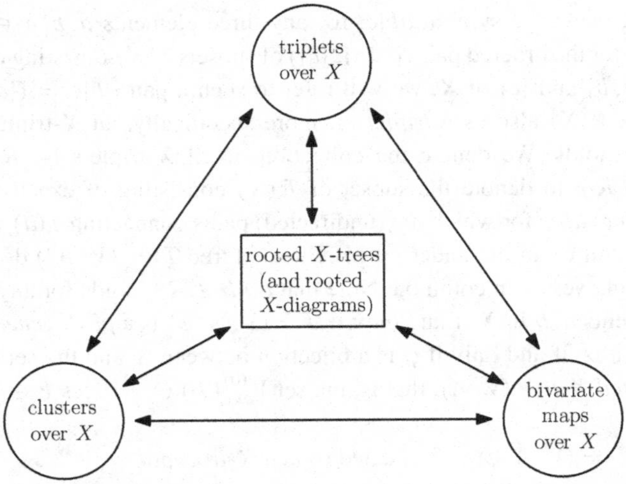

Figure 9.2 Correspondences between rooted versions of the objects pictured in Figure 9.1.

9.1 Rooted X-trees, clusters, and triplets

Denoting by $\mathcal{R}[C]$, for any subset C of X, the triplet system

$$\mathcal{R}[C] := \{ab|c : a, b \in C, c \in X - C\},$$

we begin with establishing the following simple fact:

Lemma 9.1 *Given a rooted X-tree $\mathcal{T} = (V, A, \varphi)$ and three elements $a, b, c \in X$, one has $ab|c \in \mathcal{R}_\mathcal{T}$ if and only if there exists some cluster $C \in \mathcal{C}_\mathcal{T}$ with $ab|c \in \mathcal{R}[C]$ in which case this holds in particular for the cluster $C := C_\mathcal{T}\big(\mathbf{lca}_\mathcal{T}(a, b)\big)$, while, given a cluster $C \subseteq X$, one has $C \in \mathcal{C}_\mathcal{T}$ if and only if one has $\mathcal{R}[C] \subseteq \mathcal{R}_\mathcal{T}$. That is, putting*

$$\mathcal{R}_\mathcal{C} := \bigcup_{C \in \mathcal{C}} \mathcal{R}[C]$$

for every cluster system $\mathcal{C} \subseteq \mathbb{P}(X)$ and

$$\mathcal{C}_\mathcal{R} := \{C \in \mathbb{P}(X) - \{\emptyset\} : \mathcal{R}[C] \subseteq \mathcal{R}\}$$

for every triplet system $\mathcal{R} \subseteq \mathcal{R}(X)$, one has $\mathcal{C}_{\mathcal{R}_\mathcal{T}} = \mathcal{C}_\mathcal{T}$ and $\mathcal{R}_{\mathcal{C}_\mathcal{T}} = \mathcal{R}_\mathcal{T}$ for every rooted X-tree $\mathcal{T}$ as above.

Proof Clearly, our definitions imply immediately that $a, b \in C_{\mathcal{T}}(\mathbf{lca}_{\mathcal{T}}(a, b))$ holds for all $a, b \in X$, and that $ab|c \in \mathcal{R}[C_{\mathcal{T}}(\mathbf{lca}_{\mathcal{T}}(a, b))]$ holds, for some $a, b, c \in X$ if and only if $ab|c \in \mathcal{R}_{\mathcal{T}}$ holds, implying that $\mathcal{R}_{\mathcal{T}} \subseteq \mathcal{R}_{\mathcal{C}_{\mathcal{T}}}$ holds.

Conversely, if $a, b \in C$ and $c \notin C$ holds, for some cluster $C \in \mathcal{C}_{\mathcal{T}}$ of the form $C = C(v)$ for some v in V, we must have $v \preceq_{\mathcal{T}} \mathbf{lca}_{\mathcal{T}}(a, b)$ and, therefore, $C_{\mathcal{T}}(\mathbf{lca}_{\mathcal{T}}(a, b)) \subseteq C$ and, hence, $c \notin C_{\mathcal{T}}(\mathbf{lca}_{\mathcal{T}}(a, b))$ or, equivalently, $ab|c \in \mathcal{R}_{\mathcal{T}}$ implying that also $\mathcal{R}_{\mathcal{C}_{\mathcal{T}}} \subseteq \mathcal{R}_{\mathcal{T}}$ and, hence, $\mathcal{R}_{\mathcal{T}} = \mathcal{R}_{\mathcal{C}_{\mathcal{T}}}$ as well as $C \in \mathcal{C}_{\mathcal{R}_{\mathcal{T}}}$ and, hence, $\mathcal{C}_{\mathcal{T}} \subseteq \mathcal{C}_{\mathcal{R}_{\mathcal{T}}}$ holds.

Finally, if $C \in \mathcal{C}_{\mathcal{R}_{\mathcal{T}}}$ holds, i.e., if C is a non-empty subset of X and $\mathcal{R}[C] \subseteq \mathcal{R}_{\mathcal{T}}$ holds, we must have $C = C_{\mathcal{T}}(v)$ for the last common ancestor v of all elements x in C, i.e., the first vertex in V at which the various paths from $\mathfrak{r}_{\mathcal{T}}$ to the vertices $\varphi(x)(x \in C)$ begin to diverge: Indeed, there must exist $a_0, b_0 \in C$ such the two paths from $\mathfrak{r}$ to $\varphi(a_0)$ and $\varphi(b_0)$ begin to diverge at v implying that also $v = \mathbf{lca}_{\mathcal{T}}(a_0, b_0)$ must hold, as well as $v \preceq_{\mathcal{T}} \varphi(x)$ for all $x \in C$ and, hence, $C \subseteq C_{\mathcal{T}}(v)$. Further, as $a_0b_0|c \in \mathcal{R}_{\mathcal{T}}$ and, therefore, also $c \notin C_{\mathcal{T}}(v) = C_{\mathcal{T}}(\mathbf{lca}_{\mathcal{T}}(a_0, b_0))$ holds for all $c \in X - C$, one must also have $C_{\mathcal{T}}(v) \subseteq C$ and, therefore, $C_{\mathcal{T}}(v) = C$ implying that $\mathcal{C}_{\mathcal{R}_{\mathcal{T}}} \subseteq \mathcal{C}_{\mathcal{T}}$ holds which together with the inclusion $\mathcal{C}_{\mathcal{T}} \subseteq \mathcal{C}_{\mathcal{R}_{\mathcal{T}}}$ established above implies that also $\mathcal{C}_{\mathcal{T}} = \mathcal{C}_{\mathcal{R}_{\mathcal{T}}}$ holds, as claimed. □

In analogy to the Galois connection between split and quartet systems discussed in Chapter 8, associating the triplet system $\mathcal{R}_{\mathcal{C}}$ to a cluster system $\mathcal{C}$ and, conversely, the cluster system $\mathcal{C}_{\mathcal{R}}$ to a triplet system $\mathcal{R}$ also yields a Galois connection between triplet and cluster systems. We leave it to the reader to explore further details relating to this Galois connection and, rather, provide now — in analogy to the corresponding consistency results in Chapter 3 — the following simple intrinsic characterizations of the cluster and triplet systems associated to rooted X-trees:

Theorem 9.2 (i) *Given a cluster system $\mathcal{C} \subseteq \mathbb{P}(X)$, there exists a rooted X-tree $\mathcal{T} = (V, A, \varphi)$ with $\mathcal{C}_{\mathcal{T}} = \mathcal{C}$ if and only if $X \in \mathcal{C}$ holds and $\mathcal{C}$ is a hierarchy, in which case $\mathcal{T}$ is uniquely determined, up to canonical isomorphism, by $\mathcal{C}$.*

(ii) *Similarly, given a triplet system $\mathcal{R} \subseteq \mathcal{R}(X)$, there exists a rooted X-tree $\mathcal{T} = (V, A, \varphi)$ with $\mathcal{R}_{\mathcal{T}} = \mathcal{R}$ if and only if $\mathcal{R}$ satisfies the following two conditions:*

($\mathcal{R}$1) *For any three distinct elements $a, b, c \in X$, at most one of the triplets $ab|c$, $bc|a$ and $ca|b$ is contained in $\mathcal{R}$.*

($\mathcal{R}$2) *For any four elements $a, b, c, d \in X$, $ab|c \in \mathcal{R}$ implies $ad|c \in \mathcal{R}$ or $ab|d \in \mathcal{R}$.*

And again, if this holds, $\mathcal{T}$ is uniquely determined, up to canonical isomorphism, by $\mathcal{R}$.

Proof (i) If $\mathcal{C} = \mathcal{C}_{\mathcal{T}}$ holds for some rooted X-tree $\mathcal{T} = (V, A, \varphi)$, we must have $X = C(\mathfrak{r}_{\mathcal{T}}) \in \mathcal{C}$. Moreover, $\mathcal{C}$ must be a hierarchy as $\emptyset \neq C(v)$ holds for every $v \in V$ in view of (9.1), and as $x \in X$, $v_1, v_2 \in V$, and $x \in C(v_1) \cap C(v_2)$ implies that both, v_1 and v_2, are vertices on the path $\mathbf{p}_{T}(\varphi(x), \mathfrak{r}_{\mathcal{T}})$ from $\varphi(x)$ to $\mathfrak{r}_{\mathcal{T}}$. So, one of them, say v_1, must be passed first by that path in which case $C(v_1) \subseteq C(v_2)$ must hold.

Conversely, if $\mathcal{C} \subseteq \mathbb{P}(X)$ is a hierarchy that contains X, the associated *Hasse graph* $H(\mathcal{C}) := \big(\mathcal{C}, A(\mathcal{C})\big)$ with vertex set $\mathcal{C}$ and arc set

$$\begin{aligned} A(\mathcal{C}) &:= \big\{(C_1, C_2) \in \mathcal{C} \times \mathcal{C} : |\{C \in \mathcal{C} : C_1 \supseteq C \supseteq C_1\}| = 2\big\} \\ &= \big\{(C_1, C_2) \in \mathcal{C} \times \mathcal{C} : C_1 \supsetneq C_2 \text{ and } \emptyset = \{C \in \mathcal{C} : C_1 \supsetneq C \supsetneq C_2\}\big\} \end{aligned}$$

is a rooted tree relative to the root $\mathfrak{r} := X$, and we have $C \preceq_{H(\mathcal{C})} C'$ for some $C, C' \in \mathcal{C}$ relative to that tree if and only if $C \supseteq C'$ holds (see Figure 9.5 for an example of a Hasse graph).

Furthermore, defining a labeling map $\varphi_{\mathcal{C}} : X \to \mathcal{C}$ by associating, to any $x \in X$, the unique smallest cluster $C_x := C_x^{\mathcal{C}}$ in $\mathcal{C}$ containing x, i.e., the intersection $\varphi_{\mathcal{C}}(x) := \bigcap_{C \in \mathcal{C}_x} C$ of all clusters C in $\mathcal{C}_x := \{C \in \mathcal{C} : x \in C\}$, the set of all $C \in \mathcal{C}$ containing x, we obtain a triple $\mathcal{T}(\mathcal{C}) := (\mathcal{C}, A(\mathcal{C}), \varphi_{\mathcal{C}})$ which is easily seen to form a rooted X-tree whose underlying directed tree $T(\mathcal{C})$ coincides with $H(\mathcal{C})$ and for which $\mathcal{C} = \mathcal{C}_{\mathcal{T}(\mathcal{C})}$ holds: Indeed, we have

$$C = \{x \in X : \varphi_{\mathcal{C}}(x) \subseteq C\} = \{x \in X : C \preceq_{H(\mathcal{C})} \varphi_{\mathcal{C}}(x)\}$$

for every $C \in \mathcal{C}$. Moreover, the image $\varphi_{\mathcal{C}}(X)$ contains (at least) all vertices $C \in \mathcal{C}$ of degree 1 as well as all those vertices C of degree 2 that are distinct from the root X as it consists of all $C \in \mathcal{C}$ that are distinct from the union $\bigcup_{C' \in \mathcal{C}, C' \subsetneq C} C' = \bigcup_{(C,C') \in A(\mathcal{C})} C'$ of all of its proper subclusters C' in $\mathcal{C}$. So, it contains in particular every $C \in \mathcal{C}$ for which there is only one or no cluster $C' \in \mathcal{C}$ with $(C, C') \in A(\mathcal{C})$ and, hence, also the root X in case its out-degree in $T(\mathcal{C})$ is 1.

Finally, given any rooted X-tree $\mathcal{T} = (V, A, \varphi)$ with $\mathcal{C}(\mathcal{T}) = \mathcal{C}$, the map $\psi : V \to \mathcal{C} : v \mapsto C_{\mathcal{T}}(v)$ must indeed be a canonical isomorphism from $\mathcal{T}$ onto $\mathcal{T}(\mathcal{C})$: In view of our assumption $\mathcal{C}(\mathcal{T}) = \mathcal{C}$, it must be surjective, and it must be injective in view of (9.2) — so, it must be a graph isomorphism in view of (9.1). And $\psi\big(\varphi(x)\big) = C_{\mathcal{T}}\big(\varphi(x)\big)$ must coincide with $\varphi_{\mathcal{C}}(x) = \bigcap_{C \in \mathcal{C}_x} C$ for every $x \in X$ as $C_{\mathcal{T}}\big(\varphi(x)\big) = \{y \in X : \varphi(x) \preceq_{\mathcal{T}} \varphi(y)\}$ clearly is a cluster in $\mathcal{C}_x$ that — in view of the transitivity of the binary relation "$\preceq_{\mathcal{T}}$" — must, in turn, be contained in any cluster $C = C_{\mathcal{T}}(v) = \{y \in X : v \preceq_{\mathcal{T}} \varphi(y)\} \in \mathcal{C}_{\mathcal{T}}$ that contains x.

(ii) In view of Lemma 9.1 and (i), it suffices to note that a triplet system $\mathcal{R} \subseteq \mathcal{R}(X)$ is of the form $\mathcal{R} = \mathcal{R}_{\mathcal{C}}$ for some hierarchy $\mathcal{C} \subseteq \mathbb{P}(X)$ with $X \in \mathcal{C}$ if and only if $\mathcal{R}$ satisfies the Conditions ($\mathcal{R}1$) and ($\mathcal{R}2$).

It is obvious that ($\mathcal{R}1$) and ($\mathcal{R}2$) must hold for any triplet system $\mathcal{R} \subseteq \mathcal{R}(X)$ that is of the form $\mathcal{R} = \mathcal{R}_{\mathcal{C}}$ for some hierarchy $\mathcal{C} \subseteq \mathbb{P}(X)$: We cannot have $ab|c \in \mathcal{R}$ and $bc|a \in \mathcal{R}$ for some $a, b, c \in X$ in this case as this would imply the existence of clusters $C, C' \in \mathcal{C}$ with $ab|c \in \mathcal{R}[C]$ and $bc|a \in \mathcal{R}[C']$ which is impossible as $b \in C \cap C'$ implies that $C \subseteq C'$ or $C' \subseteq C$ must hold.

Similarly, $a, b, c, d \in X$ and $ab|c \in \mathcal{R}[C]$ for some $C \subseteq X$ implies that either $ad|c \in \mathcal{R}[C]$ (in case $d \in C$) or $ab|d \in \mathcal{R}[C]$ (in case $d \notin C$) must hold. So, also ($\mathcal{R}2$) holds for every triplet system $\mathcal{R}$ of the form $\mathcal{R} = \mathcal{R}_{\mathcal{C}}$ for some cluster system $\mathcal{C} \subseteq \mathbb{P}(X)$.

It is also obvious that $\mathcal{C}_{\mathcal{R}}$ must be a hierarchy that contains X for every triplet system $\mathcal{R} \subseteq \mathcal{R}(X)$ that satisfies the Condition ($\mathcal{R}1$) as $C, C' \in \mathcal{C}_{\mathcal{R}}$, $C \cap C' \neq \emptyset$, $C - C' \neq \emptyset$, and $C' - C \neq \emptyset$ would imply the existence of elements $b \in C \cap C'$, $a \in C - C'$, and $c \in C' - C$ with $ab|c \in \mathcal{R}$ and $bc|a \in \mathcal{R}$ in contradiction to ($\mathcal{R}1$).

So, it remains to show that $\mathcal{R} = \mathcal{R}_{\mathcal{C}_{\mathcal{R}}}$ must hold for every triplet system $\mathcal{R} \subseteq \mathcal{R}(X)$ that satisfies the conditions ($\mathcal{R}1$) and ($\mathcal{R}2$): By definition, we have $\mathcal{R}_{\mathcal{C}_{\mathcal{R}}} \subseteq \mathcal{R}$. Conversely, if $ab|c \in \mathcal{R}$ holds for some triplet $ab|c \in \mathcal{R}(X)$, we have $a, b \in C_{\mathcal{R}}(a, b) := X - \{x \in X : ab|x \in \mathcal{R}\}$ and $c \notin C_{\mathcal{R}}(a, b)$. And we have $C_{\mathcal{R}}(a, b) \in \mathcal{C}_{\mathcal{R}}$ as $a', b' \in C_{\mathcal{R}}(a, b)$ and $c' \notin C_{\mathcal{R}}(a, b)$ implies $ab|c' \in \mathcal{R}$ as well as, in view of ($\mathcal{R}2$) applied to $ab|c'$ and a' or b' respectively, also $aa'|c' \in \mathcal{R}$ and $ab'|c' \in \mathcal{R}$. Thus, ($\mathcal{R}2$) applied to $aa'|c'$ and b' implies that also $aa'|b' \in \mathcal{R}$ or $a'b'|c' \in \mathcal{R}$ must hold while ($\mathcal{R}2$) applied to $ab'|c'$ and a' implies that $ab'|a' \in \mathcal{R}$ or $a'b'|c' \in \mathcal{R}$ must hold. So, as $aa'|b' \in \mathcal{R}$ and $ab'|a' \in \mathcal{R}$ cannot hold simultaneously in view of ($\mathcal{R}1$), $a'b'|c' \in \mathcal{R}$ must indeed hold for all $a', b' \in C_{\mathcal{R}}(a, b)$ and $c' \notin C_{\mathcal{R}}(a, b)$ implying that also $C_{\mathcal{R}}(a, b) \in \mathcal{C}_{\mathcal{R}}$ and, hence, $ab|c \in \mathcal{R}_{\mathcal{C}_{\mathcal{R}}}$ holds for any $ab|c \in \mathcal{R}$. So, we must have $\mathcal{R} = \mathcal{R}_{\mathcal{C}_{\mathcal{R}}}$, as claimed. □

Corollary 9.3 *Any triplet system $\mathcal{R} \subseteq \mathcal{R}(X)$ that satisfies the conditions ($\mathcal{R}1$) and ($\mathcal{R}2$) also satisfies the conditions*

($\mathcal{R}3$) *$ab|d, bc|d \in \mathcal{R}$ implies $ac|d \in \mathcal{R}$.*
($\mathcal{R}4$) *$ab|c, cd|a \in \mathcal{R}$ implies $ab|d \in \mathcal{R}$.*

Proof ($\mathcal{R}3$): Choose some clusters $C, C' \in \mathcal{C}_{\mathcal{R}}$ with $a, b \in C$, $d \notin C$ and $b, c \in C'$, $d \notin C'$. Note that $b \in C \cap C'$ implies that either $C \subseteq C'$ and, then, also $a, c \in C'$ and $d \notin C'$ or that $C' \subseteq C$ and, then, also $a, c \in C$ and $d \notin C$ holds. So, $ac|d \in \mathcal{R}$ must hold in any case.

($\mathcal{R}4$): Choose some clusters $C, C' \in \mathcal{C}_{\mathcal{R}}$ with $a, b \in C, c \notin C$ and $c, d \in C'$ and $a \notin C'$. Then, we must have $d \notin C$ and, therefore, $ab|d \in \mathcal{R}$ as claimed as $a, b \in C$ and $c \notin C$ together with $c, d \in C'$ and $a \notin C'$ implies that neither $C \subsetneq C'$ nor $C' \subsetneq C$ holds and, therefore, $C \cap C' = \emptyset$. □

It is clearly of quite some interest to further investigate cluster and triplet systems with all sorts of specific properties as this might help to design better algorithms for deriving trees or networks from real biological data. We will come back to this point later in this section.

9.2 Dated rooted X-trees and hierarchical dissimilarities

Suppose now that we are given a dated rooted X-tree $\mathcal{T} = (V, A, \varphi, \tau)$. It follows immediately from its definition that the map $D := D_{\mathcal{T}}$ defined in the last section is a *dissimilarity*, that is, $D(x, x) \leq D(x, y) = D(y, x)$ holds for all $x, y \in X$. It is also easy to check that, in addition, D satisfies the following 3-*point condition* for all $x, y, z \in X$ (cf. Figure 9.3(b)):

$$D(x, y) \leq \max\{D(x, z), D(y, z)\}. \tag{9.3}$$

Indeed, assume — without loss of generality — that, say, $D(y, z) \leq D(x, z)$ holds. Clearly, this implies that down the line from the last common ancestor $\mathbf{lca}_{\mathcal{T}}(x, z)$ of x and z to z we must somewhere meet the last common ancestor $\mathbf{lca}_{\mathcal{T}}(y, z)$ of y and z. So, $\mathbf{lca}_{\mathcal{T}}(x, z)$ must also be an ancestor of y implying that also $D(x, y) = \tau(\mathbf{lca}_{\mathcal{T}}(x, y)) \leq \tau(\mathbf{lca}_{\mathcal{T}}(x, z)) = D(x, z)$ must hold in this case.

A symmetric bivariate map $D \in \mathbb{R}^{X \times X}$ that satisfies this condition is necessarily a dissimilarity (putting $y := x$ in (9.3) yields that $D(x, x) \leq D(x, z)$ must hold for all $x, z \in X$) and will, henceforth, also be called a *hierarchical dissimilarity*. And if D is simultaneously a metric, it will also be called an *ultrametric*. Clearly, a hierarchical dissimilarity D is a metric — and, thus, an ultrametric — if and only if $D(x, x) = 0$ holds for all $x \in X$.

We will show now that any dated rooted X-tree $\mathcal{T}$ is, in fact, encoded by the associated hierarchical dissimilarity $D_{\mathcal{T}}$ (cf. [21]):

Theorem 9.4 *A dissimilarity $D : X \times X \to \mathbb{R}$ is of the form $D = D_{\mathcal{T}}$ for some dated rooted X-tree $\mathcal{T}$ if and only if it is a hierarchical dissimilarity. Furthermore, the tree $\mathcal{T}$ is uniquely determined, up to canonical isomorphism, by the map D.*

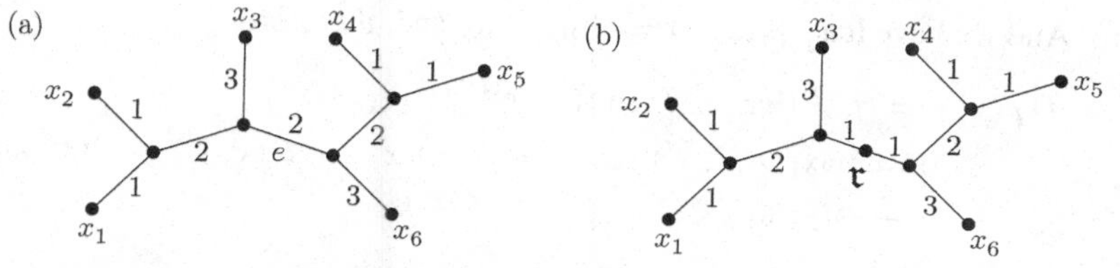

Figure 9.3 (a) An example of an edge-weighted phylogenetic $\{x_1, \ldots, x_6\}$-tree $\mathcal{T}$. (b) The metric D induced by the tree $\mathcal{T}$ in (a) satisfies the 3-point condition (9.3) since the new vertex $\mathfrak{r}$ that subdivides edge e has the same distance to all leaves in $\mathcal{T}$.

Proof It has been noted already above that every dissimilarity D of the form $D = D_{\mathcal{T}}$ for some dated rooted X-tree $\mathcal{T}$ is a hierarchical dissimilarity. To show that, conversely, every hierarchical dissimilarity is also of this form, assume that D is a hierarchical dissimilarity and proceed as follows:

(i) put $X_\alpha = X_\alpha^{(D)} := \{x \in X : D(x, x) \leq \alpha\}$ for every $\alpha \in \mathbb{R}$,
(ii) note that the collection $\mathcal{C} = \mathcal{C}(D)$ consisting of all subsets C of X of the form $C = C_\alpha(x) = C_\alpha^{(D)}(x) := \{y \in X : D(x, y) \leq \alpha\}$ for some $\alpha \in \mathbb{R}$ and $x \in X_\alpha$ forms a hierarchy that contains X — the latter resulting for $\alpha := \max D := \max\{D(x, y) : x, y \in X\}$,
(iii) define a map $\tau^{(D)} : \mathcal{C}(D) \to \mathbb{R}$ by putting
$$\tau^{(D)}(C) := \max\{D(x, y) : x, y \in C\}$$
for every $C \in \mathcal{C}$, and
(iv) note finally that $\varphi_{\mathcal{C}}(x) = C_{D(x,x)}(x)$ holds for every $x \in X$.

Thus, we may consider the dated rooted tree
$$\mathcal{T}^{(D)} := \big(\mathcal{C}(D), A\big(\mathcal{C}(D)\big), \varphi^{(D)}, \tau^{(D)}\big)$$
with vertex set $\mathcal{C}(D)$, edge set $A\big(\mathcal{C}(D)\big)$, labeling map $\varphi^{(D)} := \varphi_{\mathcal{C}(D)}$, and dating map $\tau^{(D)}$ as defined above for which $D = D_{\mathcal{T}^{(D)}}$ holds:

(i) Indeed, the triple $\big(\mathcal{C}(D), A\big(\mathcal{C}(D)\big), \varphi_{\mathcal{C}(D)}\big)$ is clearly one — and hence, up to canonical isomorphism, the *unique* — rooted X-tree $\mathcal{T}$ with $\mathcal{C}(D) = \mathcal{C}_{\mathcal{T}}$.
(ii) Furthermore, $\tau^{(D)}$ is strictly monotonically decreasing as $C_1 \supsetneq C_2$ holds for every arc $(C_1, C_2) \in A\big(\mathcal{C}(D)\big)$ and, therefore, $\tau^{(D)}(C_1) = \max\{D(x, y) : x, y \in C_1\} > \tau^{(D)}(C_2) = \max\{D(x, y) : x, y \in C_2\}$ because $D(x, y') > D(x, y)$ must hold for all $x, y \in C_2$ and $y' \in C_1 - C_2$ (as $D(x, y') \leq D(x, y)$ and $x, y \in C_2$ would imply $y' \in C_2$).

(iii) And we have $\mathbf{lca}_{\mathcal{T}^{(D)}}(x, y) = C_{D(x,y)}(x)$ and, therefore,

$$\begin{aligned} D_{\mathcal{T}}(x, y) &= \tau^{(D)}\big(\mathbf{lca}_{\mathcal{T}^{(D)}}(x, y)\big) = \tau^{(D)}\big(C_{D(x,y)}(x)\big) \\ &= \max\{D(x', y') : x', y' \in X;\ D(x, x'), D(x, y') \leq D(x, y)\} \\ &= D(x, y) \end{aligned}$$

for all $x, y \in X$.

And finally, mapping each vertex v of an arbitrary dated rooted X-tree $\mathcal{T}$ with $D = D_{\mathcal{T}}$ onto the subset $C_{\mathcal{T}}(v)$ of X, it is also easily verified that $C_{\mathcal{T}}(v) \in \mathcal{C}(D)$ holds and that this map defines a canonical isomorphism from $\mathcal{T}$ onto $\mathcal{T}^{(D)}$. We leave details to the reader. □

It is also worth noting that, given any dissimilarity $D : X \times X \to \mathbb{R}$ defined on X, there exists a unique largest hierarchical dissimilarity $D^{(0)}$ defined on X with $D^{(0)} \leq D$ (i.e., $D^{(0)}(x, y) \leq D(x, y)$ for all x, y) — also called the *subdominant hierarchical dissimilarity* for D — which can be defined by first denoting by $\mathcal{D}_{\leq D}$ the collection of all hierarchical dissimilarities D' defined on X with $D' \leq D$ and then putting

$$D^{(0)}(x, y) := \sup\{D'(x, y) : D' \in \mathcal{D}_{\leq D}\}$$

for all $x, y \in X$: Indeed, one has

$$\begin{aligned} D^{(0)}(x, x) &= \sup\{D'(x, x) : D' \in \mathcal{D}_{\leq D}\} \\ &\leq \sup\{D'(x, y) : D' \in \mathcal{D}_{\leq D}\} = D^{(0)}(x, y) \end{aligned}$$

and

$$\begin{aligned} D^{(0)}(x, y) &= \sup\{D'(x, y) : D' \in \mathcal{D}_{\leq D}\} \\ &\leq \sup\{\max\{D'(x, z), D'(z, y)\} : D' \in \mathcal{D}_{\leq D}\} \\ &= \max\{\sup\{D'(x, z) : D' \in \mathcal{D}_{\leq D}\}, \sup\{D'(z, y) : D' \in \mathcal{D}_{\leq D}\}\} \\ &= \max\{D^{(0)}(x, z), D^{(0)}(z, y)\} \end{aligned}$$

for all $x, y, z \in X$.

Moreover, it is easy to see that $D^{(0)}(x, y)$ coincides, for all $x, y \in X$, with the minimum, over all sequences $x = x_0, x_1, \ldots, x_\ell = y$ of distinct points in X, of the numbers $\max\{D(x_{i-1}, x_i) : i = 1, \ldots, \ell\}$. Thus, the hierarchy $\mathcal{C}(D^{(0)})$ contains a subset A of X if and only if there exists some real number α such that the set A forms a connected component of the graph $\big(X, \big\{\{x, y\} : D(x, y) \leq \alpha\big\}\big)$.

In particular, $\mathcal{C}(D^{(0)})$ contains the collection $\mathcal{A}(D)$ of *Apresjan clusters* for D [4] consisting of all non-empty subsets A of X for which $D(a, a') < D(a, b)$

holds for all $a, a' \in A$ and $b \in X - A$ (which is clearly a hierarchy as $A_1, A_2 \in \mathcal{A}(D)$, $a \in A_1 \cap A_2$, $a' \in A_1 - A_2$, and $a'' \in A_2 - A_1$ would imply the contradiction $D(a, a') < D(a, a'') < D(a, a')$): Indeed, assuming that $X \neq A \in \mathcal{A}(D)$ holds, choose $a_0 \in A$ and $b_0 \in X - A$ with $D(a_0, b_0) \leq D(a, b)$ for all $a \in A$ and $b \in X - A$ and note that $D(a, a_0) < D(a_0, b_0) \leq D^{(0)}(a, b)$ and, hence, $D^{(0)}(a, a') \leq \max\{D(a, a_0), D(a', a_0)\} < D(a_0, b_0) \leq D^{(0)}(a, b)$ must hold for all $a, a' \in A$ and $b \in X - A$.

Remarkably, to construct $\mathcal{A}(D)$ for a dissimilarity $D \in \mathbb{R}^{X \times X}$, we only need to compare "distances" between pairs a, b and a', b' of elements in X for which the intersection $\{a, b\} \cap \{a', b'\}$ is non-empty. More specifically, ranking the elements of X relative to some fixed element $a \in X$ (and D) by defining the *rank*

$$rk_a(x) = rk_a^D(x) := |\{b \in X : D(a, b) \leq D(a, x)\}|,$$

it is easily seen that $C \in \mathcal{A}(D)$ holds for a non-empty subset C of X if and only if $rk_a(x) \leq |C|$ holds for all $a, x \in C$. One may say that the resulting function $rk_a : X \to \mathbb{N} : x \mapsto rk_a(x)$ represents, for any $a \in X$, the "world view" of a relative to D that may also be viewed as a "phylogenetic signature" of a relative to D. For more details on algorithms for computing the subdominant hierarchical dissimilarities and these rank functions see e.g., [47, 53, 119]. In the following sections, we will discuss some further properties of cluster systems, triplet systems, and dissimilarities and relate them to split systems, quartet systems, and metrics using the "Farris transform".

9.3 Affine versus projective clustering and the combinatorial Farris transform

In case X is a set of species, taxonomists commonly make use of the canonical one-to-one correspondence between rooted and augmented finite trees mentioned in Chapter 1 that easily extends to X-trees as follows. First, choose any element $*$ not contained in X and put $X^* := X + *$. Then, one may associate, to each (dated) rooted X-tree $\mathcal{T} = (V, A, \varphi)$ (or $\mathcal{T} = (V, A, \varphi, \tau)$) the X^*-tree $\mathcal{T}^\circ = (V, A^\circ, \varphi^\circ)$ where φ° extends the labeling map φ to X^* by putting $\varphi^\circ(*) := \mathfrak{r}_{\mathcal{T}}$ (or the edge-weighted X^*-tree $\mathcal{T}^\circ = (V, A^\circ, \varphi^\circ, \omega^\circ)$ with φ° as above and ω° denoting that edge-weighting of A° that maps every edge $e = \{u, v\} \in A^\circ$ onto the (necessarily positive!) real number $|\tau(u) - \tau(v)|$). Once again, it is well-known folklore and easy to see that this sets up a canonical one-to-one correspondence between (edge-weighted) X^*-trees and (dated) rooted X-trees.

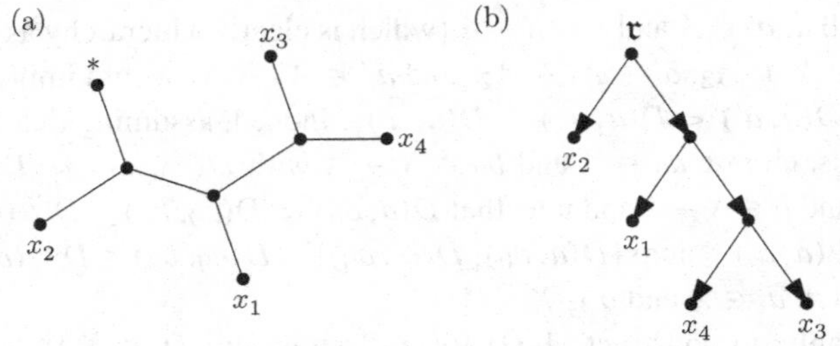

Figure 9.4 (a) A phylogenetic X^*-tree $\mathcal{T}^*$ for $X := \{x_1, x_2, x_3, x_4\}$. (b) The rooted X-tree $\mathcal{T}$ obtained by cutting off the pendant edge of $\mathcal{T}^*$ containing the leaf $*$, declaring the interior vertex incident with that edge to represent the root of $\mathcal{T}$, and orienting the remaining edges accordingly.

So, taxonomists may first choose an "outgroup" $*$ for X that can be any species that is distantly, but not too distantly related to the species in X (e.g., a frog or a bird, but not an insect or a snail for a diverse group of mammals) and then use any tree-(re-)construction algorithm to obtain a good X^*-tree $\mathcal{T}^* = (V^*, E^*, \varphi^*)$ for the extended set X^*. If the outgroup $*$ was chosen appropriately, the vertex $\varphi^*(*)$ will be a leaf in $\mathcal{T}^*$ that is distinct from the vertices $\varphi^*(x)$ for every $x \in X$. So, one may then either consider the (unique!) rooted X-tree $\mathcal{T} = (V, A, \varphi)$ with $\mathcal{T}^\circ = \mathcal{T}^*$.

Alternatively, if — for example — one does not want a frog or a bird to play the role of a root for a group of mammals, one first modifies the X^*-tree $\mathcal{T}^*$ as follows: One deletes the vertex $\varphi^*(*)$ and the pendant edge it is attached to. Then, one modifies the value $\varphi^*(*)$ the map φ^* attains on the element $*$ by putting $\varphi^*(*) := v_{\mathcal{T}^*}\big(\varphi^*(*)\big)$. And then, one applies the above construction of rooted X-trees from X^*-trees to this modified tree to obtain the desired rooted X-tree. That is, one cuts off the pendant edge of $\mathcal{T}^*$ containing the leaf $\varphi^*(*)$ of $\mathcal{T}^*$ representing the outgroup "$*$", declares the interior vertex incident with that edge to represent the root of the tree, and orients the remaining edges of $\mathcal{T}^*$ "away" from that root. For example, the rooted $\{x_1, x_2, x_3, x_4\}$-tree $\mathcal{T}$ in Figure 9.4(b) is obtained from the phylogenetic $\{x_1, x_2, x_3, x_4, *\}$-tree $\mathcal{T}^*$ depicted in Figure 9.4(a) in this way.

And if there is any good edge-weighting for the X^*-tree (V^*, E^*, φ^*), then there is — as indicated above — a corresponding dating map for the rooted X-tree one has obtained. In addition, checking whether performing this procedure for a number of various distinct outgroups $*$ yields isomorphic or, at least, sufficiently similar (dated) rooted X-trees can help to check your construction's reliability.

In a more abstract setting, this construction suggests we consider a more general interplay between two different types of clustering models [51]: In the *affine* clustering model, the information that is being sought is gathered in terms of non-empty subsets or *clusters* of X whereas, in the *projective* model, we look for suitable *splits* of X. As in classical geometry, it turns out that, while the affine model is often more easily grasped and seems to reflect the intuitive understanding of clustering better, results and proofs often have a more elegant appearance when presented using the projective model (see, for example, [33, 35]).

As suggested by the example above, the affine and projective clustering models can be related by some sort of combinatorial Farris transform in the following way: Given a set X and a cluster system $\mathcal{C} \subseteq \mathbb{P}(X)$, we augment X by a new element "$*$" not contained in X (in analogy to adding a "point at infinity") and define the split system $\Sigma^*(\mathcal{C})$ on $X^* := X + *$ by

$$\Sigma^*(\mathcal{C}) := \{C|(X^* - C) : C \in \mathcal{C}\}.$$

Similarly, for any split system $\Sigma \subseteq \Sigma(X^*)$, we define the *combinatorial Farris transform* $\mathcal{C}_*(\Sigma) \subseteq \mathbb{P}(X)$ of Σ by

$$\mathcal{C}_*(\Sigma) := \{\overline{S}(*) : S \in \Sigma\}.$$

It is obvious that $\Sigma^*(\mathcal{C}_*(\Sigma)) = \Sigma$ holds for any split system $\Sigma \subseteq \Sigma(X^*)$, and $\mathcal{C}_*(\Sigma^*(\mathcal{C})) = \mathcal{C}$ for any cluster system $\mathcal{C} \subseteq \mathbb{P}(X)$.

The following proposition asserts that the combinatorial Farris transform of a split system $\Sigma \subseteq \Sigma(X^*)$ is a hierarchy if and only if Σ is a compatible split system:

Proposition 9.5 *A split system $\Sigma \subseteq \Sigma(X^*)$ is compatible if and only if its combinatorial Farris transform $\mathcal{C}_*(\Sigma)$ is a hierarchy.*

Proof Suppose that $\mathcal{C}_*(\Sigma)$ is a hierarchy and consider two arbitrary splits $S, S' \in \Sigma$. If we have $\overline{S}(*) \cap \overline{S'}(*) = \emptyset$, then S and S' are compatible. Otherwise, since $\mathcal{C}_*(\Sigma)$ is a hierarchy, we may assume, without loss of generality, that $\overline{S}(*) \subseteq \overline{S'}(*)$ holds implying as well that S and S' are compatible in view of $\overline{S}(*) \cap S'(x) = \emptyset$.

Conversely, suppose that Σ is compatible and consider two splits $S, S' \in \Sigma$. As $* \in S(*) \cap S'(*)$ always holds, we have either (i) $\overline{S}(*) \cap \overline{S'}(*) = \emptyset$, or (ii) $\overline{S}(*) \cap S'(*) = \emptyset$, and, therefore, $\overline{S}(*) \subseteq \overline{S'}(*)$, or (iii) $S(*) \cap \overline{S'}(*) = \emptyset$ and, therefore, $\overline{S'}(*) \subseteq \overline{S}(*)$ implying that $C^*(\Sigma)$ is indeed a hierarchy, as claimed. □

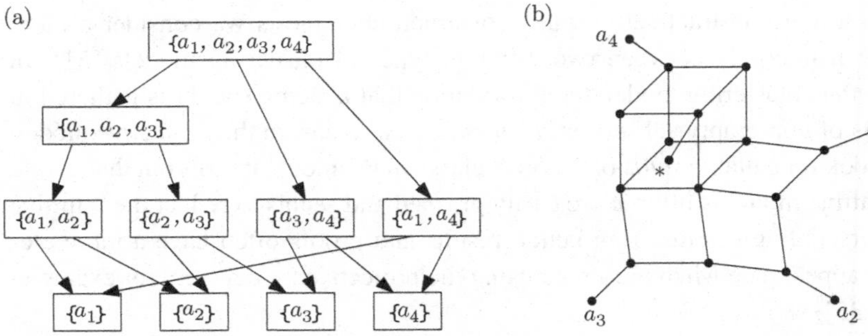

Figure 9.5 (a) The Hasse graph $H(\mathcal{C})$ for the cluster system $\mathcal{C}$ over $X = \{a_1, \ldots, a_4\}$ containing all singletons and the subsets $\{a_1, a_2, a_3, a_4\}$, $\{a_1, a_2, a_3, \}$, $\{a_1, a_2\}$, $\{a_2, a_3\}$, $\{a_3, a_4\}$, and $\{a_1, a_4\}$. (b) The Buneman graph of the associated split system $\Sigma^*(\mathcal{C})$.

In Chapter 4, we saw that we can represent any split system $\Sigma \subseteq \Sigma(X)$ by an X-labeled Buneman graph $(\mathcal{B}(\Sigma), \varphi_\Sigma)$, and that, in case Σ is compatible, this graph represents, up to canonical isomorphism, the unique X-tree $\mathcal{T}$ with $\Sigma = \Sigma_\mathcal{T}$. Remarkably, one can view the Hasse graph of an arbitrary cluster system $\mathcal{C}$ as an "affine" analogon of the Buneman graph: Figure 9.5 presents a particular instance, and the reader is invited to work out the relationship between Hasse and Buneman graphs for arbitrary cluster systems and their associated split systems in general.

Further properties of the Hasse graph associated to a cluster system and its relationship to X-trees are studied in [19]. As with the Buneman graph of split systems that are not compatible, the Hasse graphs $H(\mathcal{C})$ of a cluster system $\mathcal{C}$ that is not a hierarchy can be very complex. To overcome this difficulty, there has been a great deal of activity recently in developing alternative ways to associate an X-labeled network to a cluster system, see for example [100].

Next, consider a phylogenetic X^*-tree $\mathcal{T}^*$ and the associated phylogenetic rooted X-tree $\mathcal{T}$ obtained from $\mathcal{T}^*$ as described above. It is easy to see that, putting

$$\mathcal{R}(\mathcal{Q}) := \{ab|c \in \mathcal{R}(X) : ab|c* \in \mathcal{Q}\} \cup \{aa|b : a, b \in X, a \neq b\} \subseteq \mathcal{R}(X)$$

for every quartet system $\mathcal{Q} \subseteq \mathcal{Q}(X^*)$ and

$$\mathcal{Q}(\mathcal{R}) := \{ab|cd \in \mathcal{Q}(X) : ab|c \in \mathcal{R} \text{ and } ab|d \in \mathcal{R}\} \cup$$
$$\{ab|c* \in \mathcal{Q}(X^*) : ab|c \in \mathcal{R}\} \subseteq \mathcal{Q}(X^*)$$

for every triplet system $\mathcal{R} \subseteq \mathcal{R}(X)$, one has $\mathcal{R}_{\mathcal{T}} = \mathcal{R}(\mathcal{Q}_{\mathcal{T}^*})$ and $\mathcal{Q}_{\mathcal{T}^*} = \mathcal{Q}(\mathcal{R}_{\mathcal{T}})$ for the triplet system $\mathcal{R}_{\mathcal{T}}$ associated to $\mathcal{T}$ and the quartet system $\mathcal{Q}_{\mathcal{T}^*}$ associated to $\mathcal{T}^*$.

In consequence, we have

Theorem 9.6 *Given a triplet system $\mathcal{R} \subseteq \mathcal{R}(X)$ and a quartet system $\mathcal{Q} \subseteq \mathcal{Q}(X^*)$, the following assertions are equivalent:*

(i) *There exists a phylogenetic rooted X-tree $\mathcal{T}$ with $\mathcal{R} = \mathcal{R}_{\mathcal{T}}$ and $\mathcal{Q} = \mathcal{Q}_{\mathcal{T}^\circ}$ where $\mathcal{T}^\circ$, of course, denotes the X^*-tree associated with $\mathcal{T}$.*

(ii) *$\mathcal{Q}$ is thin, transitive, and saturated, and one has*

$$\mathcal{R} = \{ab|c \in \mathcal{R}(X) : ab|c* \in \mathcal{Q}\} \cup \{aa|b : a, b \in X, a \neq b\}.$$

(iii) *$\mathcal{R}$ satisfies the conditions ($\mathcal{R}1$) and ($\mathcal{R}2$) and one has*

$$\mathcal{Q} = \{ab|cd \in \mathcal{Q}(X) : ab|c \in \mathcal{R} \text{ and } ab|d \in \mathcal{R}\} \cup \{ab|c* \in \mathcal{Q}(X^*) : ab|c \in \mathcal{R}\}.$$

In particular, a quartet system $\mathcal{Q} \subseteq \mathcal{Q}(X^*)$ is thin, transitive, and saturated if and only if the triplet system $\mathcal{R}(\mathcal{Q}) \subseteq \mathcal{R}(X)$ satisfies the conditions ($\mathcal{R}1$) and ($\mathcal{R}2$) in which case also $\mathcal{Q} = \mathcal{Q}\big(\mathcal{R}(\mathcal{Q})\big)$ holds.

9.4 Hierarchical dissimilarities, hyperbolic maps, and their Farris transform

In this section, we will investigate a certain construction that will allow us to relate metrics and hierarchical dissimilarities by manipulating bivariate maps D using univariate maps f in a rather natural and simple way: Given any symmetric bivariate map $D \in \mathbb{R}^{X \times X}$ and any map $f \in \mathbb{R}^X$, we define the *Farris transform* D^f of D relative to f to be the map

$$D^f : X \times X \to \mathbb{R} : (x, y) \mapsto D(x, y) - f(x) - f(y). \tag{9.4}$$

Such transforms were originally proposed by J. S. Farris *et al.* in [75, p.181] (see also [73, 74]), and were dubbed Farris transforms in [6] (see also [10, 63]). Clearly, D^f is a dissimilarity if and only if $f(y) - f(x) \leq D(x, y) - D(x, x)$ holds for all $x, y \in X$. Moreover, given D, there exists some such map f if and only if $D(x, x) + D(y, y) \leq 2D(x, y)$ holds for all $x, y \in X$: Indeed, if such a map f exists, one must have $0 = \big(f(y) - f(x)\big) + \big(f(x) - f(y)\big) \leq \big(D(x, y) - D(x, x)\big) + \big(D(x, y) - D(y, y)\big)$ and, therefore,

$D(x,x)+D(y,y) \le 2D(x,y)$ as claimed. And if this holds for all $x, y \in X$, one has $f(y) - f(x) \le D(x,y) - D(x,x)$ for the map

$$f := f_D : X \to \mathbb{R} : x \mapsto \frac{D(x,x)}{2}. \tag{9.5}$$

To motivate this definition, first put $X^* := X + *$ for some element "$*$" not in X, as above. Then, given a dated rooted X-tree $\mathcal{T} = (V, A, \varphi, \tau)$, consider the associated hierarchical dissimilarity $D_{\mathcal{T}}$ and the metric $D_{\mathcal{T}^\circ}$ defined on X^* that is associated with the corresponding edge-weighted X^*-tree $\mathcal{T}^\circ$. Clearly, given any two elements $x, y \in X$, one has

$$\begin{aligned} & D_{\mathcal{T}^\circ}(x,y) + 2\tau(\mathfrak{r}_{\mathcal{T}}) \\ &= \tau\big(\mathbf{lca}_{\mathcal{T}}(x,y)\big) - \tau\big(\varphi(x)\big) + \tau\big(\mathbf{lca}_{\mathcal{T}}(x,y)\big) - \tau\big(\varphi(y)\big) + 2\tau(\mathfrak{r}_{\mathcal{T}}) \\ &= [\tau(\mathfrak{r}_{\mathcal{T}}) - \tau\big(\varphi(x)\big)] + [\tau(\mathfrak{r}_{\mathcal{T}}) - \tau\big(\varphi(y)\big)] + 2\tau\big(\mathbf{lca}_{\mathcal{T}}(x,y)\big) \\ &= D_{\mathcal{T}^\circ}(*,x) + D_{\mathcal{T}^\circ}(*,y) + D_{\mathcal{T}}(x,y). \end{aligned} \tag{9.6}$$

Thus, defining the function $f = f_{\mathcal{T}} : X \to \mathbb{R}$ by putting $f_{\mathcal{T}}(x) := \tau(\mathfrak{r}_{\mathcal{T}}) - D_{\mathcal{T}^\circ}(*,x) = \tau\big(\varphi(x)\big) = \frac{1}{2}D_{\mathcal{T}}(x,x)$, we see that $D_{\mathcal{T}^\circ}|_X = D_{\mathcal{T}}^f$ or, equivalently, $D_{\mathcal{T}} = (D_{\mathcal{T}^\circ}|_X)^{-f}$ holds.

Note that these facts imply in particular that every hierarchical dissimilarity D is "hyperbolic" (that is, 0-hyperbolic as defined in Chapter 5), i.e., it satisfies the 4-point condition

$$D(x,y) + D(u,v) \le \max\{D(x,u) + D(y,v), D(x,v) + D(y,u)\} \tag{9.7}$$

for all $x, y, u, v \in X$. Indeed,

(i) this holds for the associated Farris transform D^f relative to the map $f : X \to \mathbb{R} : x \mapsto \frac{D(x,x)}{2}$ that, as we have just seen, necessarily coincides with the metric $D_{\mathcal{T}^\circ}|_X$ induced by the edge-weighted X^*-tree $\mathcal{T}^\circ$, and

(ii) a symmetric bivariate map $D \in \mathbb{R}^{X \times X}$ is hyperbolic if and only if this holds for one — or, equivalently, for every — Farris transform D^f of D, a fact that we will use below again and again without any further discussion.

More specifically, we have

Lemma 9.7 (i) *A symmetric bivariate map* $D \in \mathbb{R}^{X \times X}$ *is a hierarchical dissimilarity if and only if*

(i.1) *D is hyperbolic and*

(i.2) $\max D(x, X) := \max\{D(x,y) : y \in X\}$ *coincides with* $\max D$ *for all* $x \in X$.

(ii) *More generally, an arbitrary symmetric bivariate map* $D \in \mathbb{R}^{X \times X}$ *is hyperbolic if and only if there exists some map* $f \in \mathbb{R}^X$ *such that the Farris transform*

D^f of D relative to f is a hierarchical dissimilarity if and only if the Farris transform D^a of D relative to the map $f = f_a : X \to \mathbb{R} : x \mapsto D(x, a)$ is a hierarchical dissimilarity for one — or, equivalently, for every — element $a \in X$.

(iii) *Furthermore, if $D : X \times X \to \mathbb{R}$ is a symmetric bivariate map that is hyperbolic and if f is an arbitrary map in $\mathbb{R}^X$, then the following assertions are equivalent:*

(iii.1) *D^f is a hierarchical dissimilarity.*
(iii.2) *$D(x, y) + f(z) \leq \max\{D(x, z) + f(y), D(y, z) + f(x)\}$ holds for all $x, y, z \in X$.*
(iii.3) *There exists some real number $\gamma \in \mathbb{R}$ for which the γ-translate $f_\gamma : X \to \mathbb{R} : x \mapsto f(x) + \gamma$ of f is contained in $T(D)$.*

Proof (i) We have seen already that every hierarchical dissimilarity is hyperbolic while

$$\max_{y \in X} D(x, y) \leq \max\{D(x, x'), \max\{D(y, x') : y \in X\}\}$$

and, therefore, also $\max D(x, X) = \max D(x', X)$ must hold for all $x, x' \in X$ implying that also $\max D(x, X) = \max D$ holds for all $x \in X$, as claimed. Conversely, if D is hyperbolic and if $D(x, z'), D(y, z') \leq D(z, z')$ holds for some $x, y, z, z' \in X$ (as would be the case for any $z' \in X$ with $D(z, z') = \max D(z, X) = \max D$), then one has also $D(x, y) \leq \max\{D(x, z), D(y, z)\}$ in view of $D(x, y) + D(z, z') \leq \max\{D(x, z) + D(y, z'), D(x, z') + D(y, z)\} \leq \max\{D(x, z), D(y, z)\} + D(z, z')$.

(ii) We noted already that a symmetric bivariate map $D \in \mathbb{R}^{X \times X}$ is hyperbolic whenever there exists some map $f \in \mathbb{R}^X$ such that the Farris transform D^f of D relative to f is hyperbolic and so, in particular, in case D^f is a hierarchical dissimilarity. Thus, it suffices to note that, for all $x, y, z, a \in X$, we have

$$\begin{aligned} &\max\{D^a(x, z), D^a(y, z)\} - D^a(x, y) \qquad (9.8)\\ &= \max\begin{Bmatrix} D(x, z) - D(x, a) - D(z, a) \\ D(y, z) - D(y, a) - D(z, a) \end{Bmatrix} - \big(D(x, y) - D(x, a) - D(y, a)\big)\\ &= \max\{D(x, z) + D(y, a), D(y, z) + D(x, a)\} - D(x, y) - D(z, a) \end{aligned}$$

for the Farris transform D^a of D relative to $f = f_a : x \mapsto D(x, a)$. So, if $D : X \times X \to \mathbb{R}$ is hyperbolic, we must have

$$D^a(x, y) \leq \max\{D^a(x, z), D^a(y, z)\},$$

for every $a \in X$, i.e., D^a is hierarchical, while — conversely — if D^a is hierarchical for one $a \in X$, it must be hyperbolic and so, then, must be D.

(iii) Finally, if $D \in \mathbb{R}^{X \times X}$ is a hyperbolic symmetric bivariate map and if f is an arbitrary map in $\mathbb{R}^X$, then D^f clearly is a hierarchical dissimilarity if and only if

$$D(x, y) - f(x) - f(y) \leq \max\{D(x, z) - f(x) - f(z), D(y, z) - f(y) - f(z)\}$$

or, equivalently, $D(x, y)+f(z) \leq \max\{D(x, z)+f(y), D(y, z)+f(x)\}$ holds for all $x, y, z \in X$ or, in view of (i), just as well if and only if there exists some real number $\gamma \in \mathbb{R}$ such that $2\gamma = \max D^f(x, X) = \max_{y \in X} \big(D(x, y) - f(x) - f(y)\big)$ or, equivalently, $f(x) + \gamma = \max_{y \in X} \big(D(x, y) - (f(y) + \gamma)\big)$ holds for all $x \in X$ which, by definition of the tight span, just means that the map f_γ is contained in the tight span $T(D)$ of D. □

In this context, it is also worth noting that our observations imply the following characterization of hyperbolic maps:

Lemma 9.8 *Given a symmetric bivariate map $D \in \mathbb{R}^{X \times X}$ and a point $a \in X$ such that D satisfies the inequality* (9.7) *for $x := a$ and all $y, u, v \in X$, then D must satisfy this inequality for all x, y, u, v in X, i.e., D must be hyperbolic.*

Proof Indeed, our assumption implies in view of (9.8) that $D^a(x, y)$ does not exceed $\max\{D^a(x, z), D^a(y, z)\}$ for all $x, y, z \in X$. So, it implies that D^a is a hierarchical dissimilarity and that, in consequence, D must be hyperbolic, as claimed. □

Interestingly, there is also, as was to be expected, a close relationship between this construction and the combinatorial Farris transform considered in the previous section that relates splits and clusters:

Suppose, for example, that D is a metric on X and recall that, in Chapter 7, we defined the split system $\Sigma_D^{(a)}$, for every $a \in X$, to contain exactly all those splits S in $\Sigma(X)$ for which

$$D(x, y) + D(z, a) < D(x, z) + D(y, a) \tag{9.9}$$

and, therefore, also

$$D(x, y) + D(z, a) < \min\{D(x, z) + D(y, a), D(y, z) + D(x, a)\}$$

or equivalently — cf. (9.8) —

$$D^a(x, y) < \min\{D^a(x, z), D^a(y, z)\}$$

holds for all $x, y \in \overline{S}(a)$ and $z \in S(a)$.

Thus, the hierarchy $\mathcal{C}_a\left(\Sigma_D^{(a)}\right) = \left\{\overline{S}(a) : S \in \Sigma_D^{(a)}\right\}$ associated with the compatible split system $\Sigma_D^{(a)}$ coincides exactly with the collection of proper Apresjan clusters of the Farris transform D^a of D. In particular, $a \notin A$ holds for every proper cluster $A \in \mathcal{A}(D^a)$ and, in view of $D^a(a, x) = 0$ for all $x \in X$, even for all proper clusters $A \in \mathcal{C}(\overline{D^a})$. In view of the identity $\Sigma_D = \bigcap_{a \in X} \Sigma_D^{(a)}$, it follows immediately that, for every split $S \in \Sigma(X)$, we have $S \in \Sigma_D$ if and only if $\overline{S}(a) \in \mathcal{C}_D^{(a)}$ holds for all $a \in X$. Note also that, in a similar way, one can associate cluster systems to D (or just as well to the various Farris transforms D^f of D) that correspond to the weakly compatible split systems Σ^D (or Σ^{D^f}, respectively) introduced in Chapter 7 (see [9] for more details).

It is also worth noting that our result that every hierarchical dissimilarity is hyperbolic generalizes as follows to arbitrary symmetric bivariate maps: In analogy to the terminology introduced in Chapter 5 according to which, following [82], a symmetric bivariate map $D \in \mathbb{R}^{X \times X}$ is defined to be Δ-hyperbolic for some given real number $\Delta \in \mathbb{R}$ if

$$D(x, y) + D(u, v) \leq \Delta + \max\{D(x, u) + D(y, v), D(x, v) + D(y, u)\}$$

holds for all $x, y, u, v \in X$, we say that D is Δ-*hierarchical* if

$$D(x, y) \leq \Delta + \max\{D(x, z), D(y, z)\}$$

holds for all $x, y, z \in X$.

Clearly, a map D as above is Δ-hyperbolic for some $\Delta \in \mathbb{R}$ if and only if this holds for one or, just as well, for all Farris transforms of D. Note also that a symmetric bivariate map $D \in \mathbb{R}^{X \times X}$ can be Δ-hyperbolic or Δ-hierarchical only if $\Delta \geq 0$ holds and that, since X is finite, there always exists, for every D as above, some sufficiently large $\Delta \geq 0$ such that D is both, Δ-hyperbolic as well as Δ-hierarchical. More specifically, we may, for every bivariate map $D \in \mathbb{R}^{X \times X}$, define its *hyperbolicity index*

$$\begin{aligned}\Delta(D) &:= \inf\{\Delta \in \mathbb{R} : D \text{ is } \Delta\text{-hyperbolic}\}\\ &= \max_{x,y,u,v \in X} \{D(x, y) + D(u, v)\\ &\qquad - \max\{D(x, u) + D(y, v), D(x, v) + D(y, u)\}\}\end{aligned}$$

and the *ultrametricity index*

$$\begin{aligned}\Delta^*(D) &:= \inf\{\Delta \in \mathbb{R} : D \text{ is } \Delta\text{-hierarchical}\}\\ &= \max_{x,y,z \in X} \{D(x, y) - \max\{D(x, z), D(z, y)\}\}.\end{aligned}$$

As mentioned above, the Farris transform was traditionally used as a tool to relate rooted and unrooted trees or rather the metrics induced on X by such trees [6]. As Δ-hierarchical and Δ-hyperbolic maps generalize hierarchical and hyperbolic maps, respectively, the next theorem can be seen as a generalization of this relationship. This result is also interesting in view of the fact that the converse of the first assertion in the theorem below clearly does not hold: There exist treelike metrics D with $\Delta^*(D)$ arbitrarily large while, since D is treelike, $\Delta(D) = 0$ holds.

Theorem 9.9 *Given a symmetric bivariate map $D \in \mathbb{R}^{X\times X}$ and a number $\Delta \in \mathbb{R}$, the following hold:*

(i) *If D is Δ-hierarchical, then it is 2Δ-hyperbolic.*
(ii) *D is Δ-hyperbolic if and only if D^a is Δ-hierarchical for every $a \in X$, that is, we have $\Delta(D) = \max\{\Delta^*(D^a) : a \in X\}$.*
(iii) *If D^a is Δ-hierarchical for some $a \in X$, then D is 2Δ-hyperbolic. In particular, $\Delta(D) \leq 2\min\{\Delta^*(D^a) : a \in X\}$ always holds.*

Proof (i) Consider arbitrary elements $x_1, x_2, y_1, y_2 \in X$. Without loss of generality, assume that $D(x_1, y_1) = \min\{D(x_i, y_j) : i, j \in \{1, 2\}\}$ holds. Then, if D is Δ-hierarchical, we have

$$\begin{aligned} D(x_1, x_2) + D(y_1, y_2) &\leq (\max\{D(x_1, y_1), D(x_2, y_1)\} + \Delta) \\ &\quad + (\max\{D(x_1, y_1), D(x_1, y_2)\} + \Delta) \\ &= D(x_2, y_1) + D(x_1, y_2) + 2\Delta \\ &\leq \max\left\{\begin{matrix} D(x_2, y_1) + D(x_1, y_2) \\ D(x_1, y_1) + D(x_2, y_2) \end{matrix}\right\} + 2\Delta, \end{aligned}$$

as required.

(ii) This follows immediately from (9.8).

(iii) If D^a is Δ-hierarchical for some $a \in X$, then it is 2Δ-hyperbolic by (i), implying that D must also be 2Δ-hyperbolic, as claimed. □

9.5 Hierarchical dissimilarities, generalized metrics, and the tight-span construction

In this section, we will consider hierarchical dissimilarities and their tight span $T(D) := \{f \in \mathbb{R}^X : f(x) = \sup_{y\in X}(D(x, y) - f(y))\}$ as defined, for any symmetric bivariate map $D \in \mathbb{R}^{X\times X}$, in the first section of Chapter 5. As noted there, we have $\|f, g\| = \max\{f(x) - g(x) : x \in X\}$ and, therefore, also "$f \leq g \Rightarrow f = g$" for all $f, g \in T(D)$.

It is also easily verified that $T(D^f) = T(D) - f := \{g - f : g \in T(D)\}$ holds, for any map $f \in \mathbb{R}^X$, for the tight span $T(D^f)$ of the Farris transform D^f of a symmetric bivariate map $D \in \mathbb{R}^{X\times X}$ relative to f (as $g(x) = \sup_{y\in X}\big(D(x,y) - g(y)\big)$ holds for some $g \in \mathbb{R}^X$ and $x \in X$ if and only if $g(x) - f(x) = \sup_{y\in X}\big(D(x,y) - f(x) - f(y) - (g(y) - f(y))\big)$ holds).

Further, it follows immediately from our definitions that $\max D(x, X)$ coincides with $\max D$ for all $x \in X$ if and only if the constant map $\rho_D : X \to \mathbb{R} : x \mapsto \frac{1}{2}\max D$ is contained in $T(D)$.

Next, note that a symmetric bivariate map $D \in \mathbb{R}^{X\times X}$ is a Farris transform of a metric defined on X if and only if the following *generalized triangle inequality*

$$D(x,y) + D(z,z) \leq D(x,z) + D(z,y) \tag{9.10}$$

holds for all x, y, z in X in which case we define D to be a *generalized metric*: Indeed, given any symmetric bivariate map $D \in \mathbb{R}^{X\times X}$ and any map $f \in \mathbb{R}^X$, the Farris transform D^f of D vanishes on the diagonal if and only if f coincides with the map f_D defined in (9.5). Furthermore, the corresponding Farris transform $D^{\#} := D^{f_D}$ is a metric if and only if $D(x,y) - \frac{D(x,x)}{2} - \frac{D(y,y)}{2} \leq D(x,z) - \frac{D(x,x)}{2} - \frac{D(z,z)}{2} + D(z,y) - \frac{D(z,z)}{2} - \frac{D(y,y)}{2}$ or, equivalently, (9.10) holds. Clearly, every hyperbolic map and therefore, in particular, every hierarchical dissimilarity is a generalized metric.

Furthermore, given any generalized metric $D \in \mathbb{R}^{X\times X}$ and any element $x \in X$, the sum of f_D and the Kuratowski map $k_x^{D^{\#}}$ of x relative to $D^{\#}$ is an element of $T(D)$, i.e., we have $gk_x = gk_x^D \in T(D)$ for the map

$$gk_x : X \to \mathbb{R} : y \mapsto D^{\#}(x,y) + f_D(y) = D(x,y) - \frac{1}{2}D(x,x)$$

that we also dub the *generalized Kuratowski map* relative to the generalized metric D.

Clearly, the observations above combined with the facts derived in the first section of Chapter 5 regarding the Kuratowski embedding $X \to T(D) : x \mapsto k_x$ in case D is a metric imply that the map gk_x is the only map in $T(D)$ for which $f(x) = \frac{1}{2}D(x,x)$ holds, and that the L_∞-distance $\|gk_x, gk_y\|$ between any two such maps gk_x and gk_y coincides with $D^{f_D}(x,y) = D(x,y) - \frac{D(x,x)+D(y,y)}{2}$ for all $x, y \in X$. Hence, the map $X \to T(D) : x \mapsto gk_x$ yields, for every generalized metric $D \in \mathbb{R}^{X\times X}$, an isometric embedding of the metric space $(X, D^{\#})$ into $T(D)$ endowed with the L_∞-metric.

Next note that, given a generalized metric $D \in \mathbb{R}^{X \times X}$ with only non-negative values, one can also associate to D the map $D^0 : X \times X \to \mathbb{R}_{\geq 0}$ defined by putting

$$D^0(x, y) := \begin{cases} 0 & \text{if } x = y, \\ D(x, y) & \text{otherwise.} \end{cases}$$

Furthermore, the tight span of D^0 is closely related to that of D and $D^{\#}$ in case D is a non-negative generalized metric:

Proposition 9.10 *Assume that D is a non-negative generalized metric defined on a finite set X of cardinality at least 2. Then, the following hold:*

(i) *The tight span $T(D^0)$ of D^0 contains the tight span $T(D)$ of D.*
(ii) *A map $f \in T(D^0)$ is actually contained in $T(D)$ if and only if $f(x) \geq \frac{D(x,x)}{2}$ holds for all $x \in X$.*
(iii) *The map $T(D) \to \mathbb{R}^X : f \mapsto (f^{\#} : X \to \mathbb{R}_{\geq 0} : x \mapsto f(x) - \frac{D(x,x)}{2})$ is an isometry from $T(D)$ onto $T(D^{\#})$.*
(iv) *And, conversely, the map $T(D^{\#}) \to T(D^0) : f \mapsto \left(f_{\#} : X \to \mathbb{R}_{\geq 0} : x \mapsto f(x) + \frac{D(x,x)}{2} \right)$ is an isometry from $T(D^{\#})$ into $T(D^0)$ whose image is the subspace $T(D)$ of $T(D^0)$.*

Proof (i) Note first that $f \in T(D^0)$ holds for every $f \in T(D)$ as $f \in T(D)$ implies $f(x) \geq \frac{D(x,x)}{2} \geq 0$ and, therefore, also

$$0 \leq f(x) = \max\{D(x, y) - f(y) : y \in X\} = \max\{D^0(x, y) - f(y) : y \in X\}$$

as

$$\begin{aligned} f(x) &= \max\{D(x, y) - f(y) : y \in X\} \\ &= \max\{D(x, x) - f(x), \max\{D(x, y) - f(y) : y \in X - x\}\} \\ &= \max\{D^0(x, y) - f(y) : y \in X\} \end{aligned}$$

must hold except perhaps in case $f(x) = D(x, x) - f(x) \geq \max\{D(x, y) - f(y) : y \in X\}$. However, as noted above, $f \in T(D)$ and $f(x) = \frac{D(x,x)}{2}$ implies $f(y) = g_{k_x}(y) = D(x, y) - \frac{D(x,x)}{2}$ for all $y \in X$ and, therefore, also $f(x) = \max\{D^0(x, y) - f(y) : y \in X\}$ in case $f(x) = D(x, x) - f(x)$ provided X contains at least two distinct elements.

(ii) Indeed, if $f \in P(D^0)$ holds for $f \in \mathbb{R}^X$, we have $f \in P(D)$ and, therefore, $f(x) \geq \max\{D(x, y) - f(y) : y \in X\}$ if and only if $f(x) + f(x) \geq D(x, x)$ holds for all $x \in X$ as $f(x) + f(y) \geq D^0(x, y) = D(x, y)$ holds clearly for any two distinct elements $x, y \in X$. And we must have $f(x) = \max\{D(x,y) - f(y) : y \in X\}$ for every $f \in T(D^0)$ with $f(x) \geq \frac{D(x,x)}{2}$ for all $x \in X$ as $f(x) = \max\{D^0(x,y) - f(y) : y \in X\}$ implies $f(x) \leq \max\{D(x,y) - f(y) : y \in X\} \leq f(x)$.

(iii) Defining $f^\# \in \mathbb{R}^X$ for every $f \in \mathbb{R}^X$ by putting $f^\#(x) := f(x) - \frac{D(x,x)}{2}$ for all $x \in X$ as above, it suffices to note that

$$f(x) = \max\{D(x,y) - f(y) : y \in X\}$$

holds, for some $f \in \mathbb{R}^X$, for all $x \in X$ if and only if

$$f(x) - \frac{D(x,x)}{2} = \max_{y \in X}\left(\left(D(x,y) - \frac{D(x,x)}{2} - \frac{D(y,y)}{2}\right) - \left(f(y) - \frac{D(y,y)}{2}\right)\right)$$

or, equivalently, $f^\#(x) = \max\{D^\#(x,y) - f^\#(y) : y \in X\}$ holds for all $x \in X$.

Furthermore, we have $\|f, g\| = \|f^\#, g^\#\|$ for all $f, g \in \mathbb{R}^X$ as $f^\#(x) - g^\#(x) = f(x) - g(x)$ holds, for all $f, g \in \mathbb{R}^X$, for every $x \in X$.

(iv) This is an obvious consequence of the assertions (i) and (iii). □

It is also easy to see that $T(D_0)$ can be described, as a metric space, quite easily in terms of its subspace $T(D)$: Indeed, $T(D_0)$ can be obtained from $T(D)$ as follows:

(i) adjoin, for each $x \in X$ with $D(x,x) > 0$, a closed interval of length $D(x,x)/2$ attached at the point gk_x, and define the metric on $T(D^0)$ to be the unique largest metric on this *amalgamation* that induces
(ii) the L_∞-metric on $T(D)$
(iii) and the standard metric on all of these intervals.

Put differently, $T(D^0)$ is the push-out of

(i) the isometry $X \to T(D^\#) : x \mapsto gk_x$ of the metric space $(X, D^\#)$ into its tight span $T(D^\#)$ and
(ii) the isometry $X \to \bigcup_{x \in X}[0, D(x,x)/2] : x \mapsto (0)_{x \in X}$ of the space $(X, D^\#)$ into the disjoint union $\bigcup_{x \in X}[0, D(x,x)/2]$ of all these intervals.

Here, we consider this disjoint union $\bigcup_{x \in X}[0, D(x,x)/2]$ as a metric space by defining the distance between two real numbers α and β in the same interval by $|\alpha - \beta|$ and between two real numbers α and β in two distinct intervals indexed by some x and some y in X with $x \neq y$ by $\alpha + \beta + D^\#(x,y)$. The situation is depicted in Figure 9.6. We leave the straightforward, yet a bit cumbersome verification of this claim (based on the fact that a map with $f(x) \leq \frac{D(x,x)}{2}$ for

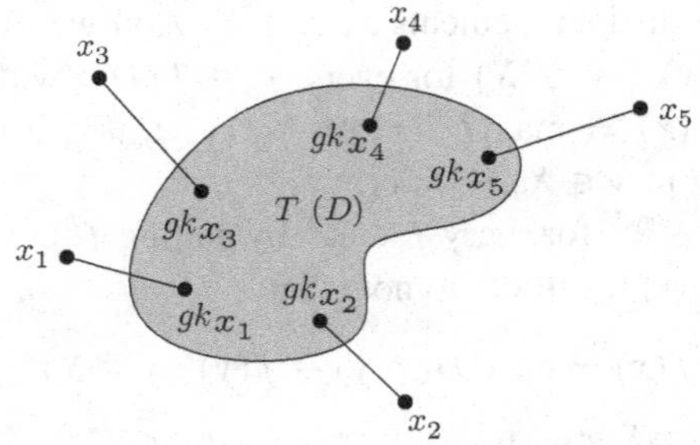

Figure 9.6 The construction of $T(D^0)$: The embedding of $(X, D^{\#})$ into $T(D^{\#})$ defined in the text maps each point x_i in X to the point gk_{x_i} in $T(D^{\#})$ to which intervals of suitable length are attached.

some $x \in X$ is contained in $T(D^0)$ if and only if $f(y) = D(x, y) - f(x)$ holds for all $y \in X$) to the interested reader.

We conclude this section by remarking that Theorem 9.4 could also be established using Theorem 9.9, the fact that both, D^0 and $D^{\#}$, are hyperbolic whenever D is and hence, in particular, in case D is a hierarchical dissimilarity, and our previous results regarding the tight span of hyperbolic metrics. Again, we leave the details as an exercise for the interested reader.

9.6 Algorithmic issues

As noted already above, to construct the Apresjan hierarchy $\mathcal{A}(D)$ for a dissimilarity $D \in \mathbb{R}^{X \times X}$, we only need to compare "distances" between pairs of elements in X of the form a, b and a, b'. This clearly has advantages not only regarding the stability of the construction and its significance (as any monotone transformation of the values of D will yield the same clusters), but also obvious algorithmic advantages.

Furthermore, the Farris transform can be applied in quite a few interesting ways to support phylogenetic analysis. For example, using the various Farris transforms of the form D^f for $f \in \mathbb{R}^X$ and, in particular, the Farris transforms of the form D^a for $a \in X$, one can derive whole families of hierarchies and then, considering any family $(\mathcal{C}_i)_{i \in I}$ indexed by a finite set I, one may use the fact that the set system

$$\mathcal{C}(I) := \bigcup_{J \subseteq I, 2|J| > |I|} \{C \subseteq X : \forall_{j \in J} C \in \mathcal{C}_j\}$$

is also a hierarchy (as any two subsets J, J' of I with $2|J|, 2|J'| > |I|$ must share at least one element j) to obtain further hierarchies that might actually be larger than any one of the constituent hierarchies used in its construction. Note also that, given any two subsets Y, Y' of X with $2|Y|, 2|Y'| \geq |X|$ and any two proper clusters $A, A' \subseteq X$ with $A \in \mathcal{C}(\overline{D^a})$ for all $a \in Y$ and $A' \in \mathcal{C}(\overline{D^{a'}})$ for all $a' \in Y'$, we must either have $Y \cap Y' \neq \emptyset$ or $A \cap A' = \emptyset$ in view of $A \cap A' \cap (Y \cup Y') = \emptyset$. So, also the set system

$$\mathcal{A}_D := \bigcup_{Y \subseteq X, 2|Y| \geq |X|} \{A \subseteq X : \forall_{a \in Y} A \in \mathcal{C}(\overline{D^a})\}$$

is a hierarchy that could be of interest in any phylogenetic analysis based on some dissimilarity D.

Furthermore, looking at the "phylogenetic signatures" rk_a^D associated, as noted above, to any dissimilarity D and element $a \in X$, one may use them to define a derived metric ∂D on X by putting

$$\partial D(x, y) := \left\| rk_x^D, rk_y^D \right\|$$

for all $x, y \in X$ where "$\| \dots \|$" can be any metric on $\mathbb{R}^X$, and consider the collection $\mathcal{A}(\partial D)$ of its Apresjan clusters: Indeed, one may even iterate this procedure as long as one wishes.

Finally, we note that, although the characterization of triplet systems induced by rooted phylogenetic trees in Theorem 9.6 is similar to the one presented for quartets in Theorem 3.7, triplet systems are in a significant way somehow simpler than quartet systems. For example, we mentioned in Chapter 6 that it is NP-complete to decide whether a quartet system is compatible. Remarkably, there is — in contrast — a polynomial time algorithm to decide whether a given triplet system $\mathcal{R} \subseteq \mathcal{R}(X)$ is *compatible*, that is, whether there exists a rooted phylogenetic X-tree $\mathcal{T}$ with $\mathcal{R} \subseteq \mathcal{R}_{\mathcal{T}}$. This algorithm, called BUILD, was presented by Aho *et al.* in [2], and we will briefly describe it here: Define for any triplet system $\mathcal{R} \subseteq \mathcal{R}(X)$ and any non-empty subset $Y \subseteq X$ the *cluster graph* $G_{(\mathcal{R},Y)} = (V_{(\mathcal{R},Y)}, E_{(\mathcal{R},Y)})$ with vertex set $V_{(\mathcal{R},Y)} := Y$ and edge set $E_{(\mathcal{R},Y)}$ consisting of those $\{u, v\} \in \binom{Y}{2}$ for which there exists some $w \in Y - \{u, v\}$ such that $uv|w \in \mathcal{R}$ holds. For example, Figure 9.7(b) presents the cluster graph $G_{(\mathcal{R},X)}$ corresponding to the triplet system $\mathcal{R}$ on $X := \{a, b, c, d, e\}$ consisting of those triplets in $\mathcal{R}(X)$ which are depicted in Figure 9.7(a).

Now, given a triplet system $\mathcal{R} \subseteq \mathcal{R}(X)$, the algorithm BUILD recursively constructs a hierarchy $\mathcal{C}$ on X or outputs that $\mathcal{R}$ is not compatible. If $n = 1$, the recursion bottoms out and the algorithm returns $\mathcal{C} := \{X\}$. If $n \geq 2$, the algorithm first computes the cluster graph $G_{(\mathcal{R},X)}$. If $G_{(\mathcal{R},X)}$ is connected, the algorithm returns "not compatible". Otherwise, it computes the connected

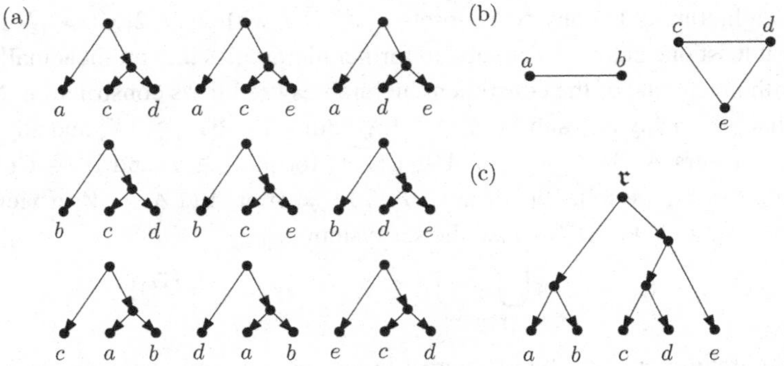

Figure 9.7 (a) A collection of nine rooted phylogenetic trees, each representing a single triplet in a triplet system $\mathcal{R}$ on $X = \{a, b, c, d, e\}$. (b) The cluster graph $G_{(\mathcal{R},X)}$. (c) A rooted phylogenetic X-tree $\mathcal{T}$ with $\mathcal{R} \subseteq \mathcal{R}_\mathcal{T}$.

components $X_1, \ldots, X_l$ of $G_{(\mathcal{R},X)}$. Then, for every $i \in \{1, \ldots, l\}$, BUILD is called recursively with triplet system $\mathcal{R}_i := \mathcal{R} \cap \mathcal{R}(X_i)$ as its input. If any one of these recursive calls yields "not compatible", the algorithm returns "not compatible". Otherwise, for every $i \in \{1, \ldots, l\}$, we obtain a hierarchy $\mathcal{C}_i$ on X_i and the algorithm returns the hierarchy $\mathcal{C} := \mathcal{C}_1 \cup \cdots \cup \mathcal{C}_l \cup \{X\}$ on X.

It is not hard to show by induction on the size of X that BUILD always returns "not compatible" if the input triplet system $\mathcal{R} \subseteq \mathcal{R}(X)$ is not compatible and that, conversely, if $\mathcal{R}$ is compatible, BUILD always returns a hierarchy $\mathcal{C}$ that contains X and, for every $x \in X$, the 1-subset $\{x\}$. In the latter case, Theorem 9.2 implies that there exists a phylogenetic rooted X-tree $\mathcal{T}$, unique up to canonical isomorphism, with $\mathcal{C}_\mathcal{T} = \mathcal{C}$ and this tree also satisfies $\mathcal{R} \subseteq \mathcal{R}_\mathcal{T}$.

To conclude this chapter, we note that there is currently a great deal of interest in developing methods for constructing "rooted directed X-labeled networks" from triplet systems, i.e., X-labeled directed graphs without directed cycles and exactly one vertex of in-degree 0. As these networks can be quite complex in nature, it can be useful to have some parameter that captures how much such a network deviates from being a rooted X-tree. One such parameter, called the *level* of the network, is the smallest integer k so that, for each "two-connected" (cf. [49]) component of the underlying undirected graph, the number of *reticulation vertices* that it contains, that is, those vertices that have more than one incoming edge, is at most k. For example, the vertex v in Figure 9.8(a) is a reticulation vertex since it has three incoming edges, namely e_1, e_2, and e_3. The other reticulation vertices in this network are u and w. As the vertices that are not leaves form a two-connected component containing u, v, and w,

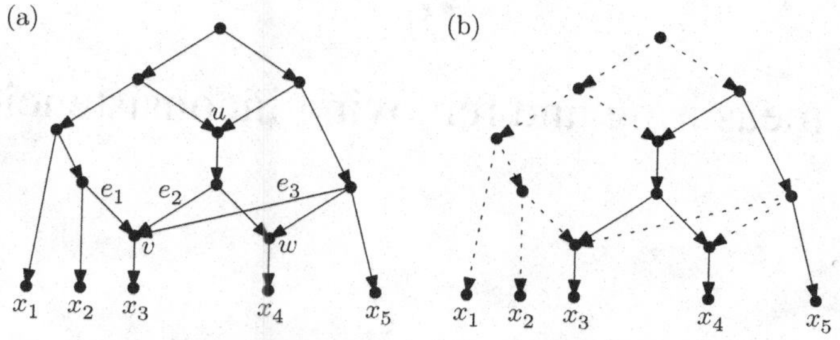

Figure 9.8 (a) An example of a rooted directed X-labeled network. The level of this network is 3. (b) The network displays the triplet $x_3x_4|x_5$, that is, removing the dotted edges and suppressing any resulting vertices of degree 2, we obtain a rooted phylogenetic tree on $\{x_3, x_4, x_5\}$ that represents precisely this triplet.

the level of the network is 3. Note that, for small $k \in \mathbb{N}$, a *level-k network*, that is, a network of level k, is close to being a rooted X-tree. Indeed, a network is a rooted X-tree if and only if it is a level-0 network.

One goal for reconstructing a rooted network from a given triplet system $\mathcal{R}$ could be to compute a network that displays as many of the triplets in $\mathcal{R}$ as possible in the sense illustrated in Figure 9.8(b). However, in contrast to rooted X-trees, it can be shown that it is NP-complete to decide even if there exists a level-1 network that displays every triplet in a given subset $\mathcal{R} \subseteq \mathcal{R}(X)$ [104, 127]. Nevertheless, several algorithms have been recently developed that try to exploit special properties of the input that are often met for triplet systems that come from biological data, such as that the triplet system is in some sense not too sparse [104, 126, 127], and this remains an exciting area of ongoing research.

10

On measuring and removing inconsistencies

In the preceding chapters, we have seen how to encode trees in terms of splits, metrics, and quartets, and we have studied the relationship between these various types of encodings. Most of our discussions were concerned with what we might call "ideal data", that is, data that correspond perfectly to a tree. However, as one might imagine, real biological data rarely come in this form. In consequence, many methods have been devised for tree and network reconstruction that try to find a tree or network that "best" fits the data, usually relative to some measure of optimality (see e.g., [76]).

It is beyond the scope of this book to review all of these methods here. Instead, we will briefly illustrate in this chapter how measuring *inconsistencies* in our three basic structures, split systems, metrics, and quartet systems, can be of some use in guiding the process of tree and network reconstruction.

To avoid uninteresting discussions of small cases, we will assume throughout this chapter that X is a finite set of cardinality $n \geq 3$.

10.1 k-compatibility

We have seen in Chapter 3 that compatible split systems are precisely those split systems that encode X-trees. Given that split systems arising from real data will rarely be compatible, we will now consider a simple way to measure how far a given split system is from being compatible. This yields some interesting mathematical results and, as we shall see, can also provide a useful tool for analyzing biological data.

More specifically, given any integer $k \geq 1$, we dub a split system $\Sigma \subseteq \Sigma(X)$ *k-compatible* if every totally incompatible subsystem $\Sigma' \subseteq \Sigma$ has cardinality at most k. Note that a split system is compatible if and only if it is 1-compatible and that every 2-compatible split system is weakly compatible, while, e.g., $\Sigma := \Sigma(\langle 4\rangle)$ is 3-, but not weakly compatible. And it is an interesting exercise related to the theorems of Sperner and Dilworth to determine the smallest number $k = k(n)$ such that $\Sigma(\langle n\rangle)$ is k-compatible.

Intriguingly, k-compatible split systems Σ — or, rather, the associated collection $\bigcup \Sigma$ consisting of all subsets of X that are part of Σ — were originally studied by A. Lomonosov, M. Karzanov, and others in the context of certain multi-commodity flow problems in [78, 110] and other papers where such collections were dubbed $(k-1)$-*cross-free set systems*.

A question that arises naturally when studying k-compatible split systems and which was raised already by M. Karzanov is to ask for their size: Let

$$\text{karz}_k(n) := \max\{|\Sigma| : \Sigma \subseteq \Sigma(\langle n\rangle) \text{ and } \Sigma \text{ is } k\text{-compatible}\},$$

denote the cardinality of the largest possible k-compatible split system contained in $\Sigma(\langle n\rangle)$ — or as well in the set $\Sigma(X)$ of X-splits for any set X of cardinality n. Then, how does $\text{karz}_k(n)$ scale with n and k? In [107], Karzanov and Lomonosov conjectured that $\text{karz}_k(n)$ is $O(kn)$. It is obvious that this bound holds in case $kn \geq |\Sigma(\langle n\rangle)| = 2^{n-1} - 1$. However, as Lomonosov originally pointed out, it is not hard to show that $\text{karz}_k(n)$ is $O(kn \log n)$:

Lemma 10.1 *For any two positive integers k, n, we have*

$$\text{karz}_k(n) \leq n + \sum_{i=2}^{\lfloor \frac{n-1}{2} \rfloor} \left\lfloor \frac{kn}{i} \right\rfloor + \begin{cases} k & \text{if } n \text{ is even,} \\ 0 & \text{otherwise.} \end{cases}$$

Proof Suppose that $\Sigma \subseteq \Sigma(\langle n\rangle)$ is a k-compatible split system and note that — in view of the fact that any two distinct splits of the form $S = A|B$ and $S' = A'|B'$ with $A' \cap A \neq \emptyset$ and $|A| = |A'| = \|S\| = \|S'\|$ must be incompatible — the cardinality of the various sets

$$\Sigma(x, i) := \{S \in \Sigma : |S(x)| = i\} \quad (x \in X, i \in \mathbb{N})$$

can never exceed k (nor 1 in case $i = 1$). Furthermore, the cardinality of $\Sigma_{\|S\|=i} := \{S \in \Sigma : \|S\| = i\}$ coincides with $\frac{\sum_{x\in X} |\Sigma(x,i)|}{i}$ in case $1 \leq i \leq \frac{n-1}{2}$ while, in case n is even, $\Sigma_{\|S\|=\frac{n}{2}}$ coincides with $\Sigma(x, i)$ for every $x \in X$. This implies

$$|\Sigma| = \sum_{i=1}^{\lfloor \frac{n}{2} \rfloor} |\Sigma_{\|S\|=i}|$$

$$= |\Sigma_{\|S\|=1}| + \sum_{i=2}^{\left\lfloor \frac{n-1}{2} \right\rfloor} \frac{\sum_{x \in X} |\Sigma(x,i)|}{i} + |\Sigma_{\|S\|=\frac{n}{2}}|$$

$$\leq n + \sum_{i=2}^{\left\lfloor \frac{n-1}{2} \right\rfloor} \left\lfloor \frac{kn}{i} \right\rfloor + \begin{cases} k & \text{if } n \text{ is even,} \\ 0 & \text{otherwise,} \end{cases}$$

as claimed. □

Remarkably, absolutely precise bounds are known for small k. For example, it follows directly from our results in Chapters 2 and 3 (in particular Theorem 3.3) that $\mathrm{karz}_1(n) = 2n - 3$ holds for all $n \geq 2$. In case $k = 2$, it is also known that $\mathrm{karz}_2(n) = 4n - 10$ holds and that, in fact, every 2-compatible split system $\Sigma \subseteq \Sigma(X)$ with $|\Sigma| = 4n - 10$ must be circular [67]. The proof of these facts is quite involved. But restricting attention to circular split systems, a simple proof can be given that nicely illustrates the flavor of some of the pertinent arguments:

Theorem 10.2 *The cardinality of any 2-compatible circular split system defined on an n-set X never exceeds* $4n - 10$.

Proof As any circular split system $\Sigma \subseteq \Sigma(X)$ contains at most $\binom{n}{2}$ splits, we must have $|\Sigma| \leq \binom{n}{2}$ and therefore $|\Sigma| \leq 4n - 10$ $(= \binom{n}{2})$ in case $n = 4$ and $n = 5$. Further, as any circular split system on X contains at most n splits of a fixed size i and $n < 2n - 5$ holds for all $n \geq 6$, we have $|\Sigma| < 4n - 10$ for any circular split system $\Sigma \subseteq \Sigma(X)$ that contains only splits of size at most 2.

Next, put $[i, j] := \{i, i+1, \ldots, j\}$ for all $i, j \in \mathbb{N}_{\geq 0}$ with $i \leq j$, and let $\Sigma_\circ(\langle n \rangle)$ denote the (canonical maximal) circular split system $\Sigma_\circ(\langle n \rangle)$ that consists of all splits $S \in \Sigma(\langle n \rangle)$ of the form $S = S(i|j) := [i, j-1]|(\langle n \rangle - [i, j-1])$ for some $i, j \in \langle n \rangle$ with $i < j$. In addition, assume that $\Sigma \subseteq \Sigma_\circ(\langle n \rangle)$ is a 2-compatible split system that (i) contains at least one split of size 3 or larger and (ii) has maximal cardinality among all such split systems. Let $i := \min\{\|S\| : S \in \Sigma, \|S\| \geq 3\}$ denote the minimal size attained by all splits in Σ of size at least 3 and assume also that, without loss of generality, the split $S := S(1|i+1) = [1, i]|[i+1, n]$ is contained in Σ.

We claim that $i = 3$ must hold: Otherwise, consider the split $S_0 := S(2|i+1)$ (see, for example, Figure 10.1 for the case $n = 8$ and $i = 4$) and note that, by construction, S_0 cannot be contained in Σ and that, therefore, a pair of splits must exist in Σ that are incompatible with S_0 as well as with each other. In particular, they also must have size at least 3 (as any two splits in $\Sigma_\circ(\langle n \rangle)$ that

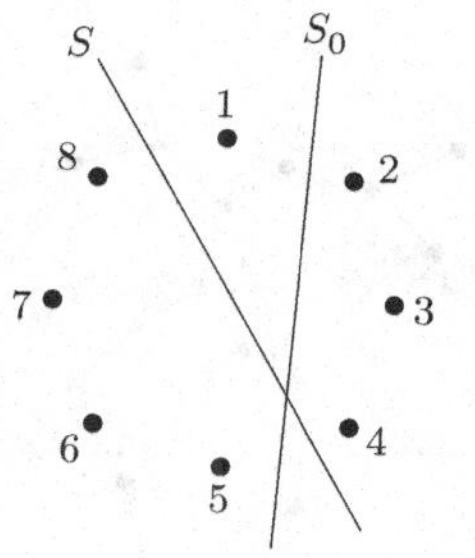

Figure 10.1 An example of a configuration considered in the proof of Theorem 10.2.

are incompatible with a split of size 2 in $\Sigma_\circ(\langle n\rangle)$ must be compatible with each other.)

At the same time, at least one of them must be compatible with S. But this implies that this split must be of the form $S(1|l)$ for some $l \le i$ and, hence, of size $l-1 < i$ in contradiction to $\|S(1|l)\| \ge \min\{\|S\| : S \in \Sigma, \|S\| \ge 3\} = i$. This establishes our claim.

So, $i = 3$ and $S = S(1|4) = [1,3]|[4,n]$ must hold. Define $X' := X \setminus \{2\}$ and $\Sigma' := \Sigma|_{X'}$. Note that Σ' is 2-compatible and circular. Hence, induction implies that $|\Sigma'| \le 4n - 14$ must hold, and it remains to show that $|\Sigma| \le |\Sigma'| + 4$ must also hold.

To this end, note first that there is exactly one split in $\Sigma_\circ(\langle n\rangle)$, namely $S_0 := S(2|3)$, whose restriction to X' does not yield a split of X'. Thus, it suffices to show that there exist at most three pairs of splits $S', S'' \in \Sigma$ such that the restriction of S' and S'' to X' yield the same split of X', namely the pair $S', S'' = S(1|2), S(1|3)$, the pair $S', S'' = S(2|4), S(3|4)$, and at most one pair of splits $S', S'' \in \Sigma$ of the form $S', S'' = S(2|i), S(3|i)$ for some $i \in [5,n]$: Indeed, any three splits of the form $S(1|i_1)$, $S(2|i_2)$, and $S(3|i_3)$ with $4 \le i_1 < i_2 < i_3 \le n$ must be pairwise incompatible. So, in view of $S = S(1|4) \in \Sigma$, there can be only one pair of splits $S', S'' \in \Sigma$ of the form $S', S'' = S(2|i)$, $S(3|i)$ for some $i \in [5,n]$, as claimed. □

In view of the fact that 2-compatible split systems with maximal cardinality must be circular, one might wonder whether this is also the case for $k \ge 3$. However, this is not the case. For example, already for $k = 3$, the union of the split system $\Sigma_\circ(\langle 7\rangle)$ and the three 2-splits $S_{\{1,3\}}$, $S_{\{4,6\}}$, and $S_{\{5,7\}}$ on the set $\langle 7\rangle$ is a 3-compatible split system with 24 splits, while any circular split system on a set with seven elements contains at most $\binom{7}{2} = 21$ splits — so,

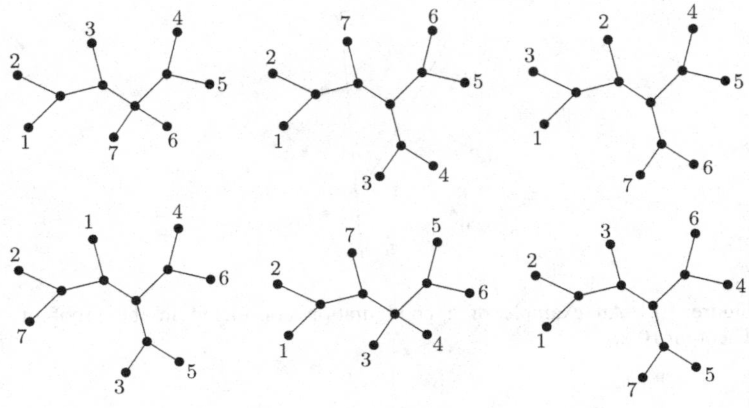

Figure 10.2 A family $\mathcal{P}$ of six phylogenetic trees on $X = \langle 7 \rangle$.

adding just any single one of those three splits in $\Sigma(\langle n \rangle) - \Sigma_\circ(\langle n \rangle)$ would produce a counterexample. Note, however, that it was shown in [66] (see also [38]) that, for $2k + 1 \leq n$, every k-compatible and circular split system on an n-set of maximal size has cardinality exactly $2nk - \binom{2k+1}{2}$ (while, for $2k + 1 \geq n$, every circular split system is k-compatible).

We now illustrate how considerations such as these may be used in practical applications to generate split networks. Suppose that $\mathcal{P}$ is a family of X-trees. Consider the split system

$$\Sigma_{>\frac{1}{2}}(\mathcal{P}) := \left\{ S \in \Sigma(X) : |\{\mathcal{T} \in \mathcal{P} : S \in \Sigma_{\mathcal{T}}\}| > \tfrac{1}{2}|\mathcal{P}| \right\},$$

that is, the set of those splits that are contained in $\Sigma_{\mathcal{T}}$ for more than half of the trees $\mathcal{T} \in \mathcal{P}$. Note that $\Sigma_{>\frac{1}{2}}(\mathcal{P})$ is compatible since, by the Pigeon Hole Principle, there must exist, for every pair of splits $S, S' \in \Sigma_{>\frac{1}{2}}(\mathcal{P})$, a tree $\mathcal{T} \in \mathcal{P}$ with $S, S' \in \Sigma_{\mathcal{T}}$. So, in particular, S and S' must be compatible. In consequence, there exists, up to canonical isomorphism, a unique X-tree $\mathcal{T}^*$ with $\Sigma_{>\frac{1}{2}}(\mathcal{P}) = \Sigma_{\mathcal{T}^*}$ that is commonly known as the *majority rule consensus tree* of $\mathcal{P}$ [112]. For example, the majority rule consensus tree for the family $\mathcal{P}$ of phylogenetic trees in Figure 10.2 is depicted in Figure 10.3(a). Note that there are many other ways to compute a consensus tree of a family of X-trees — see [31] for a recent overview.

Now, for every integer $k \geq 1$, consider the split system

$$\Sigma_{>\frac{1}{k+1}}(\mathcal{P}) := \left\{ S \in \Sigma(X) : |\{\mathcal{T} \in \mathcal{P} : S \in \Sigma_{\mathcal{T}}\}| > \frac{1}{k+1}|\mathcal{P}| \right\}.$$

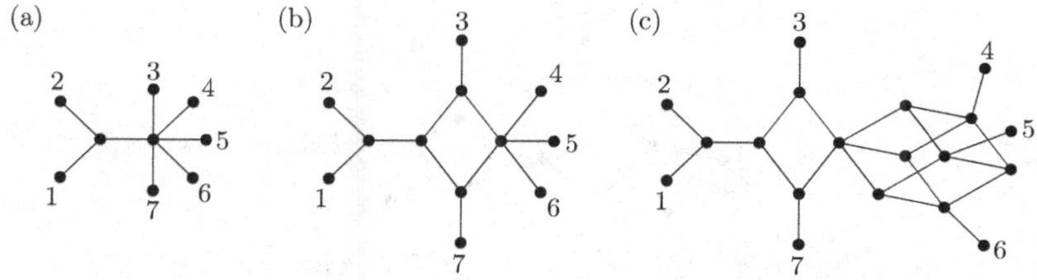

Figure 10.3 (a)–(c) Consensus networks for the trees in Figure 10.2 for $k = 1, 2, 3$, respectively. (a) The majority rule consensus tree. (b) The Buneman graph representing $\Sigma_{>\frac{1}{3}}(\mathcal{P})$. (c) The Buneman graph representing $\Sigma_{>\frac{1}{4}}(\mathcal{P})$.

Invoking the Pigeon Hole Principle again, it follows that $\Sigma_{>\frac{1}{k+1}}(\mathcal{P})$ must be k-compatible. So, the Buneman graph $\mathcal{B}(\Sigma_{>\frac{1}{k+1}})$ can be thought of as a *consensus network* of $\mathcal{P}$ [96]. Since $\Sigma_{>\frac{1}{k+1}}(\mathcal{P})$ is k-compatible, it follows (cf. Observation (B*-i) and (B*-iv) in Chapter 4) that every subgraph of $\mathcal{B}(\Sigma_{>\frac{1}{k+1}})$ that is a hypercube has dimension at most k, and that the number of edges of $\mathcal{B}(\Sigma_{>\frac{1}{k+1}})$ is comparatively small for small k. In particular, $\mathcal{B}(\Sigma_{>\frac{1}{k+1}})$ can be quite instructive and easy to draw for small k (see, for example, Figure 10.3(b) and (c)).

Note that related approaches to find consensus networks and, more generally, *super-networks* from collections $\mathcal{P}$ of *partial trees* on X (that is, collections $\mathcal{P}$ such that each $\mathcal{T} \in \mathcal{P}$ is an X'-tree for some subset $X' \subseteq X$) have been developed in, for example, [95, 99].

A particularly simple approach works by *greedy* elimination of non-fitting splits: Given an arbitrary split system $\Sigma \subseteq \Sigma(X)$, one may, starting with $\Sigma_0 := \Sigma$, construct a sequence $(\Sigma_i)_{i=0,1,\dots}$ of smaller and smaller split systems Σ_i by eliminating — in a recursive step-by-step fashion — always one of those splits $S \in \Sigma_i$ for which the number of splits $S' \in \Sigma_i$ that are incompatible with S is as large as possible (or this holds for the sum of the weights of all those splits in case we are dealing with a weighted split system, or — in case we want to obtain, more generally, a k-compatible subsystem of Σ for some fixed $k \geq 1$ — for the number of $(k+1)$-subsets of pairwise incompatible splits containing S, or for the sum over the weights of all splits distinct from S in all such subsets, or ...). That is, for each $i \geq 0$, one puts $\Sigma_{i+1} := \Sigma_i - S$ where $S \in \Sigma_i$ is chosen as explained, and stops when no pair of incompatible splits or, more generally, no $(k+1)$-subset of pairwise incompatible splits, is left.

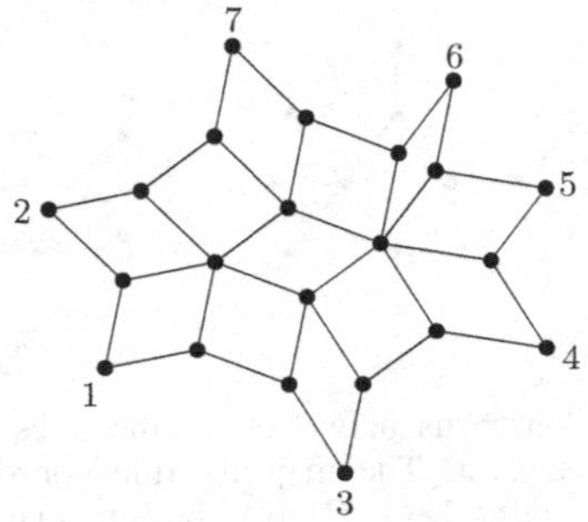

Figure 10.4 For the split system Σ induced by the six trees depicted in Figure 10.2, a split graph displaying a 2-compatible subset of Σ obtained by greedy elimination as described in the text.

To illustrate this construction, consider the split system $\Sigma := \bigcup_{T \in \mathcal{P}} \Sigma_T$ on $X := \langle 7 \rangle$ where $\mathcal{P}$ is the collection of phylogenetic X-trees depicted in Figure 10.2. Clearly, $\Sigma^* := \Sigma \cap \Sigma^*(X)$ consists of the ten 2-splits $S_{\{1,2\}}$, $S_{\{1,3\}}$, $S_{\{2,7\}}$, $S_{\{3,4\}}$, $S_{\{3,5\}}$, $S_{\{4,5\}}$, $S_{\{4,6\}}$, $S_{\{5,6\}}$, $S_{\{5,7\}}$, and $S_{\{6,7\}}$ and the two 3-splits $S_{\{1,2,3\}}$, $S_{\{1,2,7\}}$. We want to construct a 2-compatible subset of $\Sigma_0 := \Sigma^*$ and, therefore, fix $k := 2$: Only the two splits $S_{\{3,5\}}$ and $S_{\{5,7\}}$ are incompatible to six other splits in Σ_0 while all others are incompatible to at most five. So, $S_{\{3,5\}}$ may be eliminated first. Next, exactly the 2-splits $S_{\{5,7\}}$ and $S_{\{6,7\}}$ in $\Sigma_1 := \Sigma^* - S_{\{3,5\}}$ are incompatible with precisely five other splits in Σ_1. So, we may next form $\Sigma_2 := \Sigma_1 - S_{\{5,7\}}$. Now, exactly the three splits $S_{\{3,4\}}$, $S_{\{4,6\}}$, and $S_{\{6,7\}}$ are incompatible with precisely four other splits in Σ_2. So, we may next eliminate either one. Yet, only eliminating $S_{\{4,6\}}$ yields a 2-compatible split system, *viz.* the split system

$$\Sigma_3 := \Sigma_2 - S_{\{4,6\}} = \{S_A : A \in \mathcal{A}\}$$

with $\mathcal{A} := \{\{1,2\}, \{1,3\}, \{2,7\}, \{3,4\}, \{4,5\}, \{5,6\}, \{6,7\}, \{1,2,3\}, \{1,2,7\}\}$ that actually happens to be a cyclic split system relative to the linear order of X for which $2 < 1 < 3 < 4 < 5 < 6 < 7$ holds. The Buneman graph $\mathcal{B}(\Sigma_3)$ is depicted in Figure 10.4.

Finally, it should be mentioned that there is still another fascinating way of transforming a given weighted split system $\mu_{obs} : \Sigma(X) \to \mathbb{R}$ of experimentally observed split weights in a specific way into another weighted split system μ_{caus} that might provide a better representation of the underlying "causative" phylogenetic branching process than μ_{obs}, called *Hadamard conjugation*, that was developed more than 20 years ago by Mike Hendy and his colleagues from the New Zealand school of phylogenetics (cf. [91, 92, 93]).

To describe Hadamard conjugation, note first that, for each subset Σ of $\Sigma(X)$, we can define an equivalence relation $\sim_\Sigma$ on X by putting

$$x \sim_\Sigma y \iff |\{S \in \Sigma : S(x) \neq S(y)\}| \equiv 0 \pmod 2.$$

This is indeed an equivalence relation as — putting $\Delta_{x|y}(\Sigma) := \Delta(\phi_x^\Sigma, \phi_y^\Sigma) = \{S \in \Sigma : S(x) \neq S(y)\}$ — one has

$$\Delta_{x|z}(\Sigma) = \Delta_{x|y}(\Sigma) \,\Delta\, \Delta_{y|z}(\Sigma)$$

and, therefore,

$$\begin{aligned} |\Delta_{x|z}(\Sigma)| &= |\Delta_{x|y}(\Sigma) \,\Delta\, \Delta_{y|z}(\Sigma)| \\ &= |\Delta_{x|y}(\Sigma)| + |\Delta_{y|z}(\Sigma)| - 2\,|\Delta_{x|y}(\Sigma) \cap \Delta_{y|z}(\Sigma)| \\ &\equiv |\Delta_{x|y}(\Sigma)| + |\Delta_{y|z}(\Sigma)| \pmod 2 \end{aligned}$$

for all $x, y, z \in X$. Furthermore, the set $S_\Sigma := X/\sim_\Sigma$ of $\sim_\Sigma$-equivalence classes is a partition of X into either one or two non-empty subsets. So, S_Σ is either also an X-split or the trivial partition of X into exactly one subset, *viz.* the set X. Furthermore, given any split $S \in \Sigma(X)$, let Σ_S denote the collection of all subsets Σ of $\Sigma(X)$ with $S = S_\Sigma$.

Next, assume that our weighted split system $\mu_{obs} : \Sigma(X) \to \mathbb{R}$ is actually derived from, say, an X-indexed family $\mathfrak{S} = (s(x))_{x \in X}$ of aligned $\{\pm 1\}$-sequences $s(x) = (s_1(x), s_2(x), \ldots, s_\ell(x))$ of length $\ell \geq 1$ by putting

$$\begin{aligned} \mu_{obs} &:= \mu_{\mathfrak{S}} : \Sigma(X) \to \mathbb{R} : S \mapsto \mu_{\mathfrak{S}}(S) \\ &:= \frac{|\{i \in \langle \ell \rangle : S = \{s_i^{-1}(+1), s_i^{-1}(-1)\}\}|}{\ell}, \end{aligned}$$

so that $(1 - \sum_{S \in \Sigma(X)} \mu_{obs}(S))\ell$ clearly coincides with the number of constant sites in $\mathfrak{S}$. Assume also

(i) that the family $\mathfrak{S}$ of aligned $\{\pm 1\}$-sequences is actually the result of an evolutionary replication, mutation, and selection process including, of course, back and parallel mutations to which various "proper phylogenetic splits" contributed in varying measures specified by the weighted split system μ_{caus},
(ii) and that one can apply certain standard modeling "IID" assumptions from phylogenetic analysis as formalized, e.g., in Neyman's statistical 2-state model (cf. [117]) which imply that μ_{obs} and μ_{caus} are related to one another by the equation

$$\mu_{obs}(S) = \sum_{\Sigma \in \Sigma_S} \prod_{S' \in \Sigma} \mu_{caus}(S') \prod_{S' \in \Sigma(X) - \Sigma} (1 - \mu_{caus}(S')). \qquad (10.1)$$

It is then easy to see that, considering the right-hand side of (10.1) as a polynomial P_S in the variables $\mu_{caus}(S')$ ($S' \in \Sigma(X)$) and the system of all of these polynomials as a polynomial map

$$P : \mathbb{R}^{\Sigma(X)} \to \mathbb{R}^{\Sigma(X)} : \mu \mapsto (P_S(\mu))_{S \in \Sigma(X)}$$

with

$$P_S(\mu) := \sum_{\Sigma \in \Sigma_S} \prod_{S' \in \Sigma} \mu(S') \prod_{S' \in \Sigma(X) - \Sigma} (1 - \mu(S'))$$

for all $S \in \Sigma(X)$ and $\mu \in \mathbb{R}^{\Sigma(X)}$, the Jacobian $\Sigma(X) \times \Sigma(X)$-matrix

$$\left(\frac{\partial}{\partial \mu(S')} P_S \right)_{S, S' \in \Sigma(X)},$$

evaluated at $\mu_0 : \Sigma(X) \to \mathbb{R} : S \mapsto 0$ (the "all-zero weighted split system") coincides with the identity $\Sigma(X) \times \Sigma(X)$-matrix $\left(\delta_{S,S'}\right)_{S,S' \in \Sigma(X)}$. Thus, one can apply the Inverse Function Theorem to conclude that, in a sufficiently small neighborhood U of μ_0 in the space $\mathbb{R}^{\Sigma(X)}$ of all (positively or negatively) weighted split systems over X, there must exist a unique analytic (and actually algebraic) map $P' : U \to \mathbb{R}^{\Sigma(X)}$ with $P(P'(\mu)) = P'(P(\mu)) = \mu$ for all maps $\mu \in U$ and, thus, with $P'(\mu_{obs}) = \mu_{caus}$ provided μ_{obs} is sufficiently close to μ_0, that is, not too many mutations have completely wiped the phylogenetic signal in μ_{obs}.

Of course, there is no guarantee that, given an experimentally observed weighted split system μ_{obs}, the associated transformed weighted split system $P'(\mu_{obs})$ has compatible support, yet the hope is that, at least, the splits with large $P'(\mu_{obs})$-value are compatible. In consequence, one can conclude that the "IID" assumptions of sequence evolution on which Neyman's 2-state model is based do not apply to the sequence sites in question whenever there are too many pairwise incompatible splits with large $P'(\mu_{obs})$-value — an equally valuable, and often not totally unexpected, insight.

10.2 Δ-hierarchical approximations

Given a metric D on X, one approach often taken in phylogenetics is to try to find an edge-weighted X-tree $\mathcal{T} = (V, E, \omega, \varphi)$ such that the induced metric $D_{\mathcal{T}}$ *approximates* D as closely as possible. More specifically, denoting, for any two subsets $\mathcal{D}_1, \mathcal{D}_2$ of $\mathbb{R}^{X \times X}$, by $\|\mathcal{D}_1, \mathcal{D}_2\|_\infty$ the infimum of the L_∞ distances

$$\|D_1, D_2\|_\infty := \sup\{|D_1(x, y) - D_2(x, y)| : x, y \in X\}$$

over all $D_1 \in \mathcal{D}_1$ and $D_2 \in \mathcal{D}_2$, we can look for a metric D_0 in the set $\mathbb{T}(X)$ of all metrics in $\mathbb{R}^{X \times X}$ that satisfy the 4-point condition for which $\|D, D_0\|_\infty$ coincides with

$$\|D, \mathbb{T}(X)\|_\infty := \inf\{\|D, D'\|_\infty : D' \in \mathbb{T}(X)\},$$

an infimum that is easily seen to be attained as $\mathbb{T}(X)$ is a closed subset of $\mathbb{R}^{X \times X}$ and $|D'(x, y)|$ is, for any bivariate map $D' \in \mathbb{R}^{X \times X}$ with $\|D, D'\|_\infty \leq \alpha$ for some $\alpha \in \mathbb{R}$, bounded from above by $\max\{D(x, y) : x, y \in X\} + \alpha$. However, given D and some $\alpha \in \mathbb{R}$, even deciding whether there exists a metric $D' \in \mathbb{T}(X)$ with $\|D, D'\|_\infty \leq \alpha$ is NP-complete [1], and this holds also for many other variants of this problem, even if α is assumed to be an integer (see, for example, [46, 72]).

In view of this, researchers have focused on constructing treelike metrics D_0 such that $\|D, D_0\|_\infty \leq \rho$ holds for some real number ρ that is defined independently of the actual construction of D_0. Intriguingly, this task was first studied by M. Gromov in the context of his investigations of *hyperbolic groups* [82]. The bound was given in terms of the hyperbolicity index $\Delta(D)$ introduced in Chapter 9 that can be regarded as a measure of inconsistency in that it measures how far D is from being a treelike metric (or, equivalently, from satisfying the 4-point condition). More specifically, M. Gromov noted that there is some constant $c > 0$ such that $\|D, \mathbb{T}(X)\|_\infty \leq c\,\Delta(D)(\log_2 n)^2$ holds for every metric D, and he showed how to explicitly construct a treelike metric D' with $\|D, D'\|_\infty \leq c\Delta(D)(\log_2 n)^2$. This upper bound was improved in [29] to

$$\|D, \mathbb{T}(X)\|_\infty \leq \Delta(D)\lceil \log_2 n \rceil. \tag{10.2}$$

And it was shown in [1] that the same ideas can be used to compute a treelike metric D' with $\|D, D'\|_\infty \leq 3\|D, \mathbb{T}(X)\|_\infty$ in polynomial time.

As we shall see below (Theorem 10.4(iii)), a somewhat more general result analogous to (10.2) can be derived even when approximating arbitrary symmetric bivariate maps by symmetric Δ-hierarchical maps. Indeed, let $\mathcal{D}(X|\Delta\texttt{-hier})$ denote the set of all Δ-hierarchical maps in $\mathbb{R}^{X \times X}$ and note first that, given any symmetric map $D : X \times X \to \mathbb{R}$ and any real number $\Delta \geq 0$, there exists — as in case $\Delta = 0$ — a *subdominant Δ-hierarchical dissimilarity* for D, that is, a unique largest Δ-hierarchical dissimilarity $D^{(\Delta)}$ in

$$\mathcal{D}_{\leq D}(X|\Delta\texttt{-hier}) := \{D' \in \mathcal{D}(X|\Delta\texttt{-hier}) : D' \leq D\},$$

which can be shown just as in case $\Delta = 0$.

Note that $D_1^{(\Delta)} \leq D_2 \leq D_1$ implies $\mathcal{D}_{\leq D_1}(X|\Delta\text{-}\mathtt{hier}) = \mathcal{D}_{\leq D_2}(X|\Delta\text{-}\mathtt{hier})$ and, hence, $D_1^{(\Delta)} = D_2^{(\Delta)}$ for all $D_1, D_2 \in \mathbb{R}^{X \times X}$.

To see how well $D^{(\Delta)}$ approximates D, define the map $\partial_\Delta D \in \mathbb{R}^{X \times X}$ by

$$\partial_\Delta D(x, y) := \min \left\{ \Delta + \min_{z \in X} \max\{D(x, z), D(y, z)\}, D(x, y) \right\}$$

for all $x, y \in X$, define the maps $\partial_\Delta^k D$ for all $k \in \mathbb{N}_{\geq 0}$ recursively by $\partial_\Delta^0 D := D$ and $\partial_\Delta^{k+1} D := \partial_\Delta(\partial_\Delta^k D)$ for all $k \in \mathbb{N}_{\geq 0}$, and put

$$\partial_\Delta^\infty D(x, y) := \inf\{\partial_\Delta^k D(x, y) : k \in \mathbb{N}_{\geq 0}\}$$

for all $x, y \in X$. Note that it follows immediately from the above definitions that

(i) $\partial_\Delta^k D \leq \partial_\Delta^{k-1} D$ holds for all $k > 0$,
(ii) $D = \partial_\Delta D$ holds if and only if one has $\Delta \geq \Delta^*(D)$, and
(iii) one has $D' \leq \partial_\Delta D$ for all maps $D' \in \mathcal{D}_{\leq D}(X|\Delta\text{-}\mathtt{hier})$,

implying that also $(\partial_\Delta D)^{(\Delta)} = D^{(\Delta)}$ and, therefore, also $D^{(\Delta)} \leq \partial_\Delta^\infty D \leq D$ must hold.

In addition, consider, for all $k \in \mathbb{N}_{\geq 1}$, the $(k+1)$-*variate map*

$$D^{[k]} : X^{\{0,1,2,\ldots,k\}} \to \mathbb{R} : (x_0, x_1, \ldots, x_k) \mapsto \max\{D(x_{\nu-1}, x_\nu) : \nu = 1, \ldots, k\},$$

and note that

$$D^{[k+l]}(x_0, x_1, \ldots, x_{k+l}) = \max\{D^{[k]}(x_0, x_1, \ldots, x_k), D^{[l]}(x_k, x_{k+1}, \ldots, x_{k+l})\}$$

holds for all $k, l \geq 1$ and $x_0, x_1, \ldots, x_{k+l} \in X$, and that

$$D(x_0, x_k) \leq D^{[k]}(x_0, x_1, \ldots, x_k)$$

holds, for all $k \geq 1$ and $x_0, x_1, \ldots, x_k$ in X in case D is a hierarchical dissimilarity.

Furthermore, we have:

Lemma 10.3 *With Δ, k and $x_0, x_1, x_2, \ldots, x_k \in X$ as above, one has*

$$D(x_0, x_k) \leq D^{[k]}(x_0, x_1, x_2, \ldots, x_{k-1}, x_k) + \Delta \lceil \log_2 k \rceil$$

for every $D \in \mathcal{D}(X|\Delta\text{-}\mathtt{hier})$.

Proof We use induction on k. For $k=1$, we clearly have $D(x_1,x_2)=D^{[1]}(x_1,x_2)$ for all $x_1,x_2\in X$.

Now, assume $k>1$, put $l:=2^{(\lceil\log_2 k\rceil-1)}$, so that $l=2^{(\lceil\log_2 k\rceil-1)}<k=2^{\log_2 k}\le 2^{\lceil\log_2 k\rceil}=2l$ holds, and note that, by induction,

$$\begin{aligned}
D(x_0,x_k) &\le \max\{D(x_0,x_l),D(x_l,x_k)\}+\Delta\\
&\le \max\left\{\begin{matrix} D^{[l]}(x_0,\dots,x_l)+\Delta\lceil\log_2 l\rceil \\ D^{[k-l]}(x_l,\dots,x_k)+\Delta\lceil\log_2(k-l)\rceil\end{matrix}\right\}+\Delta\\
&\le \max\{D^{[l]}(x_0,\dots,x_l),D^{[k-l]}(x_l,\dots,x_k)\}+\Delta(\lceil\log_2(l)\rceil+1)\\
&= D^{[k]}(x_0,\dots,x_k)+\Delta((\lceil\log_2 k\rceil-1)+1)\\
&= D^{[k]}(x_0,\dots,x_k)+\Delta(\lceil\log_2 k\rceil)
\end{aligned}$$

holds indeed for all $x_0,x_1,\dots,x_k\in X$ for every Δ-hierarchical map D. □

We now state and prove the main result of this section:

Theorem 10.4 *Given any symmetric bivariate map $D\in\mathbb{R}^{X\times X}$ and any non-negative real number Δ, one has $D^{(\Delta)}=\partial_\Delta^\infty D$. Moreover, one has*

(i) *$D^{(\Delta)}(a,x)=D(a,a)$ for some $a\in X$ and all $x\in X$ whenever $D(a,x)=D(a,a)$ holds for this element $a\in X$ and all $x\in X$,*
(ii) *$\|D,D^{(\Delta)}\|_\infty\le\max\{0,\Delta^*(D)\lceil\log_2(n-1)\rceil-\Delta\}$ and, therefore, also $\|D,\mathcal{D}(X|\Delta\text{-hier})\|\le\frac{1}{2}\max\{0,\Delta^*(D)\lceil\log_2(n-1)\rceil-\Delta)\}$.*

Proof We have already seen that $D^{(\Delta)}\le\partial_\Delta^\infty D\le D$ always holds. Hence, in order to show that $D^{(\Delta)}=\partial_\Delta^\infty D$ holds, it remains to show that $\partial_\Delta^\infty D$ is Δ-hierarchical. But this follows immediately in view of the fact that

$$\begin{aligned}
\partial_\Delta^{k+1}D(x,y) &= \min\left\{\Delta+\min_{z\in X}\max\{\partial_\Delta^k D(x,z),\partial_\Delta^k D(y,z)\},\partial_\Delta^k D(x,y)\right\}\\
&\le \Delta+\max\{\partial_\Delta^k D(x,z_0),\partial_\Delta^k D(y,z_0)\}
\end{aligned}$$

holds for all $x,y,z_0\in X$ and all $k\ge 0$.

(i) Now suppose that $D(a,x)=D(a,a)$ holds, for some $a\in X$, for all $x\in X$. Define the symmetric bivariate map D' on X by putting, for all $x,y\in X$, $D'(x,y):=D(a,a)$ if $a\in\{x,y\}$ and $D'(x,y):=\min\{D(z,z'):z,z'\in X\}$ otherwise. By definition, one has $D'\le D$. And D' is a hierarchical dissimilarity: Indeed, one has $D'(x,y)\le\max\{D'(x,z),D'(y,z)\}$ for all $x,y,z\in X$ with $a\notin\{x,y\}$, and this inequality holds also in case, say,

$a = x$ as this implies $D'(x, y) = D(a, a) \leq \max\{D'(x, z), D'(y, z)\} = \max\{D'(a, a), D'(y, z)\}$. But then, as $D' \leq D^{(\Delta)} \leq D$ must hold, we get $D(a, a) = D'(a, x) \leq D^{(\Delta)}(a, x) \leq D(a, x) = D(a, a)$ and, hence, $D^{(\Delta)}(a, x) = D(a, a)$ for all $x \in X$, as required.

(ii) First, we consider the case $\Delta = 0$: Note that $D^{(0)}(x, y)$ coincides, for any two elements $x, y \in X$, with the infimum over all terms of the form $D^{[k]}(x, x_1, \ldots, x_{k-1}, y)$ where k runs through all integers in $\mathbb{N}_{\geq 1}$ and the $x_1, \ldots, x_{k-1}$ over all $(k-1)$-tuples of distinct elements in $X - \{x, y\}$: Indeed, given any $(k+1)$-tuple $x_0, x_1, \ldots, x_{k-1}, x_k$ of elements in $X - \{x, y\}$ with $x_i = x_j$ for some $i, j \in \{0, 1, \ldots, k\}$ with $0 < j - i \leq k - 1$, the set $\{(x_{\nu-1}, x_\nu) : \nu = 1, \ldots, k\}$ contains the two subsets $\{(x_{\nu-1}, x_\nu) : \nu = 1, \ldots, i\}$ and $\{(x_{\nu-1}, x_\nu) : \nu = j + 2, \ldots, k\}$ and the pair $(x_j, x_{j+1}) = (x_i, x_{j+1})$ implying that $D^{[k]}(x_0, x_1, \ldots, x_{k-1}, x_k) \geq D^{[i+k-j]}(x_0, \ldots, x_i, x_{j+1}, \ldots, x_{k-1}, x_k)$ holds which readily implies our claim.

In consequence, there exist, for any two distinct elements x, y in X, some integer $k \leq n - 1$ and $k - 1$ *distinct* elements $x_1, \ldots, x_{k-1} \in X - \{x, y\}$ with $D^{(0)}(x, y) = D^{[k]}(x, x_1, \ldots, x_{k-1}, y)$. By Lemma 10.3, this implies that

$$\begin{aligned} D(x, y) &\leq D^{[k]}(x, x_1, \ldots, x_{k-1}, y) + \Delta^*(D)\lceil \log_2(k) \rceil \\ &= D^{(0)}(x, y) + \Delta^*(D)\lceil \log_2(k) \rceil \\ &\leq D^{(0)}(x, y) + \Delta^*(D)\lceil \log_2(n-1) \rceil \end{aligned}$$

holds.

Moreover, we have

$$\begin{aligned} D(x, x) &\leq \min\{\max\{D(x, z), D(z, x)\} : z \in X\} + \Delta^*(D) \\ &= \min\{D^{[2]} D(x, z, x) : z \in X\} + \Delta^*(D) \\ &= \inf\{D^{[k]}(x, x_1, \ldots, x_{k-1}, x) : k \geq 2,\ x_1, \ldots, x_{k-1} \in X\} + \Delta^*(D) \\ &= D^{(0)}(x, x) + \Delta^*(D) \leq D^{(0)}(x, x) + \Delta^*(D)\lceil \log_2(n-1) \rceil \end{aligned}$$

for every $x \in X$. Hence, we have $\| D, D^{(0)} \|_\infty \leq \max\{0, \Delta^*(D)\lceil \log_2(n-1) \rceil\}$ in case $\Delta = 0$, as claimed.

It remains to show that (ii) holds also for every positive Δ. In case $\Delta \geq \Delta^*(D)$, we have $D = D^{(\Delta)}$ — so, all our claims hold for essentially trivial reasons. Otherwise, consider the symmetric bivariate map D' on X defined by putting

$$D'(x, y) := \min\{D^{(0)}(x, y) + \Delta, D(x, y)\}$$

for all $x, y \in X$. Clearly, $D^{(0)} \le D' \le D$ holds, and it is also easy to see that D' is Δ-hierarchical:

Indeed, one has $D'(x, y) \le \max\{D'(x, z), D'(y, z)\} + \Delta$ for all $x, y, z \in X$ as $D'(x, y) \le D^{(0)}(x, y) + \Delta$ implies

$$D'(x, y) \le \max\{D^{(0)}(x, z), D^{(0)}(z, y)\} + \Delta \le \max\{D'(x, z), D'(y, z)\} + \Delta.$$

Moreover, for all $x, y \in X$ with $D'(x, y) = D(x, y)$, we clearly have

$$D(x, y) \le D'(x, y) + \Delta^*(D)\lceil \log_2(n-1)\rceil - \Delta$$

in view of $\Delta \le \Delta^*(D)$. And for all $x, y \in X$ with $D'(x, y) = D^{(0)}(x, y) + \Delta$, we have

$$\begin{aligned} D(x, y) &\le D^{(0)}(x, y) + \Delta^*(D)\lceil \log_2(n-1)\rceil \\ &= D'(x, y) + \Delta^*(D)\lceil \log_2(n-1)\rceil - \Delta. \end{aligned}$$

Thus, $D' \le D^{(\Delta)} \le D$ must hold and, therefore, also $\|D, D^{(\Delta)}\|_\infty \le \|D, D'\|_\infty \le \Delta^*(D)\lceil \log_2(n-1)\rceil - \Delta$, as required.

Finally, considering, for any $\Delta \ge 0$, the symmetric Δ-hierarchical map $D'' \in \mathbb{R}^{X \times X}$ obtained by lifting the map $D^{(\Delta)} \le D$ by half its distance to D, i.e., by putting

$$D''(x, y) := D^{(\Delta)}(x, y) + \frac{1}{2}\|D, D^{(\Delta)}\|_\infty$$

for all $x, y \in X$, we have

$$\begin{aligned} D(x, y) - D''(x, y) &= D(x, y) - D^{(\Delta)}(x, y) - \frac{1}{2}\|D, D^{(\Delta)}\|_\infty \\ &\le \|D, D^{(\Delta)}\|_\infty - \frac{1}{2}\|D, D^{(\Delta)}\|_\infty \\ &= \frac{1}{2}\|D, D^{(\Delta)}\|_\infty \end{aligned}$$

and

$$\begin{aligned} D''(x, y) - D(x, y) &= D^{(\Delta)}(x, y) + \frac{1}{2}\|D, D^{(\Delta)}\|_\infty - D(x, y) \\ &\le \frac{1}{2}\|D, D^{(\Delta)}\|_\infty \end{aligned}$$

and, therefore,

$$\|D, D''\|_\infty \le \frac{1}{2}\|D, D^{(\Delta)}\|_\infty,$$

implying that

$$\begin{aligned}\|D, \mathcal{D}(X|\Delta\texttt{-hier})\|_\infty &\leq \|D, D''\|_\infty \leq \frac{1}{2}\|D, D^{(\Delta)}\|_\infty \\ &\leq \frac{1}{2}\max\{0, \Delta^*(D)\lceil \log_2(n-1)\rceil - \Delta\}\end{aligned}$$

must also hold, as claimed. □

In [56], simulations were performed to test how well Gromov's bound presented in Theorem 10.4(ii) performs in practice. It was found that the bound was far from being tight in general. Indeed, the results suggested that, for a randomly generated symmetric bivariate map D and $\Delta = 0$, the quantity $\|D, D^{(0)}\|_\infty$ rarely exceeds $2\Delta^*(D)$ independently of the cardinality of X. Even so, we conclude by noting that, as was shown also in [56], defining the map $D : \langle n\rangle \times \langle n\rangle \to \mathbb{R}_{\geq 0}$ by choosing an arbitrary positive constant C and putting

$$D(x, y) := \begin{cases} 1 + C\lceil \log_2 |x-y|\rceil & \text{if } x \neq y, \\ 0 & \text{else,} \end{cases}$$

for all $x, y \in \langle n\rangle$, the equality $\|D, D^{(0)}\|_\infty = \Delta^*(D)\lceil \log_2(n-1)\rceil$ holds, that is, there are bivariate maps D for which Gromov's bound is actually tight. It could therefore be of interest to better understand such worst case examples, with the view to possibly obtaining improved versions of Gromov's bound for less "extreme" cases.

10.3 Quartet-Joining and QNet

As mentioned in Chapter 6, it is NP-complete to decide whether, for a given quartet system $\mathcal{Q} \subseteq \mathcal{Q}(X)$, there exists a phylogenetic X-tree $\mathcal{T}$ that displays $\mathcal{Q}$ [123]. So, how could we nonetheless generate a tree or a network from an arbitrary "generic" quartet system derived from biological data? Various methods have been proposed including, for example *Quartet-Puzzling* [124], the *$\mathcal{Q}$*-method* [22], and an integer linear programming approach [132]. Here, we describe a method called *Quartet-Joining* (cf. [89]) that follows an *agglomerative approach* to construct trees. It is analogous to the *Neighbor-Joining* algorithm [120] which is one of the most popular methods to generate phylogenetic trees from metrics.

The key idea in Quartet-Joining is, relative to a weighted quartet system $\mu : \mathcal{Q}(X) \to \mathbb{R}_{\geq 0}$ on a set X, to use a scoring function σ_μ to decide when a pair of elements in X can be regarded as "neighbors" relative to μ. In particular,

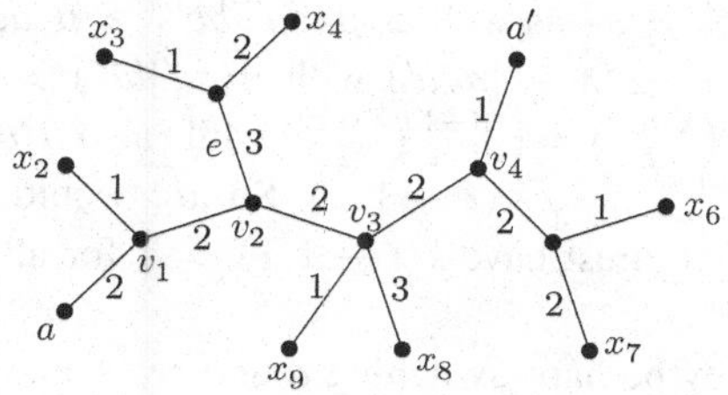

Figure 10.5 An example used in the proof of Lemma 10.5.

in case $\mu = \mu_{\mathcal{T}}$ is the weighted quartet system corresponding to an edge-weighted X-tree $\mathcal{T}$, the scoring function should select a pair of elements that form a cherry. More specifically, for any two distinct elements $a, a' \in X$, put

$$\sigma_\mu(a, a') := \sum_{\{b,b'\} \in \binom{X - \{a,a'\}}{2}} \mu(aa'|bb').$$

Then, as we now show, $\sigma_\mu(a, a')$ is maximized precisely when a and a' form a cherry in the tree $\mathcal{T}$ in case we have $\mu = \mu_{\mathcal{T}}$:

Lemma 10.5 *Let $\mu = \mu_{\mathcal{T}}$ be the weighted quartet system that is induced by some edge-weighted simple phylogenetic X-tree $\mathcal{T}$. Then, any two distinct elements $a, a' \in X$ that maximize σ_μ form a cherry in $\mathcal{T}$.*

Proof If a, a' did not form a cherry in $\mathcal{T}$, the length ℓ of the path $\mathbf{p} := \mathbf{p}_{\mathcal{T}}(a, a') = v_0 := a, v_1, \ldots, v_\ell := a'$ from a to a' would be larger than 2. For each $i = 1, \ldots, \ell - 1$ and each edge $e \in E_i := E_{v_i} \backslash E(\mathbf{p})$ (that is, distinct from $\{v_{i-1}, v_i\}$ and $\{v_i, v_{i+1}\}$), let $X(i, e)$ denote the set of all $x \in X$ with $e = e^{v_i \to x}$, and let $k(i, e)$ denote the cardinality of this set. In Figure 10.5, an example with $\ell = 5$ is depicted where we have, e.g., $X(2, e) = \{x_3, x_4\}$. Note that $\mu(aa'|bb') > 0$ holds for two distinct elements $b, b' \in X - \{a, a'\}$ if and only if there exists some $i = i_{bb'} \in \{1, \ldots, \ell - 1\}$ and some edge $e = e_{bb'} \in E_i$ with $b, b' \in X(i, e)$ in which case $\mu(aa'|bb') = D_{\mathcal{T}}(med_{\mathcal{T}}(b, b', v_i), v_i)$ holds. In other words, we have

$$\sigma_\mu(a, a') = \sum_{i=1}^{\ell-1} \sum_{e \in E_i} \sum_{\{b,b'\} \in \binom{X(i,e)}{2}} \mu(aa'|bb').$$

If there were to exist some $e_1 \in E_1$ with $k(1, e_1) = 1$, the two elements a, a'' with a'' denoting the unique element in $X(1, e_1)$ would form a cherry and one would have $b, b' \neq a''$ as well as $\mu(aa''|bb') \geq \mu(aa'|bb')$ for any two distinct elements $b, b' \in X - \{a, a'\}$ with $\mu(aa'|bb') > 0$ or, equivalently, with $\{b, b'\} \in \bigcup_{i=1}^{\ell-1} \binom{X(i,e)}{2} = \bigcup_{i=2}^{\ell-1} \binom{X(i,e)}{2}$ while also $\mu(aa''|a'c) > 0$ would hold for every $c \in \bigcup_{e \in E_{\ell-1}} X(\ell - 1, e)$. So, σ_μ would not be maximized by a, a'. Similarly, we must have $k(\ell - 1, e) > 1$ for all $e \in E_{\ell-1}$ if a, a' maximize σ_μ.

Now, assume — by perhaps switching a and a' — that $\min_{e \in E_1} k(1, e) \leq \min_{e \in E_{\ell-1}} k(\ell - 1, e)$ holds and choose some $e_1 \in E_1$ with $\min_{e \in E_1} k(1, e) = k(1, e_1)$. Furthermore, note that — in view of $k(1, e_1) > 1$ — one can find two distinct elements c, c' in $X(1, e_1)$ for which the distance $D_{\mathcal{T}}(med_{\mathcal{T}}(c, c', v_1), v_1) = \mu(aa'|cc'') =: r$ between v_1 and the median of c, c', and v_1 is as large as possible implying that, in particular, c, c' must form a cherry in $\mathcal{T}$.

We claim that $\sigma_\mu(a, a') < \sigma_\mu(c, c')$ must hold for these two elements in $X(1, e_1)$: Indeed, note that $|\{a, a', b, b', c, c'\}| = 6$ and

$$r + \mu(aa'|bb') \leq \mu(cc'|bb')$$

holds for any two distinct elements $b, b' \in Y := X - (\{a, a'\} \cup X(1, e_1))$ while $\mu(aa'|bb') \leq r$ must hold for any two distinct elements b, b' in $X(1, e_1)$.

Thus, noting that, by our choice of e_1, we must have $k(1, e_1) \leq |Y|$, our claim follows immediately from

$$\begin{aligned}
\sigma_\mu(a, a') &= \sum_{i=1}^{\ell-1} \sum_{e \in E_i} \sum_{\{b,b'\} \in \binom{X(i,e)}{2}} \mu(aa'|bb') \\
&= \sum_{\{b,b'\} \in \binom{X(1,e_1)}{2}} \mu(aa'|bb') + \sum_{\{b,b'\} \in \binom{Y}{2}} \mu(aa'|bb') \\
&\leq r \binom{k(1, e_1)}{2} + \sum_{\{b,b'\} \in \binom{Y}{2}} \mu(aa'|bb') \leq \sum_{\{b,b'\} \in \binom{Y}{2}} (r + \mu(aa'|bb')) \\
&\leq \sum_{\{b,b'\} \in \binom{Y}{2}} \mu(cc'|bb') < \mu(cc'|aa') + \sum_{\{b,b'\} \in \binom{Y}{2}} \mu(cc'|bb') \\
&\leq \sigma_\mu(c, c').
\end{aligned}$$

□

We now describe the Quartet-Joining algorithm. The input to the algorithm is a weighted quartet system $\mu : \mathcal{Q}(X) \to \mathbb{R}_{\geq 0}$ on a set X with $n \geq 4$ elements. Quartet-Joining performs $n-3$ agglomerations (see Figure 10.6): We start with a star tree $\mathcal{T}_0$ for $X_0 := X$ (Figure 10.6(a)). Then, using the scoring function σ_μ, two distinct elements $x, x' \in X_0$ with maximal score $\sigma_\mu(x, x')$ are selected

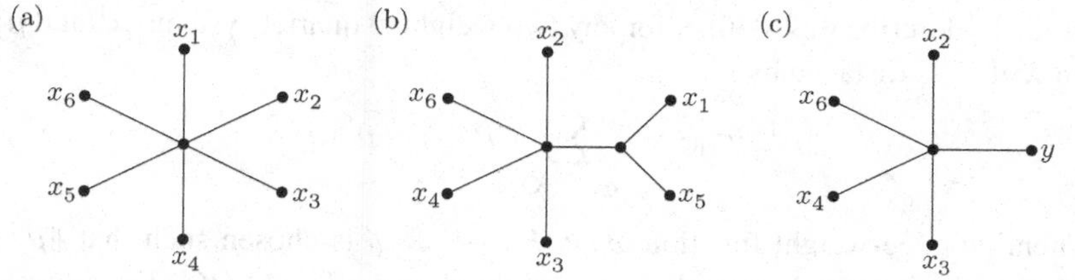

Figure 10.6 An example of an agglomeration performed by Quartet-Joining. (a) A star tree for $X = \{x_1, \ldots, x_6\}$. (b) Elements x_1 and x_5 are selected to form a cherry. (c) The cherry is replaced by a single vertex labeled by a new element y.

to form a cherry in the resulting phylogenetic tree (Figure 10.6(b)). The cherry is then replaced by a single vertex that is labeled by a new element y not yet involved in our set-up, resulting in a star tree $\mathcal{T}_1$ for $X_1 := (X_0 - \{x, x'\}) + y$ (Figure 10.6(c)). Then, based on $\mu_0 := \mu$, a weighted quartet system μ_1 on X_1 is computed and the whole process is repeated for μ_1 and X_1. This results in a sequence $X_0, X_1, \ldots, X_i, \ldots, X_{n-3}$ of sets of cardinality $|X_i| = n - i$ and star trees $\mathcal{T}_0, \ldots, \mathcal{T}_{n-3}$ for X_i and corresponding weighted quartet systems μ_i. The output phylogenetic tree $\mathcal{T}_\mu$ on X is then obtained by reversing the replacements of cherries, starting with $\mathcal{T}_{n-3}$.

It remains to describe how precisely the weighted quartet system μ_i on X_i is computed from the weighted quartet system μ_{i-1} on X_{i-1} for $i = 1, \ldots, n-3$ after two distinct elements $x, x' \in X_{i-1}$ with maximal score $\sigma_{\mu_{i-1}}$ have been selected. Recall that, in each agglomeration, the cherry formed by x and x' is replaced by a single vertex labeled by a new element y. We put

$$\mu_i(aa'|bb') = \begin{cases} \mu_{i-1}(aa'|bb') & \text{if } y \notin \{a, a', b, b\}, \\ \frac{1}{2}(\mu_{i-1}(ax|bb') + \mu_{i-1}(ax'|bb')) & \text{if } a' = y, \end{cases}$$

for all $aa'|bb' \in \mathcal{Q}(X_i)$. It is not hard to check that, if there exists an edge-weighted X_{i-1}-tree $\mathcal{T}_{i-1}$ with $\mu_{i-1} = \mu_{\mathcal{T}_{i-1}}$, then there exists also an edge-weighted X_i-tree $\mathcal{T}_i$ with $\mu_i = \mu_{\mathcal{T}_i}$. Hence, using induction, it follows by Lemma 10.5 that Quartet-Joining, applied to the weighted quartet system $\mu = \mu_\mathcal{T}$ that is induced by some edge-weighted phylogenetic X-tree $\mathcal{T}$, constructs a binary phylogenetic X-tree $\mathcal{T}_\mu$ that is a *refinement* of $\mathcal{T}$, that is, with $\Sigma_\mathcal{T} \subseteq \Sigma_{\mathcal{T}_\mu}$.

Finally, to obtain weights for the edges of the tree $\mathcal{T}_\mu = (V_\mu, E_\mu, \varphi_\mu)$ constructed by Quartet-Joining from μ, a *non-negative least squares* approach is used. To describe this, define, for any two weighted quartet systems μ^1 and μ^2 on X the L_2-distance by

$$\|\mu^1, \mu^2\|_2 := \sum_{q \in \mathcal{Q}(X)} (\mu^1(q) - \mu^2(q))^2.$$

Then, an edge-weight function $\omega_\mu : E_\mu \rightarrow \mathbb{R}_{\geq 0}$ is chosen such that $\|\mu - \mu_{\mathcal{T}'}\|_2$ is minimized over all edge-weighted trees $\mathcal{T}'_\mu = (V_\mu, E_\mu, \omega_\mu, \varphi_\mu)$ where μ is the input quartet weight function. Note that ω_μ might assign weight 0 to some edges of $\mathcal{T}_\mu$. Those edges are then contracted in the final tree output by Quartet-Joining. Moreover, since the weight of the pendant edges does not affect the induced weighted quartet system, we put $\omega_\mu(e) := 1$ for every such edge e of $\mathcal{T}_\mu$.

It can be shown that the restriction of ω_μ to the interior edges of $\mathcal{T}_\mu$ is unique and, in case μ is induced by some edge-weighted phylogenetic X-tree $\mathcal{T}$, $\mu = \mu_{\mathcal{T}_\mu}$ holds (see [89] for further details). This implies that Quartet-Joining is *consistent*, that is, the following result holds:

Theorem 10.6 *Let μ be a weighted quartet system induced by some edge-weighted phylogenetic X-tree $\mathcal{T}$. Then Quartet-Joining applied to μ outputs an edge-weighted phylogenetic X-tree that is isomorphic to $\mathcal{T}$ (up to the weights of the pendant edges of $\mathcal{T}$).*

If the given weighted quartet system μ cannot be well represented by a single tree, we could instead try to construct a network which can help to explore the degree and location of inconsistencies in the data. One way to do this is to use an agglomerative approach to network construction similar to Neighbor-Net, called *QNet* [85] (see also [34, 35]).

In QNet, rather than constructing a phylogenetic tree, a circular split system on X is constructed, which can then be represented by a split network. To this end, a linear ordering Θ of X is constructed. This is done, as in Quartet-Joining, by using the scoring function σ_μ to select two distinct elements x, x' in each agglomeration, which are then replaced by a new element y. The linear ordering Θ is then computed by reversing these replacements one at a time. To illustrate this, suppose, by induction, $y, x_3, \ldots, x_n$ is the linear ordering of $X_1 = (X_0 - \{x, x'\}) + y$ obtained in this way. Then, there are two possible choices for Θ, namely, $\Theta_1 := x, x', x_3, \ldots, x_n$ and $\Theta_2 := x', x, x_3, \ldots, x_n$. Using a scoring function similar to σ_μ, we select the linear ordering with the larger score.

Once the linear ordering Θ has been computed, a non-negative least squares approach is used to compute weights for the splits in the unique circular split system Σ_Θ with maximal cardinality that fits Θ — just as in Quartet-Joining.

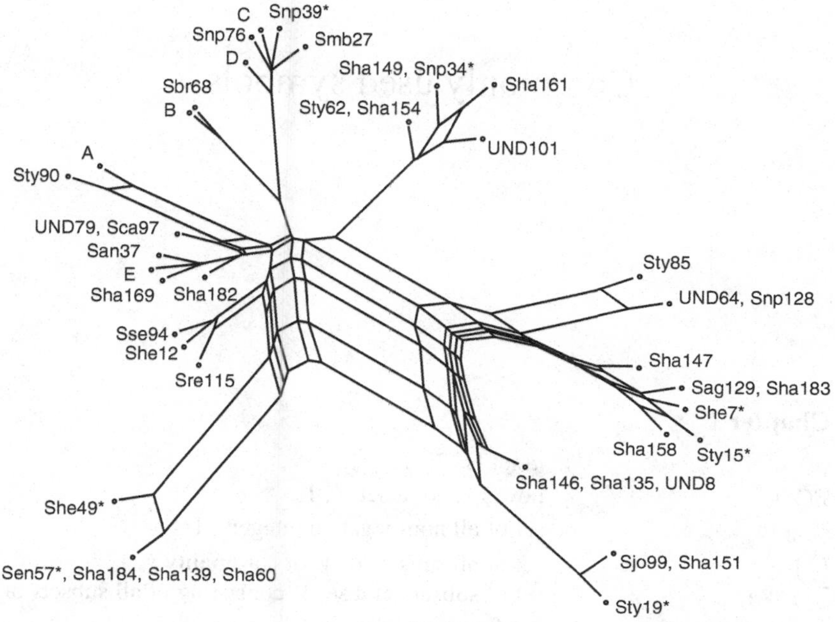

Figure 10.7 The network computed by QNet for a *Salmonella* data set (see text for more details).

One can show that the weights are unique and that QNet is consistent in the sense that, if μ is induced by some weighted circular split system ν, then the split system generated by QNet equals ν (see [89] for a proof of this fact). This implies in particular that, if μ encodes an edge-weighted X-tree, we will get back this tree. But at the same time, in case μ does not perfectly correspond to such a tree, the deviation of the network produced by QNet from being a tree also gives some indication of how far μ is from being an encoding of an X-tree.

We conclude by noting that the QNet algorithm has been implemented and run on various biological data sets. For example, in Figure 10.7, we present the network that was computed by QNet for a data set consisting of molecular sequences derived from a certain collection of *Salmonella* isolates described in [109]. It is suspected that these isolates have undergone recombination, a form of non-treelike or reticulate evolution. This is reflected in the complexity of the network. More details about this example and the QNet approach may be found in [85].

Commonly used symbols

Chapter 1

$\langle n \rangle$	n-set $\{1, 2, \ldots, n\}$, 1
$\mathbb{P}(V)$	power set of a set V, 1
$\mathbb{N}_{\geq 0}$	set of all non-negative integers, 1
$\binom{V}{k}$	set of all subsets of V of cardinality k, 1
$\mathbb{P}_{\geq k}(V)$	set of subsets of a set V consisting of all subsets of V of cardinality at least k, 1
$\mathbb{P}_{\leq k}(V)$	set of subsets of a set V consisting of all subsets of V of cardinality at most k, 1
$A + x$	set resulting from adding a single element x to a set A, 1
$A - x$	set resulting from subtracting a single element x from a set A, 1
$A - B$	set difference between two sets A and B, 1
$\max(U) = \max_{\preceq}(U)$	maximal elements in a set U (relative to a partial order $\preceq$), 2
$\min(U) = \min_{\preceq}(U)$	minimal elements in a set U (relative to a partial order $\preceq$), 2
$U_{\preceq u}$	set of all u' in a partially ordered set U with $u' \preceq u$, 2
$U_{\prec u}$	set of all u in $U_{\preceq u}$ with $u' \neq u$, 2
$\mathrm{chld}_U(u)$	maximal elements below an element u in a partially ordered set U, 2
$\bigcup \mathcal{V}$	union of all clusters in a set system $\mathcal{V}$, 2
$\bigcap \mathcal{V}$	intersection of all clusters in a set system $\mathcal{V}$, 2
$A\|B$	pair of disjoint, non-empty sets forming a (partial) split of a set X if $A \cup B = X$ ($A \cup B \subseteq X$) holds, 2
$\\|S\\|$, $\\|A\|B\\|$	size of the split $S = A\|B$, 2

$S(x)$	subset, A or B, in an X-split $S = A\|B$ that contains a given element $x \in X$, 2
$\overline{S}(x)$	subset, A or B, in an X-split $S = A\|B$ that does not contain a given element $x \in X$, 2
$A\dot{\cup}B$	disjoint union of sets A and B, 3
$\|M\|$	cardinality of the set M, 3
$G = (V, E)$	graph with vertex set V and edge set E, 4
$N_G(v) = N(v)$	set of vertices adjacent to vertex v in a graph G, 5
$E_G(v) = E(v)$	set of edges incident to a vertex v in a graph G, 5
$\deg_G(v) = \deg(v)$	degree of vertex v in a graph G, 5
$e_G(a)$	unique edge e of a graph G that is incident to a leaf a, 5
$v_G(a)$	unique vertex v of a graph G that is adjacent to a leaf a, 5
$e_G(a, b)$	edge connecting a cherry a, b to the graph G in case this edge is unique, 6
$V_{int}(G)$	set of interior vertices of a graph G, 5
$E_{int}(G)$	set of interior edges of G, 5
∂G	graph $(V_{int}(G), E_{int}(G))$ derived from a graph G, 5
$G[U]$	subgraph induced by a graph G on U, 6
$V(\mathbf{p})$	vertex set of path $\mathbf{p}$, 6
$E(\mathbf{p})$	edge set of path $\mathbf{p}$, 6
$\pi_0(G)$	set of connected components of the graph G, 7
$G(v)$	connected component containing a given vertex v of a graph G, 7
$G^{(e)}$	the graph $(V, E - e)$ that results from deleting an edge $e \in E$ from a graph $G = (V, E)$, 7
$E_G(u\|v) = E(u\|v)$	the set of all edges in a graph $G = (V, E)$ that separate the two vertices $u, v \in V$, 7
$e^{v \to u}$	unique edge in $E(v)$ that separates v from u in a tree, 8
$\mathbf{p}_T(u, v) = \mathbf{p}(u, v)$	path from u to v in a tree T, 8
$V_T[u, v]$	vertex set of the path from u to v in a tree T, 8
$E_T[u, v]$	edge set of the path from u to v in a tree T, 8
$med_T(u, v, w)$	median of the three vertices a, b, c in a tree T, 8
$A \triangle B$	symmetric difference of two sets A and B, 8
$f^{-1}(x)$	set of elements that are mapped to element x by map f, 9

$V/\sim$	set of equivalence classes of an equivalence relation $\sim$ on a set V, 9
ψG	contraction of a graph G by a contracting map ψ, 9
ψE	edge set of the contracted graph ψG, 9
ψ_e	contracting map associated to edge e, 9
ψ^e	contracting map associated to a pendant edge e, 10
G^e	the graph obtained by eliminating a pendant edge e in a graph G, 10
$A \times B$	set of all ordered pairs (a, b) with $a \in A$ and $b \in B$, 10
$V^{(i)}$	set of vertices of degree i in a graph $G = (V, E)$, 10
$T\|_U$	restriction of a tree T to a subset U of its vertex set, 11
H_k	k-dimensional hypercube, 11
$G^\circ = (V, A^\circ)$	underlying undirected graph of a directed graph G, 12
$u \preceq_{\mathcal{T}} v$	vertex u lies on the path from the root to vertex v in the rooted tree $\mathcal{T}$, 13
$\mathbb{R}$	set of real numbers, 13
$D\|_{X'}$	restriction of a metric D on X to a subset $X' \subseteq X$, 14
$[x, y]_D = [x, y]$	set of elements z with $D(x, y) = D(x, z) + D(z, y)$, 14
$G = (V, E, \omega)$	graph (V, E) with an edge-weight function ω, 15
$\mathbb{R}_{>0}$	set of positive real numbers, 15
$\mathbf{1}_E$	edge-weight function that assigns 1 to every edge of a graph, 15
$D_G = D_{(V,E,\omega)}$	metric induced by a weighted graph $G = (V, E, \omega)$, 15
D_G	metric induced by a graph $G = (V, E)$, 15
$Med_D(a, b, c)$	set of medians of the three elements a, b, c in a metric space (X, D), 16
$med_D(a, b, c)$	median of the three elements a, b, c in a metric space (X, D) in case it is unique, 16

Chapter 2

$\mathcal{T} = (V, E, \varphi)$	X-tree, 21
$\underline{\mathcal{T}}$	underlying tree of an X-tree, 21
$supp(\mathcal{T})$	support of an X-tree, 21
$\Sigma(X)$	set of all splits of X, 24
$\Sigma_{triv}(X)$	set of all trivial splits of X, 24

$\Sigma^*(X)$	set of all non-trivial splits of X, 24
S_e	split associated with an edge e in an X-tree, 24
$f^{-1}(A)$	set of all elements that are mapped to an element of the set A by a map $f : X \to A$, 24
$\Sigma_{\mathcal{T}}$	split system associated to an X-tree $\mathcal{T}$, 24
$\mathcal{C}(\Sigma)$	collection of split parts of Σ, 24
$D_{\mathcal{T}}$	metric induced by an X-tree $\mathcal{T}$, 26
$a_1a_2\|b_1b_2$	the unordered pair consisting of the two subsets $\{a_1, a_2\}$ and $\{b_1, b_2\}$, 27
$supp(q)$	support of a quartet q, 27
$\mathcal{Q}(X)$	set of all quartets on X, 27
$\mathcal{Q}_{\mathcal{T}}$	quartet system associated to the X-tree $\mathcal{T}$, 27
$\overline{\mathcal{Q}}_{\mathcal{T}}$	collection of all 4-subsets Y of X with $\mathcal{Q}(Y) \cap \mathcal{Q}_{\mathcal{T}} = \emptyset$, 28
$v_{\mathcal{T}}(Y)$	for a phylogenetic X-tree $\mathcal{T} = (V, E, \varphi)$ and a 4-subset $Y = \{a, b, c, d\}$ of X in $\overline{\mathcal{Q}}_{\mathcal{T}}$, the unique vertex $v \in V$ for which the edges $e^{v\to a}, e^{v\to b}, e^{v\to c}, e^{v\to d} \in E_v$ are all distinct, 28

Chapter 3

$\mathcal{T} = (V, E, \omega, \varphi)$	weighted X-tree, 31
$D_{\mathcal{T}}$	metric induced by a weighted X-tree $\mathcal{T}$, 31
$exc_{\mathcal{T}}(x\|X)$	eccentricity of an element x relative to the set X and the X-tree $\mathcal{T}$, 32
S_A	split $A\|X - A \in \Sigma(X)$ associated to a subset $A \subseteq X$, 40
$\nu_{\mathcal{T}}$	weighted split system induced by a weighted X-tree $\mathcal{T}$, 41
$\mathbb{T}(X)$	set of all metrics in $\mathbb{R}^{X\times X}$ that satisfy the 4-point condition, 41
$\mathcal{S}_X(\mathtt{tree})$	set of all weighted split systems ν in $\mathbb{R}_{\geq 0}^{\Sigma(X)}$ with compatible support, 41
D_ν	bivariate map associated to a weighted split system ν, 41
$\mu_{\mathcal{T}}$	weighted quartet system induced by a weighted X-tree, 46
E_q	set of edges in a phylogenetic tree that support the quartet q, 46

Chapter 4

$\mathcal{B}(\Sigma) = \big(V(\Sigma), E(\Sigma)\big)$	Buneman graph associated to a split system Σ, 51
$V(\Sigma)$	vertex set of the Buneman graph $\mathcal{B}(\Sigma)$, 51

$E(\Sigma)$	edge set of the Buneman graph $\mathcal{B}(\Sigma)$, 51
$\Delta(\phi, \psi)$	difference set $\{m \in M : \phi(m) \neq \psi(m)\}$ for two maps ϕ and ψ from a set M into a set N, 51
$V^\star(\Sigma)$	superset of $V(\Sigma)$, 51
$\mathcal{B}^\star(\Sigma)$	hypercube induced by the split system Σ, 52
ϕ^Ξ	element of $V^\star(\Sigma)$ associated to every $\phi \in V^\star(\Sigma)$ and every subset Ξ of Σ, 52
$\psi[\Xi]$	image $\{\psi(S) : S \in \Xi\}$ of a subset Ξ of Σ relative to any map ψ in $V^\star(\Sigma)$, 52
$S_{\phi,\psi}$	unique split in $\Delta(\phi, \psi)$ for two adjacent vertices in $V^\star(\Sigma)$, 53
ϕ_x	vertex in $V(\Sigma)$ that maps every $S \in \Sigma$ onto the subset $S(x)$ of X, 55
$\mathrm{res}^\star_{\Sigma\to\Sigma'}$	graph morphism that maps $V^\star(\Sigma)$ onto $V^\star(\Sigma')$ for any $\Sigma' \subseteq \Sigma$, 56
$\mathrm{res}_{\Sigma\to\Sigma'}$	graph morphism that maps $V(\Sigma)$ onto $V(\Sigma')$ for any $\Sigma' \subseteq \Sigma$, 56
ψ_Φ	element in $V^\star(\Sigma)$ associated to every $\psi \in V^\star(\Sigma)$ and every $\Phi \subseteq V^\star(\Sigma)$, 58
$\mathrm{Incpt}(\Sigma)$	family of subsets of Σ that consist of pairwise incompatible splits, 60
$\mathrm{Incpt}(\Sigma\|S)$	family of subsets of Σ that consist of pairwise incompatible splits and contain the split S, 60
$H_\phi(\psi)$	split system $\Delta_{\min}(\psi\|\phi)$ in $\mathrm{Incpt}(\Sigma)$ associated to the two vertices $\phi, \psi \in V(\Sigma)$, 61
$\mathbf{Med}_i(V')$	i-th iteration in the construction of the vertices that are generated in a median graph by V', 63
$\mathbf{Med}(V')$	the median hull of a subset V' of the point set V of a metric space (V, D), 63
κ	C-coloring of a graph, 66
$\kappa(\mathbf{p})$	set of colors of the edges of the path $\mathbf{p}$, 66
$\mathbf{P}^G_{\min}(u, v)$	set of all shortest paths from a vertex u to a vertex v in a connected graph G, 66
$\kappa(u, v)$	set of colors that appear on every shortest path from u to v in a connected graph G with edge-coloring κ, 66
(V, E, κ)	split graph consisting of a connected and bipartite graph $G = (V, E)$ together with a C-coloring κ, 66
$\mathcal{N} = (V, E, \kappa, \varphi)$	split network consisting of a split graph (V, E, κ) and a labeling map φ, 68

$\alpha^D(A:B)$	isolation index of a subset A of X versus another subset B of X relative to a metric D, 147
ν^D	weighted split system associated to D based on the isolation index, 149
Σ^D	support of ν^D, 149
$\nu^D_{x\|y}$	weighted split system associated to D, x, and y based on the isolation index, 150
$\mu(D_1\|D_2)$	metric index of a metric D_1 relative to a metric D_2, 151
$\mathbf{S}_X$	split operator for the set X, 154
$\mathbb{TSD}(X)$	space of all totally split-decomposable metrics defined on a set X, 154
D_{sp}	split-prime metric associated to a metric D, 154
$\mathbf{SP}_X$	split-prime operator for the set X, 154
$\mathbb{SP}(X)$	space of all split-prime metrics defined on a set X, 154
$\mathcal{M}(X) = \mathcal{M}(X, \mathbb{R})$	vector space of all symmetric bivariate maps from $X \times X$ into $\mathbb{R}$ that vanish on the diagonal, 161
λ_X	linear map from weighted split systems to bivariate maps on X, 161
$\overline{\mathcal{B}}(\Sigma)$	Buneman complex associated to the split system Σ, 168

Chapter 8

$\Sigma^*_{part}(X)$	collection of all partial splits $A\|B$ of a set X with $\|A\|, \|B\| \geq 2$, 171
$\mathcal{Q}(A\|B)$	quartet system induced by a partial split $A\|B$, 171
$\mathcal{Q}(\Sigma)$	quartet system induced by the split system Σ, 171
$\widehat{\mathcal{Q}}$	collection of all partial $\mathcal{Q}$-splits, that is, all partial splits in $\Sigma^*_{part}(X)$ with $\mathcal{Q}(A\|B) \subseteq \mathcal{Q}$, 172
$\Sigma(\mathcal{Q})$	collection of all splits in $\Sigma^*(X)$ that are induced by the quartet system $\mathcal{Q}$, 172
$\mathcal{Q}\|_Y$	set of all quartets in a quartet system $\mathcal{Q} \subseteq \mathcal{Q}(X)$ contained in $\mathcal{Q}(Y)$, Y a subset of X, 172
$G_{(\mathcal{Q},Y)}$	graph induced by the quartet system $\mathcal{Q}$ and the 2-subset Y, 177
$D_{\mathcal{Q}}$	metric induced by the quartet system $\mathcal{Q}$, 177
$\mathcal{Q}_D$	quartet system induced by the metric D, 179
$ab\|c\|de \in \widehat{\mathcal{Q}}$	shorthand for the assertion $abc\|de, ab\|cde \in \widehat{\mathcal{Q}}$, 181

$<_{a\to b}$	partial order associated with a transitive double cover $\mathcal{Q}$ and any two distinct elements $a, b \in X$, 181
$\sim_{a,b}$	binary relation on $X - \{a, b\}$ associated with a transitive double cover $\mathcal{Q}$, 181
$\mathcal{Q} \simeq \mathcal{Q}'$	isomorphism of quartet systems, 183
$Aut(\mathcal{Q})$	the automorphism group of a quartet system $\mathcal{Q}$, 184
$\Sigma_{strong}(\mathcal{Q})$	collection of all strong $\mathcal{Q}$-splits, 192
$Cl(G)$	cardinality of a smallest subset F of $\binom{V}{2}$ for which the graph $G\Delta F := (V, E\Delta F)$ is a disjoint union of cliques, 194
$G\Delta F$	graph obtained from a graph $G = (V, E)$ by replacing its edge set E by $E\Delta F$ of E for some subset F of $\binom{V}{2}$, 194

Chapter 9

$\mathcal{T} = (V, A, \varphi, \tau)$	dated rooted X-tree, 195
$C_{\mathcal{T}}(v)$	cluster associated to the vertex v in the rooted X-tree $\mathcal{T}$, 196
$\mathcal{C}_{\mathcal{T}}$	cluster system induced by the rooted X-tree $\mathcal{T}$, 196
$ab\|c$	triplet, 197
$\mathcal{R}_{\mathcal{T}}$	triplet system induced by the rooted X-tree $\mathcal{T}$, 197
$D_{\mathcal{T}}$	symmetric bivariate map induced by the dated rooted X-tree $\mathcal{T}$, 197
$\mathbf{lca}_{\mathcal{T}}(x, y)$	last common ancestor of the vertices labeled by x and y in the dated rooted X-tree $\mathcal{T}$, 197
$H(\mathcal{C}) = (\mathcal{C}, A(\mathcal{C}))$	Hasse graph with vertex set $\mathcal{C}$ and edge set $A(\mathcal{C})$ associated with a cluster system $\mathcal{C}$, 200
$\varphi_{\mathcal{C}}$	labeling map of the rooted X-tree associated with a hierarchy $\mathcal{C}$, 200
$C_x = C_x^{\mathcal{C}}$	smallest cluster containing an element x in a hierarchy $\mathcal{C}$, 200
$\mathcal{C}_x$	collection of all clusters in a hierarchy $\mathcal{C}$ containing an element x, 200
X_α	set of all elements $x \in X$ with $D(x, x) \leq \alpha$ (relative to some hierarchical dissimilarity D), 203
$\mathcal{C}(D)$	hierarchy associated to a hierarchical dissimilarity D, 203
$C_\alpha(x)$	set $\{y \in X : D(x, y) \leq \alpha\}$ defined for any hierarchical dissimilarity D, every $\alpha \in \mathbb{R}$, and every element $x \in X_\alpha$, 203

$\max D$	maximum value attained by a bivariate map $D : X \times X \to \mathbb{R}$, 203
$\tau^{(D)}$	dating map defined on the hierarchy $\mathcal{C}(D)$ associated to a hierarchical dissimilarity D, 203
$\mathcal{T}^{(D)}$	dated rooted X-tree associated to a hierarchical dissimilarity D, 203
$\varphi^{(D)}$	labeling map $X \to \mathcal{C}(D)$ associated to a hierarchical dissimilarity D defined on a set X, 203
$D^{(0)}$	subdominant hierarchical dissimilarity associated with a dissimilarity D defined on a set X, 204
$\mathcal{D}_{\leq D}$	collection of all hierarchical dissimilarities D' defined on X with $D' \leq D$, 204
$rk_a(x) = rk_a^D(x)$	rank of an element $x \in X$ relative to some fixed element $a \in X$ and D, 205
X^*	set X together with a new element $*$, 207
$\Sigma^*(\mathcal{C})$	split system induced by the cluster system $\mathcal{C}$, 207
$\mathcal{C}_*(\Sigma)$	cluster system induced by the split system Σ, also called the combinatorial Farris transform of Σ, 207
D^f	Farris transform of D relative to f, 209
D^a	Farris transform of D relative to the Kuratowski map k_a, 211
f_γ	γ-translate of a map $f \in \mathbb{R}^X$, 211
$\Delta(D)$	hyperbolicity index of a bivariate map $D \in \mathbb{R}^{X \times X}$, 213
$\Delta^*(D)$	ultrametricity index of a bivariate map $D \in \mathbb{R}^{X \times X}$, 213
$D^{\#}$	Farris transform of a symmetric bivariate map $D \in \mathbb{R}^{X \times X}$ that vanishes on the diagonal, 215
gk_x	generalized Kuratowski map associated to x, 215
D^0	metric D^0 associated to a non-negative generalized metric D, 216

Chapter 10

$\mathrm{karz}_k(n)$	cardinality of the largest possible k-compatible split system on a set containing n elements, 223
$\Sigma_{>\frac{1}{2}}(\mathcal{P})$	set of splits that are induced by more than half of the trees in $\mathcal{P}$, 226
$\Sigma_{>\frac{1}{k+1}}(\mathcal{P})$	set of splits that are induced by a fraction of more than $\frac{1}{k+1}$ of the trees in $\mathcal{P}$, 226

$\Vert\mathcal{D}_1, \mathcal{D}_2\Vert_\infty$	infimum of the L_∞-distances between any two maps $D_1 \in \mathcal{D}_1$ and $D_2 \in \mathcal{D}_2$, 230
$\Vert D_1, D_2\Vert_\infty$	L_∞-distance between two bivariate maps, 230
$\mathcal{D}_{\leq D}(X\vert\Delta\text{-}\mathtt{hier})$	set of all Δ-hierarchical maps below a given map $D \in \mathbb{R}^{X\times X}$, 231
$\partial_\Delta D$	certain bivariate map below the bivariate map D, 232
$D^{[k]}$	certain $k+1$-variate map associated to the bivariate map D, 232

Bibliography

[1] R. Agarwala, V. Bafna, M. Farach, B. Narayanan, M. Paterson, and M. Thorup. On the approximability of numerical taxonomy (fitting distances by tree metrics). *SIAM Journal on Computing*, **28**:1073–1085, 1999.

[2] A. Aho, Y. Sagiv, T. Szymanski, and J. Ullman. Inferring a tree from lowest common ancestors with an application to the optimization of relational expressions. *SIAM Journal on Computing*, **10**:405–421, 1981.

[3] I. Althöfer. On optimal realizations of finite metric spaces by graphs. *Discrete and Computational Geometry*, **3**:103–122, 1988.

[4] J. Apresjan. An algorithm for constructing clusters from a distance matrix. *Mashinnyi Perevod: Prikladnaja Lingvistika*, **9**:3–18, 1966.

[5] D. Avis and M. Deza. The cut cone, L^1-embeddability, complexity, and multicommodity flows. *Networks*, **21**:595–617, 1991.

[6] H.-J. Bandelt. Recognition of tree metrics. *SIAM Journal on Discrete Mathematics*, **3**:1–6, 1990.

[7] H.-J. Bandelt. Generating median graphs from Boolean matrices. In Y. Dodge, editor, *L_1-Statistical Analysis and Related Methods*, pages 305–309. North-Holland, Amsterdam, 1992.

[8] H.-J. Bandelt. Phylogenetic networks. *Verhandlungen des naturwissenschaftlichen Vereins Hamburg*, **34**:51–57, 1994.

[9] H.-J. Bandelt and A. Dress. Weak hierarchies associated with similarity measures – an additive clustering technique. *Bulletin of Mathematical Biology*, **51**:133–166, 1989.

[10] H.-J. Bandelt and A. Dress. A canonical decomposition theory for metrics on a finite set. *Advances in Mathematics*, **92**:47–105, 1992.

[11] H.-J. Bandelt and A. Dress. Split decomposition: a new and useful approach to phylogenetic analysis of distance data. *Molecular Phylogenetics and Evolution*, **1**:242–252, 1992.

[12] H.-J. Bandelt and A. Dress. A relational approach to split decomposition. In O. Opitz, B. Lausen, and R. Klar, editors, *Information and Classification – Concepts, Methods and Applications*, pages 123–131. Springer, Berlin, 1993.

[13] H.-J. Bandelt, P. Forster, and A. Rohl. Median-joining networks for inferring intraspecific phylogenies. *Molecular Biology and Evolution*, **16**:37–48, 1999.

[14] H.-J. Bandelt, P. Forster, B. C. Sykes, and M. B. Richards. Mitochondrial portraits of human population using median networks. *Genetics*, **141**:743–753, 1995.

[15] H.-J. Bandelt, K. T. Huber, and V. Moulton. Quasi-median graphs from sets of partitions. *Discrete Applied Mathematics*, **122**:23–35, 2002.

[16] H.-J. Bandelt, H. Mulder, and E. Wilkeit. Quasi-median graphs and algebras. *Journal of Graph Theory*, **18**:681–703, 1994.

[17] H.-J. Bandelt and M. van de Vel. Superextensions and the depth of median graphs. *Journal of Combinatorial Theory A*, **57**:187–202, 1991.

[18] J. Bang-Jensen and G. Gutin. *Digraphs: Theory, Algorithms and Applications*. Springer-Verlag, Berlin, 2000.

[19] M. Baroni, C. Semple, and M. Steel. A framework for representing reticulate evolution. *Annals of Combinatorics*, **8**:391–408, 2004.

[20] J. Barthélemy. From copair hypergraphs to median graphs with latent vertices. *Discrete Mathematics*, **76**:9–28, 1989.

[21] J. Barthélemy and A. Guenoche. *Trees and Proximity Representations*. John Wiley, Chichester, 1991.

[22] V. Berry and O. Gascuel. Inferring evolutionary trees with strong combinatorial evidence. *Theoretical Computer Science*, **240**:271–298, 2000.

[23] L. J. Billera, S. P. Holmes, and K. Vogtmann. Geometry of the space of phylogenetic trees. *Advances in Applied Mathematics*, **27**:733–767, 2001.

[24] O. Bininda-Edmonds, editor. *Phylogenetic Supertrees: Combining Information to Reveal the Tree of Life*. Springer, Berlin, 2004.

[25] A. Björner, M. Las Vergnas, B. Sturmfels, N. White, and G. Ziegler. *Oriented matroids*. Cambridge University Press, 1999.

[26] S. Böcker. *From subtrees to supertrees*. PhD thesis, Universität Bielefeld, 1999.

[27] S. Böcker, D. Bryant, A. Dress, and M. Steel. Algorithmic aspects of tree amalgamation. *Journal of Algorithms*, **37**:522–537, 2000.

[28] S. Böcker and A. Dress. A note on maximal hierarchies. *Advances in Mathematics*, **151**:270–282, 2000.

[29] B. Bowditch. Notes on Gromov's hyperbolicity criterion for path metric spaces. In E. Ghys *et al.*, editor, *Group Theory from a Geometric Viewpoint*, pages 64–167. World Scientific, Singapore, 1991.

[30] M. Bridson and A. Häfliger. *Metric Spaces of Non-Positive Curvature*. Springer, Berlin, 1999.

[31] D. Bryant. A classification of consensus methods for phylogenies. In M. Janowitz, F.-J. Lapointe, F. McMorris, B. Mirkin, and F. Roberts, editors, *BioConsensus*, pages 163–184. American Mathematical Society, 2003.

[32] D. Bryant and V. Berry. A structured family of clustering and tree construction methods. *Advances in Applied Mathematics*, **27**:705–732, 2001.

[33] D. Bryant and A. Dress. Linearly independent split systems. *European Journal of Combinatorics*, **28**:1814–1831, 2007.

[34] D. Bryant and V. Moulton. NeighborNet: An agglomerative method for the construction of phylogenetic networks. *Molecular Biology and Evolution*, **21**:255–265, 2004.

[35] D. Bryant, V. Moulton, and A. Spillner. Consistency of the neighbor-net algorithm. *Algorithms for Molecular Biology*, **2**(8), 2007.

[36] P. Buneman. The recovery of trees from measures of dissimilarity. In F. Hodson *et al.*, editor, *Mathematics in the Archaeological and Historical Sciences*, pages 387–395. Edinburgh University Press, 1971.

[37] P. Buneman. A note on the metric property of trees. *Journal of Combinatorial Theory, Series B*, **17**:48–50, 1974.

[38] V. Capoyleas and J. Pach. A Turán-type theorem on chords of a convex polygon. *Journal of Combinatorial Theory, Series B*, **56**:9–15, 1992.

[39] CGAL, Computational Geometry Algorithms Library. http://www.cgal.org.

[40] W. Chen, A. Dress, and W. Yu. Community structures of networks. *Mathematical in Computer Science*, **1**: 441–457, 2008.

[41] V. Chepoi and B. Fichet. A note on circular decomposable metrics. *Geometriae Dedicata*, **69**:237–240, 1998.

[42] Y. Choe, J. Koolen, K. T. Huber, V. Moulton, and Y. Won. Counting vertices and cubes in median graphs associated to circular split systems. *European Journal of Combinatorics*, **29**:443–456, 2008.

[43] M. Chrobak and L. Larmore. Generosity helps or an 11-competitive algorithm for three servers. *Journal of Algorithms*, **16**:234–263, 1994.

[44] H. Colonius and H. H. Schultze. Trees constructed from empirical relations. *Braunschweiger Berichte aus dem Institut für Psychologie*, **1**, 1977.

[45] H. Colonius and H. H. Schultze. Tree structure from proximity data. *British Journal of Mathematical and Statistical Psychology*, **34**:167–180, 1981.

[46] W. Day. Computational complexity of inferring phylogenies from dissimilarity matrices. *Bulletin of Mathematical Biology*, **29**:461–467, 1987.

[47] C. Devauchelle, A. Dress, A. Grossmann, S. Grünewald, and A. Henaut. Constructing hierarchical set systems. *Annals of Combinatorics*, **8**:441–456, 2004.

[48] M. Deza and M. Laurent. *Geometry of Cuts and Metrics*. Springer, Berlin, 1997.

[49] R. Diestel. *Graph Theory*. 3rd edn, Springer, Berlin, 2005.

[50] A. Dress. Trees, tight extensions of metric spaces, and the cohomological dimension of certain groups: A note on combinatorial properties of metric spaces. *Advances in Mathematics*, **53**:321–402, 1984.

[51] A. Dress. Towards a theory of holistic clustering. In B. Mirkin, F. R. McMorris, F. S. Roberts, and A. Rzhetsky, editors, *Mathematical Hierarchies and Biology*, pages 271–289. American Mathematical Society, 1997.

[52] A. Dress. The category of X-nets. In J. Feng, J. Jost, and M. Qian, editors, *Networks: From Biology to Theory*, pages 3–22. Springer, Berlin, 2007.

[53] A. Dress, T. Wu, and X. Xu. A note on single-linkage equivalence. *Applied Mathematics Letters*, **23**:432–435, 2010.

[54] A. Dress and P. L. Erdős. X-trees and weighted quartet systems. *Annals of Combinatorics*, **7**:155–169, 2003.

[55] A. Dress, M. Hendy, K. T. Huber, and V. Moulton. On the number of vertices and edges in the Buneman graph. *Annals of Combinatorics*, **1**:329–337, 1997.

[56] A. Dress, B. Holland, K. T. Huber, J. Koolen, V. Moulton, and J. Weyer-Menkhoff. Delta additive and delta ultra-additive maps, Gromov's trees, and the Farris transform. *Discrete Applied Mathematics*, **146**:51–73, 2005.

[57] A. Dress, K. T. Huber, J. Koolen, and V. Moulton. Blocks and cut vertices of the Buneman graph, 2011. Submitted.

[58] A. Dress, K. T. Huber, A. Lesser, and V. Moulton. Hereditarily optimal realizations of consistent metrics. *Annals of Combinatorics*, **10**:63–76, 2006.

[59] A. Dress, K. T. Huber, and V. Moulton. Some variations on a theme by Buneman. *Annals of Combinatorics*, **1**:339–352, 1997.

[60] A. Dress, K. T. Huber, and V. Moulton. A comparison between two distinct continuous models in projective cluster theory: The median and the tight-span construction. *Annals of Combinatorics*, **2**:299–311, 1998.

[61] A. Dress, K. T. Huber, and V. Moulton. Hereditarily optimal realizations: Why are they relevant in phylogenetric analysis and how does one compute them? In *Proceedings of the Euroconference, Algebraic Combinatorics and Applications*, pages 110–117. Springer, Berlin, 2001.

[62] A. Dress, K. T. Huber, and V. Moulton. An explicit computation of the injective hull of certain finite metric spaces in terms of their associated Buneman complex. *Advances in Mathematics*, **168**:1–28, 2002.

[63] A. Dress, K. T. Huber, and V. Moulton. Some uses of the Farris transform in mathematics and phylogenetics – A review. *Annals of Combinatorics*, **11**:1–37, 2007.

[64] A. Dress and D. Huson. Constructing split graphs. *IEEE Transactions on Computational Biology and Bioinformatics*, **1**:109–115, 2004.

[65] A. Dress, D. Huson, and V. Moulton. Analyzing and visualizing distance data using SplitsTree. *Discrete Applied Mathematics*, **71**:95–110, 1996.

[66] A. Dress, J. Koolen, and V. Moulton. On line arrangements in the hyperbolic plane. *European Journal of Combinatorics*, **23**:549–557, 2002.

[67] A. Dress, J. Koolen, and V. Moulton. $4n - 10$. *Annals of Combinatorics*, **8**:463–471, 2004.

[68] A. Dress and R. Scharlau. Gated sets in metric spaces. *Aequationes Mathematicae*, **34**:112–120, 1987.

[69] J. Elson, C. Herrnstadt, G. Preston, L. Thal, C. Morris, J. Edwardson, M. Beal, D. Turnbull, and N. Howell. Does the mitochondrial genome play a role in the etiology of Alzheimer's disease? *Human Genetics*, **119**:241–254, 2006.

[70] A. Erné, J. Koslowski, A. Melton, and G. Strecker. A primer on Galois connections. *Annals of the New York Academy of Sciences*, **704**:103–125, 1993.

[71] G. Estabrook. Fifty years of character compatibility concepts at work. *Journal of Systematics and Evolution*, **46**:109–129, 2008.

[72] M. Farach, S. Kannan, and T. Warnow. A robust model for finding optimal evolutionary trees. *Algorithmica*, **13**:155–179, 1995.

[73] J. Farris. On the phenetic approach to vertebrate classification. In M. Hecht, P. Goody, and B. Hecht, editors, *Major Patterns in Vertebrate Evolution*, pages 823–850. Plenum Press, New York, 1977.

[74] J. Farris. The information content of the phylogenetic system. *Systematic Zoology*, **28**:483–519, 1979.

[75] J. S. Farris, A. G. Kluge, and M. J. Eckardt. A numerical approach to phylogenetic systematics. *Systematic Zoology*, **19**:172–189, 1970.

[76] J. Felsenstein. *Inferring Phylogenies*. Sinauer Associates, Sunderland MA, 2004.

[77] S. Finnilä, I. Hassinen, L. Ala-Kokko, and K. Majamma. Phylogenetic network of the mtDNA haplogroup U in northern Finland based on sequence analysis of the complete coding region by conformation-sensitive gel electrophoresis. *American Journal of Human Genetics*, **66**:1017–1026, 2000.

[78] A. Frank, A. Karzanov, and A. Sebő. On integer multiflow maximization. *SIAM Journal on Discrete Mathematics*, **10**:158–170, 1997.

[79] M. Garey and D. Johnson. *Computers and Intractability: A Guide to the Theory of NP-Completeness*. Freeman, San Francisco, 1979.

[80] E. Gawrilow and M. Joswig. Geometric reasoning with polymake, 2005. arXiv:math.CO/0507273.

[81] M. Girvan and M. Newman. Community structure in social and biological networks. *Proceedings of the National Academy of Sciences of the United States of America*, **99**:7821–7826, 2002.

[82] M. Gromov. Hyperbolic groups. In *Essays in Group Theory*, volume 8 of *MSRI*. Springer, Berlin, 1988.

[83] M. Gromov. CAT(κ)-spaces: construction and concentration. *Journal of Mathematical Sciences*, **119**:178–200, 2004.

[84] S. Grünewald. Slim sets of binary trees. *Journal of Combinatorial Theory A*, 2011. to appear.

[85] S. Grünewald, K. Forslund, A. Dress, and V. Moulton. QNet: An agglomerative method for the construction of phylogenetic networks from weighted quartets. *Molecular Biology and Evolution*, **24**:532–538, 2007.

[86] S. Grünewald, K. T. Huber, V. Moulton, C. Semple, and A. Spillner. Characterizing weak compatibility in terms of weighted quartets. *Advances in Applied Mathematics*, **42**:329–341, 2009.

[87] S. Grünewald, K. T. Huber, V. Moulton, and C. Semple. Encoding phylogenetic trees in terms of weighted quartets. *Journal of Mathematical Biology*, **56**:465–477, 2008.

[88] S. Grünewald, J. Koolen, and W. Lee. Quartets in maximal weakly compatible split systems. *Applied Mathematics Letters*, **22**:1604–1608, 2009.

[89] S. Grünewald, V. Moulton, and A. Spillner. Consistency of the Qnet algorithm for generating planar split graphs from weighted quartets. *Discrete Applied Mathematics*, **157**:2325–2334, 2009.

[90] S. L. Hakimi and S. S. Yau. Distance matrix of a graph and its realizability. *Quarterly of Applied Mathematics*, **22**:305–317, 1964.

[91] M. D. Hendy. The relationship between simple evolutionary tree models and observable sequence data. *Systematic Zoology*, **38**:310–321, 1989.

[92] M. D. Hendy and D. Penny. A framework for the quantitative study of evolutionary trees. *Systematic Zoology*, **38**:297–309, 1989.

[93] M. D. Hendy and D. Penny. Spectral analysis of phylogenetic data. *Journal of Classification*, **10**:5–24, 1993.

[94] S. Herrmann and M. Joswig. Bounds on the f-vectors of tight spans. *Contributions to Discrete Mathematics*, **2**:161–184, 2007.

[95] B. Holland, G. Conner, K. T. Huber, and V. Moulton. Imputing supertrees and supernetworks from quartets. *Systematic Biology*, **56**:57–67, 2007.

[96] B. Holland, F. Delsuc, and V. Moulton. Visualizing conflicting evolutionary hypotheses in large collections of trees using consensus networks. *Systematic Biology*, **54**:66–76, 2005.

[97] K. T. Huber, V. Moulton, P. Lockhart, and A. Dress. Pruned median networks: a technique for reducing the complexity of median networks. *Molecular Phylogenetics and Evolution*, **19**:302–310, 2001.

[98] D. Huson. SplitsTree: analyzing and visualizing evolutionary data. *Bioinformatics*, **14**:68–73, 1998.

[99] D. Huson, T. Dezulian, T. Klöpper, and M. Steel. Phylogenetic super-networks from partial trees. *IEEE/ACM Transactions on Computational Biology and Bioinformatics*, **1**:151–158, 2004.

[100] D. Huson and R. Rupp. Summarizing multiple gene trees using cluster networks. In *Workshop on Algorithms in Bioinformatics*, volume 5251 of *LNCS*, pages 296–305. Springer, Berlin, 2008.

[101] W. Imrich and S. Klavžar. *Product Graphs: Structure and Recognition*. John Wiley, New York, 2000.

[102] W. Imrich, J. Simoes-Pereira, and C. Zamfirescu. On optimal embeddings of metrics in graphs. *Journal of Combinatorial Theory, Series B*, **36**:1–15, 1984.

[103] J. Isbell. Six theorems about metric spaces. *Commentarii Mathematici Helvetici*, **39**:65–74, 1964.

[104] J. Jansson and W.-K. Sung. Inferring a level-1 phylogenetic network from a dense set of rooted triplets. *Theoretical Computer Science*, **363**:60–68, 2006.

[105] G. Kalai. Polytope skeletons and paths. In J. Goodman and J. O'Rourke, editors, *Handbook of Discrete and Computational Geometry*, pages 455–476. Chapman & Hall/CRC Press, Boca Raton, 2004.

[106] K. Kalmanson. Edgeconvex circuits and the travelling salesman problem. *Canadian Journal of Mathematics*, **27**:1000–1010, 1975.

[107] A. Karzanov and M. Lomonosov. Systems of flows in undirected networks. In O. Larichev, editor, *Mathematical Programming*, volume 1. Institute for Systems Studies, 1978. In Russian.

[108] S. Klavžar and H. Mulder. Median graphs: characterizations, location theory and related structures. *Journal of Combinatorial Mathematics and Combinatorial Computing*, **30**:103–127, 1999.

[109] M. Kotetishvili, O. Stine, A. Kreger, J. Morris, and A. Sulakvelidze. Multilocus sequence typing for characterization of clinical and environmental *Salmonella* strains. *Journal of Clinical Microbiology*, **40**:1626–1635, 2002.

[110] M. Lomonosov. Combinatorial approaches to multiflow problems. *Discrete Applied Mathematics*, **11**:1–93, 1985.

[111] F. MacWilliams and N. Sloane. *The Theory of Error-Correcting Codes*. North-Holland, Amsterdam, 1983.

[112] T. Margush and F. McMorris. Consensus *n*-trees. *Bulletin of Mathematical Biology*, **43**:239–244, 1981.

[113] C. Meacham. Theoretical and computational considerations of the compatibility of qualitative taxonomic characters. In J. Felsenstein, editor, *Numerical Taxonomy*, pages 304–314. Springer, Berlin, 1983.

[114] V. Moulton and M. Steel. Retractions of finite distance functions onto tree metrics. *Discrete Applied Mathematics*, **91**:215–233, 1999.

[115] H. Mulder. The interval function of a graph. *Mathematical Centre Tracts*, **132**. Mathematisch Centrum, Amsterdam, 1980.

[116] M. Newman. Finding community structure in networks using the eigenvectors of matrices. *Physical Review E*, **74**, 036104, 2006.

[117] J. Neyman. Molecular studies of evolution: a source of novel statistical problems. In S. Gupta and J. Yackel, editors, *Statistical Decision Theory and Related Topics*, pages 1–27. Academic Press, New York, 1971.

[118] M. Owen and J. Provan. A fast algorithm for computing geodesic distances in tree space. *IEEE Transactions on Computational Biology and Bioinformatics*, **8**:2–13, 2011.

[119] R. Rammal, J. Angles d'Auriac, and B. Doucot. On the degree of ultrametricity. *Le Journal de Physique-Lettre*, **46**:945–952, 1985.

[120] N. Saitou and M. Nei. The neighbor-joining method: A new method for reconstructing phylogenetic trees. *Molecular Biology and Evolution*, **4**:406–425, 1987.

[121] J. M. S. Simões-Pereira. A note on the tree realizability of a distance matrix. *Journal of Combinatorial Theory*, **6**:303–310, 1969.

[122] J. M. S. Simões-Pereira and C. M. Zamfirescu. Submatrices of non-tree-realizable distance matrices. *Linear Algebra and its Applications*, **44**:1–17, 1982.

[123] M. Steel. The complexity of reconstructing trees from qualitative characters and subtrees. *Journal of Classification*, **9**:91–116, 1992.

[124] K. Strimmer and A. von Haeseler. Quartet puzzling: A quartet maximum likelihood method for reconstructing tree topologies. *Molecular Biology and Evolution*, **13**:964–969, 1996.

[125] B. Sturmfels and J. Yu. Classification of six-point metrics. *The Electronic Journal of Combinatorics*, **11**, 2004.

[126] T-H. To and M. Habib. Level-*k* phylogenetic network can be constructed from a dense triplet set in polynomial time. In *Annual Symposium on Combinatorial Pattern Matching*, LNCS. Springer, Berlin, 2009.

[127] L. van Iersel, J. Keijsper, S. Kelk, L. Stougie, F. Hagen, and T. Boekhout. Constructing level-2 phylogenetic networks from triplets. In *Annual International Conference on Research in Computational Molecular Biology*, volume 4955 of *LNCS*, pages 450–462. Springer, Berlin, 2008.

[128] J. van Lint. *Introduction to Coding Theory*. 3rd edn, Springer, Berlin, 1999.

[129] A. Verbeek. Superextensions of topological spaces. *Mathematical Centre Tracts*, **41**, 1972.

[130] R. Wetzel. *Zur Visualisierung abstrakter Ähnlichkeitsbeziehungen*. PhD thesis, Universität Bielefeld, 1995.

[131] J. Weyer-Menkhoff. *New quartet methods in phylogenetic combinatorics*. PhD thesis, Universität Bielefeld, 2003.

[132] J. Weyer-Menkhoff, C. Devauchelle, A. Grossmann, and S. Grünewald. Integer linear programming as a tool for constructing trees from quartet data. *Computational Biology and Chemistry*, **29**:196–203, 2005.

[133] E. Wilkeit. The retracts of Hamming graphs. *Discrete Mathematics*, **102**:197–218, 1992.

[134] P. Winkler. Isometric embeddings in products of complete graphs. *Discrete Applied Mathematics*, **7**:221–225, 1984.

[135] P. Winkler. The complexity of metric realisation. *SIAM Journal of Discrete Mathematics*, **1**:552–559, 1988.

[136] K. A. Zaretskii. Constructing trees from the set of distances between pendant vertices. *Uspehi Matematiceskih Nauk*, **20**:90–92, 1965.

Index